Preface

The comprehensive manual 'Cell and Tissue Culture: Laboratory Procedures' edited by A. Doyle, J.B. Griffiths and D.G. Newell, was first published in 1993, with quarterly additions and updates up to 1998. The publication has been well received by the scientific community and has now reached completion. Numerous requests have been received from a range of people, saying: 'When will a series of subset volumes be produced?' In response to this demand we have decided to look afresh at the wealth of material available in the main publication and adapt from this 'highlights', which we believe will be of particular value to targeted users. The first of these is the subset for biotechnologists, which contains selected procedures that provide essential technical information for this group of scientists. Many of the contributions have been updated from the original for this publication. It is certainly not our intention to reproduce all of the manual in this fashion but to provide core procedures for each of the specialist groups that can be identified as benefiting from them. We aim to appeal to scientists who may be new to cell culture and require the practical guidance that 'Cell and Tissue Culture: Laboratory Procedures' has to offer. There is also the added benefit of the valuable technical information being available without the major investment in the whole publication. We believe that these subsets will fulfil a need and we look forward to preparing further publications along these lines.

A. Doyle and J.B. Griffiths
Managing Editors

Safety

Neither the editors, contributors nor John Wiley & Sons Ltd. accept any responsibility or liability for loss or damage occasioned to any person or property through using the materials, instructions, methods, or ideas contained herein, or acting or refraining from acting as a result of such use. While the editors, contributors and publisher believe that the data, recipes, practical procedures and other information, as set forth in this book, are in accord with current recommendations and practice at the time of publication, they accept no legal responsibility for any errors or omissions, and make no warranty, express or implied, with respect to material contained herein. Attention to safety aspects is an integral part of all laboratory procedures and national legislations impose legal requirements on those persons planning or carrying out such procedures. It remains the responsibility of the reader to ensure that the procedures which are followed are carried out in a safe manner and that all necessary safety instructions and national regulations are implemented.

In view of ongoing research, equipment modifications and changes in governmental regulations, the reader is urged to review and evaluate the information provided by the manufacturer, for each reagent, piece of equipment or device, for any changes in the instructions or usage and for added warnings and precautions.

CHAPTER 1

THE CELL: SELECTION AND STANDARDIZATION

1.1 OVERVIEW

The biotechnologist has a wealth of systems to choose from before the decision has to be taken to establish a cell line for a specific purpose *de novo*. Naturally, the requirements (and not least the Intellectual Property considerations) will dictate the 'utility' of existing material. Even so, if the cells for exploitation or the basic tools to create the *in vitro* systems already exist, then authenticated sources of starting material are essential. This can mean obtaining cell stocks from culture collections such as ECACC, ATCC, DSMZ, Riken Cell Bank, etc. (see Appendix 3, Table A3.2). The advantage of basing a project on existing, well-characterized, authenticated and quality-controlled stocks cannot be over-emphasized. The added advantage is that much of the material available from collections is free of constraints on exploitation.

The standards for the cryopreservation, storage and routine quality control of cell stocks are widely recognized (Stacey *et al.*, 1995). Cryopreservation of a well-characterized, dependable, high-viability (achieved by controlled-rate freezing), microbial-contaminent-free cell stock is fundamental to both the academic researcher as well as the commercial producer. Existing legislation provides for a set of international standards and the US FDA/CBER 'Points to Consider' are seen as the benchmark in this field (CBER, 1993). Regulatory aspects are dealt with in more detail in section 1.3 and Chapter 6. Unfortunately, scientists embarked on a research programme leading to a cell line that might be exploitable at some later stage do not necessarily regard such guidelines as relevant to them or to the goals of their work. This is a narrow view and one that is potentially very expensive and has to be dispelled at all costs.

A further consideration is the safety aspect of handling cell lines. The minimum standard to be applied in any cell culture laboratory is Category 2 containment. Although a risk assessment is made, in most cases the extent of characterization and thus knowledge on the potential hazard of handling particular material is unknown. This is especially true with respect to the presence of adventitious agents (e.g. viruses) in cell lines. There may be particular concern in the handling of patient material with regard to hepatitis/HIV/HTLV status and a balanced view on risk has to be taken. The topic is a large one (Stacey *et al.*, 1998) and suffice to say that once minimum standards are set they can be all-embracing for every cell type handled.

Of particular importance in the routine handling of cell cultures is the mycoplasma contamination status of cell lines. If present, the concentration of mycoplasmas in the culture supernatant can be in the region of 10^6–10^8 mycoplasmas ml^{-1}. Unlike bacterial and fungal contaminants, they do not necessarily manifest themselves in terms of pH change and/or turbidity and they can be present in low

---

Cell and Tissue Culture: Laboratory Procedures in Biotechnology, edited by A. Doyle and J.B. Griffiths.
© 1998 John Wiley & Sons Ltd.

numbers. Mycoplasmas elicit numerous deleterious effects and their presence is incompatible with standardized systems. Routinely, broth and agar culture or Hoechst DNA stain are the methods of choice for detection, although increasingly polymerase chain reaction (PCR) methods are becoming available (Doyle & Bolton, 1994). Tests have to be part of a regular routine and not just seen as 'one-off' procedures at the start of a piece of work. Elimination of contamination is possible but costly in time and resources, and is not always successful, so it is better to check early rather than later; this re-emphasizes the importance of authenticated cell banks to return to in case of contamination.

Finally, it must be emphasized that no amount of testing can replace the day-to-day vigilance of laboratory workers routinely handling cells. Any alteration in normal growth pattern or morphology should not be ignored because this may well indicate a fundamental problem well in advance of other more formal testing parameters.

REFERENCES

Centre for Biologics Evaluation and Research (CBER) (1993), *Points to Consider in Characterization of Cell Lines used to Produce Biologicals*. US Food and Drugs Administration, Bethesda.

Doyle A & Bolton BJ (1994), The quality control of cell lines and the prevention, detection and cure of contamination. In: *Basic Cell Culture: a Practical Approach*, pp. 243–271. IRL Press, Oxford.

Stacey G, Doyle A & Hambleton P (eds) (1998) *Safety in Tissue Culture*. Kluwer, London.

Stacey, GN, Parodi, B & Doyle, AJ (1995) The European Tissue Culture Society (ETCS) initiative on quality control of cell lines. *Experiments in Clinical Cancer Research*: 4: 210–211.

1.2 CELL LINES FOR BIOTECHNOLOGISTS

INTRODUCTION

Animal cell lines have been used extensively for the production of a variety of therapeutic and prophylactic protein products including hormones, cytokines, enzymes, antibodies and vaccines. They offer the advantage of reproducibility and convenience over primary cell cultures and animal models as well as the potential for large-scale production. In addition, animal cells are generally capable of secreting functionally active proteins correctly folded and with correct post-translational modifications, unlike bacterial or yeast systems. In the production of recombinant proteins, fidelity in glycosylation of the product can be an important consideration influencing its secretion, degradation and biological activity. Comparisons have been made between glycosylation of recombinant proteins using insect, bacterial and mammalian expression systems, which have highlighted differences between these and the human glycosylation profile (James *et al.*, 1995). The adaptation of many cell lines to growth in serum/protein-free media has facilitated not only the downstream processing of the secreted product but also minimizes the potential risk of viral and mycoplasma contaminants, which can be inadvertently added with animal sera or animal-derived proteins such as growth factors.

Furthermore, mammalian cell lines transfected with a variety of expression systems have been widely used for the expression of recombinant proteins of commercial and therapeutic importance, some of which will be addressed here (see Table 1.2.1).

Certain cell lines require licensing agreements for their use in commercial production, although for research and development applications this is not generally necessary.

CELL LINE CHO dhfr⁻

The Chinese hamster ovary (CHO) cell line has become popular for the expression of human recombinant glycoproteins because the glycosylation enzymes present resemble those found in human cell lines (Lee *et al.*, 1989, Jenkins & Curling, 1994). CHO cells are well characterized, relatively stable, are able to produce heterologous proteins efficiently and can be grown in both attached (although not in serum-free media) and suspension conditions. Their growth as aggregates in suspension permits rapid sedimentation, facilitating media changes

---

Cell and Tissue Culture: Laboratory Procedures in Biotechnology, edited by A. Doyle and J.B. Griffiths.
© 1998 John Wiley & Sons Ltd.

for large-scale production. However, a disadvantage of this is that cell death can occur at the centre of larger aggregates (Litwin, 1991).

CHO dhfr⁻ cells, which lack the enzyme dihydrofolate reductase (DHFR), provide an expression system that can be co-transfected with a gene for a particular protein product together with the DHFR gene (Urlaub et al., 1986). The DHFR gene is amplified using methotrexate, which also amplifies the co-transfected gene of interest flanking the DHFR gene. CHO cells have been used for the production of recombinant proteins, including follicle-stimulating hormone (FSH), factor VIII, interferon-γ (IFN-γ), IFN-β, HIV1-gp120 envelope glycoprotein, recombinant tissue plasminogen activator (rtPA), herpes simplex virus (HSV) gB2, parathyroid hormone, antithrombin III, *E. coli* XGPRT, hepatitis B surface antigen, human erythropoietin, human purine nucleoside phosphorylase, SV40 small t-antigen, transforming growth factor-α (TGF-α) and TGF-β, superoxide dismutase and Epstein-Barr virus (EBV) glycoproteins.

Culture conditions

Suspension cultures can be achieved using F12 medium with 10% foetal bovine serum (FBS) and gassing the culture vessels continuously with 5% CO_2. The doubling time under these conditions is typically 18 h versus 14 h for monolayer cultures. The growth requirements of CHO dhfr⁻ cells are hypoxanthine (or adenine), glycine and thymidine in addition to the usual requirement of CHO cells for proline.

Restrictions

A WHO bank of CHO dhfr⁻ cells is deposited at the European Collection of Cell Cultures (ECACC) and permission can be obtained for their use in commercial production by a licence agreement through Professor Chasin, Department of Biological Sciences, Sherman Fairchild Centre for the Life Sciences, Columbia University, New York, NY 10027. The licence is subject to a one-time payment.

CELL LINE Sf9

Sf9 cells are a clonal derivative of a pupal ovarian line of the Fall Armyworm, *Spodoptera frugiperda* (Smith et al., 1985). The high susceptibility of Sf9 insect cells to baculovirus infection, their ability to grow rapidly at 26°C and high percentage protein secretion makes these cells a favourable system for the production of recombinant proteins using genetically manipulated baculovirus expression vector systems, such as that based on *Autographa californica* Nuclear Polyhedrosis Virus (AcMNPV). Baculovirus expression systems are shown to express functional proteins successfully at high yield (Luckow & Summers, 1988).

The spherical morphology of these cells makes them more resistant to mechanical damage, which is an important consideration in suspension culture. In addition, the ability to grow rapidly confers the advantage of reduced incubation time, lower maintenance and a higher yield of cells per unit of nutrient. Sf9 cells have been used in the production of recombinant bovine β-lactoglobulin, antistasin, β-adrenergic receptor and insecticides.

The cell line has been adapted for growth in serum-free media using spinner flasks and stirred tanks and cultures can be scaled up readily to produce high yields of recombinant protein (Jain *et al.*, 1991). In batch cultures, infection of the cells with baculovirus can be monitored easily by measuring the increase in cell volume and decrease in cell viability. Several days after infection the cells lyse, releasing the recombinant protein product. Product titre is influenced by the oxygen requirement of the insect cells.

Culture conditions

Growth is achieved at 27°C in TC-100 medium (Gibco BRL) with 10% heat-inactivated FBS or IPL-41 basal medium (J.R. Scientific, CA) with 2% FBS plus 3.3 g l^{-1} each of yeastolate and lactalbumin hydrolysate (Difco), suspension cultures being supplemented with 1–2 g l^{-1} pluronic F-68 (BASF Corp., NJ). Sf9 cells can also be cultivated in serum-free media such as Excell-400 (J.R. Scientific, CA) and SF900.

Restrictions

There is a licensing agreement through Professor Summers University of Texas (baculovirus vector system).

CELL LINE Schneider-2

This cell line is derived from the late embryonic state of *Drosophila melanogaster* and is used to produce recombinant proteins, e.g. human α_1-antitrypsin. Schneider-2 cells are adaptable to large-scale cultivation in stirred tank bioreactors and show many of the properties of Sf9 cells.

Culture conditions

Cells are cultured in M3 medium with 5% FBS at 27°C. There is no CO_2 requirement for buffering the medium.

CELL LINES COS 1/COS 7

These African green monkey kidney cell lines, derived from the CV-1 simian cell line, are transformed with SV40 and are fibroblast-like in morphology (Gluzman, 1981). Origin-defective mutants of SV40 encoding wild-type T-antigen were used stably to transform monkey kidney cells. The transformed COS cells, containing T antigen, are able to support the replication of pure populations of recombinant

SV40 to high copy number with deletions in the early region, SV40 vector replication reaching a maximum after about 2 days.

COS cells have been used for the production of recombinant proteins, including HSV-1 glycoprotein B, ricin B chain, placental alkaline phosphatase, thrombomodulin, CD7, von Willebrand factor, human dihydropteridine reductase, human β-glucuronidase, interleukin 5 and human interferon-$β_1$.

Culture conditions

Cells are cultured in Dulbecco's modified Eagle's medium (DMEM) with 10% FBS. Cells must not be allowed to reach confluence, in order to prevent syncitia formation.

Restrictions

There is a licensing agreement through MRC Technology Transfer Group, 20 Park Crescent, London, UK.

CELL LINE NIH/3T3

This is a continuous cell line established from NIH Swiss mouse embryo cultures (Torado & Green, 1963; Torado et al., 1965). The NIH/3T3 cell line is susceptible to sarcoma virus focus formation and leukaemia virus and is a valuable cell line for DNA transfection studies. It has been used for the expression of recombinant proteins, including hepatitis B surface antigen, phenylalanine hydroxylase, rat growth hormone, recombinant fibronectins, human class II MHC antigens, human insulin receptor and insulin-like growth factor. The cells are used extensively in pharmacotoxicology studies.

Culture conditions

Cells are cultured in DMEM with 10% FBS. The cells are highly contact inhibited and are sensitive to serum batches.

CELL LINE HeLa

This is a widely studied cell line derived from a human cervix adenocarcinoma (Gey et al., 1952). The cells are epithelial-like in morphology and are susceptible to polio virus type 1 and adenovirus type 3. HeLa cells are used for the expression of recombinant proteins, including mouse metallothionein 1 gene, human Cu/Zn superoxide dismutase and hepatitis B surface antigen. They have been widely used as an *in vitro* model system because of the ease with which they can be cultivated but one drawback of this is that the cell line has been responsible for widespread contamination of other cell lines (Nelson-Rees et al., 1981).

Culture conditions

Cells are cultured in Eagle's MEM (EBSS) supplemented with glutamine, non-essential amino acids and 10% FBS.

CELL LINE J558L

This is a mouse BALB/c myeloma cell line secreting IgA (Halpern & Coffman, 1972; Lundblad et al., 1972; Oi et al., 1983). The cell line is used in transfection studies.

Culture conditions

Cells are cultured in DMEM with 10% FBS.

CELL LINE Vero

The Vero cell line was derived from the kidney of a normal African green monkey and is susceptible to a wide range of viruses, including polio, rubella, arboviruses and reoviruses (Simizu & Terasima, 1988). The Vero monkey kidney cell line has been used for the industrial production of viral vaccines in preference to primary monkey kidney cells because of availability and the reduced risk of contamination by endogenous viruses. Vero cells have been used for the production of the polio (Sabin) vaccine and shown to have identical vaccine characteristics to the primary monkey kidney cells (Montagnon et al., 1991). Being a strictly anchorage-dependent cell line with fibroblast morphology, Vero cells have been adapted to culture in continuous perfusion systems, enabling high cell densities to be attained. Cell density at the time of virus inoculation has been found to influence the production of poliovirus D-antigen, with continuous perfusion producing the highest cell densities (Van der Meer et al., 1993).

A variant cell line adapted to grow in serum-free conditions is also available.

Culture conditions

Originally cultures were established in Earle's BSS containing 0.5% lactalbumin hydrolysate, 0.1% yeast extract, 0.1% polyvinylpyrrolidone and 2–5% FBS but they can be maintained successfully in Eagle's MEM (EBSS) with added 7.5% bovine serum and 2.5% FBS or Medium 199 supplemented with 5% FBS. Vero cells are adaptable to batch and continuous perfusion culture.

It should be noted that the World Health Organization has sponsored the creation of a fully characterized cell bank for vaccine manufacture purposes. Samples are available from ECACC and the American Type Culture Collection (ATCC).

MYELOMA CELL LINES

Mouse myelomas such as NS0 (a subclone of NS-1) (Galfre & Milstein, 1981) and Sp2/0-Ag14 (a re-clone of Sp2/HL-Ag derived from Sp2/HLGK) (Shulman et al., 1978), which do not express immunoglobulin (Ig), are popular fusion partners for Ig-secreting spleen cells to obtain hybrids secreting only the specific antibody (hybridoma). The myeloma cells of Sp2/0-Ag14 appear to fuse preferentially with the dividing cells of the spleen, which, of the B cells, are principally the Ig-secreting, plaque-forming cells.

Culture conditions

Cells are cultured in RPMI 1640 with glutamine and 10% FBS. The cells are resistant to 10 mM 8-azaguanine and die in the presence of HAT medium.

Restrictions

Restrictions are dependent upon cell line. NS0 in particular requires the completion of a licensing agreement through MRC Technology Transfer Group, 20 Park Crescent, London, UK.

HYBRIDOMAS

The initial development of monoclonal antibodies by Köhler and Milstein, produced by immortalization of mouse cell lines secreting specific antibodies (Köhler and Milstein, 1975), has led to the production of hybridomas, principally of mouse and rat origin, with a wide range of specificities and important applications for commercial, diagnostic and therapeutic purposes. A number of monoclonal antibodies have been licensed or are undergoing clinical trials for therapeutic application, e.g. Campath-1 raised to the lymphocyte CD52 antigen and used for the prevention of graft versus host disease, antibodies to control cancer and septic shock (PMA, 1988). Although the majority of monoclonal antibodies produced are of mouse or rat origin, human monoclonal antibodies have been obtained by immortalization of lymphocytes immunized *in vitro* and include antibodies to HIV-1 envelope glycoprotein, hepatitis B surface antigen, cytomegalovirus and rubella.

Important considerations to reduce costs and downstream processing include the necessity for high concentrations of antibody during production and high cell concentrations, which can be attained using perfusion systems (Wang et al., 1993; Shen et al., 1994). Hybridomas can be produced either in suspension, e.g. in roller bottles, stirred reactors or airlift reactors, or immobilized using a suitable matrix, e.g. hollow fibres, microbeads or microcapsules. All of these systems are adaptable to large-scale production. As with all large-scale production of cell lines, serum/protein-free media are preferable because they minimize batch variation and facilitate antibody production (Glassy et al., 1988). An important factor in the growth of cells in reduced protein media is the increased susceptibility to cell death and hence the increased risk of antibody degradation by endogenous proteases.

Culture conditions

Culture conditions are dependent upon the hybridoma but they can usually be grown in RPMI 1640 or DMEM supplemented with 10% FBS. Some cultures may need to be started on mouse macrophage feeder layers.

CELL LINE MRC-5

Human diploid cells of cell line MRC–5 are derived from normal foetal lung tissue and have a fibroblast morphology (Jacobs *et al.*, 1970). These cells will typically undergo 60–80 population doublings before dying out after 50+ passages. In production, population doubling levels up to around 28–30 are generally used, although this could probably be extended to 40 (Wood & Minor, 1990). Towards the end of their life span MRC-5 cells show increased population doubling times and the presence of abnormal cells by light microscopy, although the cells are genetically stable up to senescence (Wood & Minor, 1990). These cells are used for the production of viral vaccines, including rhinoviruses, Sabin poliovirus, measles-AIK strain, mumps, rubella, rabies, vaccinia and varicella. They show a similar virus susceptibility to another human diploid cell strain, WI-38 (see below). In comparative studies with WI-38, the MRC-5 cell line was found to replicate more rapidly and was less sensitive to adverse environmental factors.

Culture conditions

Cells are cultured in Eagle's MEM containing Earle's BSS or Hank's BSS and with 10% FBS. The life span could be increased with different media formulations, e.g. CMRL 1969> MEM> BME.

CELL LINE WI-38

Human diploid cell line WI-38 is derived from embryonic lung tissue with fibroblast morphology (Hayflick & Moorhead, 1961). The cell line has a broad range of virus susceptibility and is used for the production of human virus vaccines: rhinoviruses, measles, mumps, rubella, polio, rabies and yellow fever. The cells are anchorage dependent but will form a multilayered culture when held for long periods at 37°C with periodic pH adjustments. The cells have a life span of 50±10 population doublings, with a doubling time of 24h.

Culture conditions

Cells are cultured in Eagle's basal medium with 10% FBS.

CELL LINE Namalwa

Human B-lymphoblastoid cell line Namalwa is derived from a Burkitt's lymphoma tumour biopsy and can be grown for an indefinitely long period of time (Klein et al., 1979; Biliau et al., 1973). Although they contain Epstein-Barr (EB) nuclear antigen, Namalwa cells do not release EB virus. The cell line is used for the production of IFN-α. Namalwa cells can be induced to produce IFN by adding Sendai virus to cells pretreated with 2 mM sodium butyrate (Baker et al., 1980; Johnston, 1980) and can be primed with Sendai virus in order to increase the levels of IFN-α produced.

Culture conditions

Cells are cultured in RPMI 1640 with 5–15% FBS at 37°C and pH 6.8–7.0, although for large-scale production reduced FBS concentrations are utilized, using a substitute based on polyethylene glycol-pretreated bovine serum with a water-soluble peptide digest of animal tissue (Primatone RL) and adding pluronic F-68 to increase the viscosity of the medium (Mizrahi et al., 1980). Cultures have been scaled up for growth in large fermenters (8000 l) (Klein et al., 1979; Finter, 1987; Musgrave et al., 1993) and the cells can be grown in submerged cultures enabling easy scale-up.

Restrictions

The Namalwa cell line is cited in a US and/or other patents and must not be used to infringe patent claims.

CELL LINE BHK-21

This is a Syrian hamster cell line derived from the kidneys of 1-day-old hamsters. The cells have a fibroblast-like morphology and are used for viral replication studies, including poliovirus, rabies (Pay et al., 1985), rubella, foot and mouth disease virus (Radlett et al., 1985), VSV, HSV, adenovirus (Ad) 25 and arbovirus. Successful cultivation at scales up to 8000 l has been achieved with maximum cell density attained by minimum air sparging sufficient to satisfy the oxygen demand of the cells.

Culture conditions

Eagle's basal medium supplemented with 2% tryptose phosphate broth (TPB) and other supplementation, including additional glutamine, glucose, vitamins, lactalbumin hydrolysate and pluronic F-68; cells are also grown in DMEM supplemented with 5% tryptose phosphate broth and 5–10% FBS. Airlift fermenters (Katinger & Scheirer, 1979) and stirred vessels are used for large-scale production.

CELL LINE MDCK

This is a Cocker spaniel kidney cell line with epithelial morphology (Madin & Darby, 1958). It is used for the study of SVEV, VSV, vaccinia, Coxsackie B5, infectious canine hepatitis, adenoviruses and reoviruses.

Culture conditions

Cells are cultured in Eagle's MEM with 10% FBS.

CELL LINE GH3

The GH3 cell line was derived from the pituitary tumour of a 7-month-old female Wistar-Furth rat (Tashjian *et al.*, 1968) and is used for the study of hormones. The GH3 clone produces growth hormone at a greater rate than GH1 cells, a clone obtained from the same primary culture, and also produces prolactin. Hydrocortisone can be used to stimulate hormone production and also inhibits the production of prolactin (Bancroft *et al.*, 1969). The cells are epithelial-like in morphology but can be adapted to grow in suspension culture using Eagle's MEM, under which conditions the cells continue to produce growth hormone and prolactin (Bancroft & Tashjian, 1971). The cells are susceptible to herpes simplex and vesicular stomatitis (Indiana strain) viruses but not to poliovirus type 1.

Culture conditions

Cells are cultured in F10 supplemented with 15% horse serum and 2.5% FBS. The cells are adaptable to growth in suspension culture in spinner flasks using Eagle's MEM.

CELL LINE 293

The 293 cell line is derived from human embryonal kidney transformed with sheared human Ad5 DNA and has an epithelial morphology (Graham *et al.*, 1977; Harrison *et al.*, 1977). The cells are used for transformation studies and gene therapy applications, being sensitive to human adenovirus and adenovirus DNA. Recombinant human adenovirus vectors are being used increasingly for many applications, including vaccination and gene therapy; 293 cells can be used to isolate transformation-defective host-range mutants of Ad5 and for titrating human adenoviruses.

Culture conditions

Cells are cultured in Eagle's MEM supplemented with 10% horse serum or FBS. The cells detach at room temperature and may take days to reattach.

Table 1.2.1 Principal characteristics and applications of selected cell lines commonly used in biotechnology

Cell line	Characteristics	Applications	Refs
CHO dhfr–	Chinese hamster ovary cells; epithelial morphology	Recombinant protein production – FSH, factor VIII, IFN-γ, HIV1-gp120, rtPA, HSV gB2, parathyroid hormone, antithrombin III, M-CSF	Urlaub et al. (1986)
Sf9	Cells from pupal ovarian tissue of *Spodoptera frugiperda*; spherical morphology; rapid growth; high level of protein expression	Recombinant proteins using baculovirus expression systems – antistasin, β-adrenergic receptor; insecticides	Smith et al. (1985)
Schneider-2	*Drosophila melanogaster* cell line	Recombinant proteins – α_1-antitrypsin	
COS 1/ COS 7	African green monkey kidney; fibroblast morphology	To support the growth of recombinant SV40 viruses	Gluzman (1981)
NIH/3T3	Mouse embryo, contact inhibited; fibroblast morphology	IGF–1 production	Torado & Green (1963); Torado et al. (1965)
HeLa	Human cervix carcinoma; epithelial morphology	Virus studies	Gey et al. (1952)
J558L	Mouse myeloma	Recombinant antibody production – CD4-Ig	Halpern & Coffman (1972); Lundblad et al. (1972); Oi et al. (1983)
Vero	Monkey kidney cells; anchorage-dependent fibroblast morphology	Viral vaccines – polio, haemorrhagic viruses	Simizu & Terasima (1988)
Myeloma	Immortal cell lines, e.g. NS0, Sp2/0-Ag14	Used to produce hybridomas and recombinant antibodies	Shulman et al. (1978); Galfre & Milstein (1981)
Hybridoma	Stable hybrid lines retaining characteristics of immortal myeloma and differentiated plasma cell fusion partners	Monoclonal antibody production	Köhler & Milstein (1975)
MRC-5	Human foetal lung; fibroblast morphology	Virus studies – susceptibility similar to WI-38	Jacobs et al. (1970)
WI-38	Human foetal lung cells; fibroblast morphology; 1° diploid cells; anchorage dependent	Viral vaccines – rhinoviruses	Hayflick & Moorhead (1961) (1962)
Namalwa	Human tumour cell line	Interferon-α production	
BHK 21	Syrian hamster kidney fibroblast	Viral vaccines – foot and mouth disease (FMD), enzymes	Pay et al. (1985); Radlett et al. (1985)
MDCK	Dog cocker spaniel kidney; epithelial morphology	Animal viruses – SVEV, VSV, vaccinia, adeno and reoviruses, infectious canine hepatitis virus,	Madin & Darby (1958)
GH3	Wistar-Furth rat pituitary tumour	Produces growth hormone and prolactin	Tashjian et al. (1968)
293	Transformed primary human embryonal kidney cell line	Isolation of transformation-defective host-range mutants of Ad5	Graham et al. (1977); Harrison et al. (1977)
ΨCRE/ ΨCRIP	Mouse NIH-3T3 cells co-transformed with Moloney murine leukaemia genome	Recombinant retroviral packaging cell lines used for the development of human gene replacement therapy.	Danos & Mulligan (1988)

CELL LINE ΨCRE/ΨCRIP

Mouse NIH 3T3 cells are co-transformed with Moloney murine leukaemia genome to give rise to packaging cell lines used in the generation of helper-free recombinant retroviruses with amphotropic and ecotropic host ranges (Danos & Mulligan, 1988).

The cell lines were produced by co-transformation with two mutant Moloney murine leukaemia virus-derived proviral genomes introduced sequentially into NIH 3T3 cells and carrying complementary mutations in the gag-pol or env regions. Each genome contained a deletion of the Ψ sequence essential for the efficient encapsidation of retroviral genomes into virus particles and additional modifications to the 3' end of the provirus. This practically eliminates the potential transfer and/or recombination events that could lead to the production of helper virus (wild-type replication competent virus) by viral producers derived from these cell lines. These properties of the ΨCRE and ΨCRIP packaging cell lines make them particularly useful for *in vivo* gene transfer studies aimed at cell lineage analysis and development of human gene replacement therapy. Both ΨCRIP and ΨCRE can be used to isolate clones that stably produce high titres (10^6 cfu ml^{-1}) of recombinant retroviruses with amphotropic and ecotropic host ranges, respectively.

Culture conditions

Cells are cultured in DMEM supplemented with 2 mM glutamine and 10% FBS or calf serum.

Restrictions

These cell lines are cited in a US patent and must not be used to infringe patent claims.

REFERENCES

Baker PN, Morser J & Burke DC (1980) Effects of sodium butyrate on a human lymphoblastoid cell line (Namalwa) and its interferon production. *Journal of Interferon Research* 1 (1): 71–77.

Bancroft FC & Tashjian AH Jr (1971) Growth in suspension culture of rat pituitary cells which produce growth hormone and prolactin. *Experimental Cell Research* 64(1): 125–128.

Bancroft FC, Levine L & Tashjian AH Jr (1969) Control of growth hormone production by a clonal strain of rat pituitary cells. Stimulation by hydrocortisone. *Journal of Cell Biology* 43: 432–441.

Biliau A, Joniau M & De Somer P (1973) Mass production of human interferon in diploid cells stimulated by poly-I:C. *Journal of General Virology* 19: 1–8.

Danos O & Mulligan RC (1988) Safe and efficient generation of recombinant retroviruses with amphotropic and ecotropic host ranges. *Proceedings of the National Academy of Sciences of the USA* 85: 6460–6464.

Finter NB (1987) Human cells as a source of interferons for clinical use. *Journal of*

Interferon Research 7 (5): 497–500.

Galfre G & Milstein C (1981) Preparation of monoclonal antibodies: strategies and procedures. *Methods in Enzymology* 73B: 3–46.

Gey GO, Coffman WD & Kubicek MT (1952) Tissue culture studies of the proliferative capacity of cervical carcinoma and normal epithelium. *Cancer Research* 12 (4): 264–265.

Glassy MC, Tharakan JP & Chau PC (1988) Serum-free media in hybridoma culture and monoclonal antibody production. *Biotechnology and Bioengineering* 32 (8): 1015–1028.

Gluzman Y (1981) SV40-transformed Simian cells support the replication of early SV40 mutants. *Cell* 23: 175–182.

Graham FL, Smiley J, Russel WC & Nairn R (1977) Characteristics of a human cell line transformed by DNA from human adenovirus type 5. *Journal of General Virology* 36: 59–72.

Halpern MS & Coffman RL (1972) Polymer formation and J chain synthesis in mouse plastocytomas. *Journal of Immunology* 109: 674–680.

Harrison T, Graham F & Williams (1977) Host-range mutants of adenovirus type 5 defective for growth in HeLa cells. *Virology* 77: 319–329.

Hayflick L & Moorhead PS (1961) The serial cultivation of human diploid cell strains. *Experimental Cell Research* 25: 585–621.

Jacobs JP, Jones CM & Baille JP (1970) Characteristics of a human diploid cell designated MRC-5. *Nature* 227: 168–170.

Jain D, Ramasubramanyan K, Gould S, Lenny A, Candelore M, Tota M, Strader C, Alves K, Cuca G, Tung JS, Hunt G, Junker B, Buckland BC & Silberklang M (1991) Large-scale recombinant protein production using insect cell baculovirus expression vector system: antistasin and β-adrenergic receptor. In: Spier RE, Griffiths JB & Meignier B (eds) *Production of Biologicals from Animal Cells in Culture*, pp. 345–350. Butterworth–Heinemann, Oxford.

James DC, Freedman RB, Hoare M, Ogonah OW, Rooney BC, Larionov OA, Dobrovolsky VN, Lagutin OV & Jenkins N (1995) N-Glycosylation of recombinant human interferon-γ produced in different animal expression systems. *Bio/Technology* 13: 592–596.

Jenkins N & Curling EM (1994) Glycosylation of recombinant proteins: problems and prospects. *Enzyme and Microbial Technology* 16: 345–364.

Johnston MD (1980) Enhanced production of interferon from human lymphoblastoid (Namalwa) cells pre-treated with sodium butyrate. *Journal of General Virology* 50: 191–194.

Katinger HWD & Scheirer W (1979) Mass cultivation of mammalian cells in an airlift fermenter. *Developments in Biological Standardization* 42: 111.

Klein F, Ricketts RT, Jones WI, De Armon IA, Temple MJ, Zoon KC & Bridgen PJ (1979) Large scale production and concentration of human lymphoid interferon. *Antimicrobial Agents and Chemotherapy* 15 (3): 420–427.

Köhler G & Milstein C (1975) Continuous cultures of fused cells secreting antibody of predefined specificity. *Nature* 256: 495–497.

Lee EU, Roth J & Paulson JC (1989) Alteration of terminal glycosylation sequences on N-linked oligosaccharides of Chinese hamster ovary cells by expression of beta-galactosidase alpha 2,6-sialyltransferase. *Journal of Biological Chemistry* 264: 13848–13855.

Litwin J (1991) The growth of CHO and BHK cells as suspended aggregates in serum-free medium. In Spier RE, Griffiths JB & Meignier B (eds) *Production of Biologicals from Animal Cells in Culture*, pp. 429–433. Butterworth–Heinemann, Oxford.

Luckow VA & Summers MD (1988) Trends in the development of baculovirus expression vectors (review). *Biotechnology* 6: 47–55.

Lundblad A, Steller, Kabat EA, Hirst JW, Weigert MG and Cohn M (1972) Immunochemical studies on mouse myeloma proteins with specificity for dextran or levan. *Immunochemistry* 9: 535–544.

Madin SH & Darby NB Jr (1958) Established kidney cell lines of normal adult bovine and ovine origin. *Proceedings of the Society for Experimental and Biological Medicine* 98: 574–576.

Mizrahi A, Reuveny S, Traub A & Minai M (1980) Large scale production of human lymphoblastoid (Namalva) interferon. 1. Production of crude interferon. *Biotechnology Letters* 2 (6): 267–271.

Montagnon B, Fanget B, Vinas R, Peyron L, Vincent-Falquet JC & Caudrelier P (1991) Oral polio vaccine (Sabin) produced on large scale Vero culture. In: Spier RE, Griffiths JB & Meignier B (eds) *Production of Biologicals from Animal Cells in Culture*, pp. 695–705. Butterworth–Heinemann, Oxford.

Musgrave SC, Douglas Y, Layton G, Merrett J, Scott MF & Caulcott CA (1993) Optimisation of alpha-interferon expression in Namalwa cells. In: Kaminogawa S, Ametani A & Hachimura S (eds) *Animal Cell Technology: Basic and Applied Aspects*, Kluwer Academic, Dordrecht, The Netherlands. pp. 223–230.

Nelson-Rees WA, Daniels DW & Flandermeyer RR (1981) Cross-contamination of cells in culture. *Science* 212: 446–452.

Oi VT, Morrison SL, Herzenberg LA & Berg P (1983) Immunoglobulin gene expression in transformed lymphoid cells. *Proceedings of the National Academy of Sciences of the USA* 80: 825–829.

Pay TWF, Boge A, Menard FJR & Radlett PJ (1985) Production of rabies vaccine by an industrial scale BHK 21 suspension cell process. *Developments in Biological Standardization* 60: 171–174.

Pharmaceutical Manufacturers Association (PMA) (1988) Biotechnology products in the pipeline. *Bio/Technology* 6: 1004–1007.

Radlett PJ, Pay TWF & Garland AJM (1985) The use of BHK suspension cells for the commercial production of foot and mouth disease vaccines over a twenty year period. *Developments in Biological Standardization* 60: 163–170.

Shen B, Greenfield P & Reid S (1994) Calcium alginate immobilised hybridomas grown using a fluidised-bed perfusion system with a protein-free medium. *Cytotechnology* 14 (2): 109–114.

Shulman M, Wilde CD & Kohler G (1978) A better cell line for making hybridomas secreting specific antibodies. *Nature* 276: 269–270.

Simizu B & Terasima T (eds) (1988) *Vero Cells – Origins, Properties and Biomedical Applications*. Department of Microbiology School of Medicine Chiba University, Chiba, Japan.

Smith GE, Ju G, Ericson BL, Moschera J, Lahm H-W, Chizzonite R & Summers MD (1985) Modification and secretion of human interleukin 2 produced in insect cells by a baculovirus expression vector. *Proceedings of the National Academy of Sciences of the USA* 82: 8404–8408.

Tashjian AH Jr, Yasumura Y, Levine L, Sato GH & Parker ML (1968) Establishment of clonal strains of rat pituitary tumour cells that secrete growth hormone. *Endocrinology* 82: 342–352.

Torado GJ & Green H, (1963) Quantitative studies of the growth of mouse embryo cells in culture and their development into established lines. *Journal of Cell Biology* 17: 299–313.

Torado GJ, Habel K and Green H (1965) Antigenic and cultural properties of cells doubly transformed by polyoma virus and SV40. *Virology* 27: 179–185.

Urlaub G, Mitchell PJ, Kas E, Chasin LA, Funanage VL, Myoda TT & Hamlin J (1986) Effect of gamma rays at the dihydrofolate reductase locus: deletions and inversions. *Somatic Cell and Molecular Genetics* 12 (6): 555–666.

Van der Meer RX, Philippi MC, Romein B, Van der Velden de Groot CAM & Beuvery EC (1993) Towards a strategy for high density cultures of Vero cells in stirred tank reactors. In: Kaminogawa S, Ametani A & Hachimura S (eds) *Animal Cell Technology: Basic and Applied Aspects*, pp. 335–340. Kluwer Academic, Dordrecht, The Netherlands.

Wang G, Zhang W, Jacklin C, Eppstein L and Freedman D (1993) High cell density perfusion culture of hybridoma cells for production of monoclonal antibodies in the Collagen packed bed reactor. In: Kaminogawa S, Ametani A & Hachimura S (eds) *Animal Cell Technology: Basic and Applied Aspects*, pp. 463–70 . Kluwer Academic, Dordrecht, The Netherlands.

Wood DJ & Minor PD (1990) Use of human diploid cells in vaccine production. *Biologicals* 18(2): 143–146.

1.3 MASTER AND WORKING CELL BANKS

The large-scale culture of animal cells is now a recognized approach for the manufacture of biologicals. The regulatory aspects of such work are designed to ensure conformity and safety of such products and to a large degree dictate the way in which cultures are prepared, stored and quality controlled prior to and during the course of manufacture (Center for Biologics Evaluation and Research, 1993). In order to guarantee conformity in cells used for each production run, it is essential to establish a 'seed stock' comprising ampoules of identical cells prepared from the same original culture. This seed stock concept or seed lot system is by no means new; however, its application in cell and tissue culture is unfortunately not universal. The principles required for the preparation and handling of cell stocks in a manufacturing environment are equally relevant in the research laboratory to ensure correct identity, conformity and consistency of cell stocks. Also of general importance is the necessity to minimize the number of passages (and thus population doublings or PDs) occurring before the preparation of stocks. This avoids genetic drift and the loss of differentiated characteristics that may occur during extended serial subculture. The principles of animal cell banking (World Health Organization, 1987) were originally developed for the preparation of stocks of human diploid cells used in vaccine manufacture. It is important to realize that the guidelines established for this work are relevant to all cell stocks used for a wide range of purposes, including diagnostic procedures, research and commercial production. This section aims to give an overview of the general considerations in the production of master and working banks of animal cells.

The ability to create a quality-assured master cell bank (MCB) (Hayflick 1989; Montagnon 1989) with inbuilt consistency and reproducibility guarantees a reserve of original starting material whatever problems may occur with stocks derived at a later date. Once thorough quality control of the MCB is complete, one MCB ampoule is used to set up a larger working cell bank (WCB). (A list of tests for quality control is given in Table 1.3.1., see also sections 1.6–1.9.) In this way all subsequent WCBs derived from the MCB should be identical in terms of PDs. This procedure achieves long-term uniformity of stocks, which cannot be guaranteed by cells derived from improper banking procedures that cause cells used for production to be taken progressively further from the cells of origin (Figure 1.3.1). Thus successive banks may show the effects of cellular aging and, in the event of contamination by other animal cells or microorganisms, the original pure culture will not be retrievable.

Cell and Tissue Culture: Laboratory Procedures in Biotechnology, edited by A. Doyle and J.B. Griffiths.
© 1998 John Wiley & Sons Ltd.

MASTER AND WORKING CELL BANKS

Table 1.3.1 Quality control tests applied to master and working cell banks

Quality control tests applied to master cell banks
　　Viability (at 0 and 24 h)
　　Sterility test – absence of bacteria and fungi
　　Karyology
　　DNA analysis (e.g. DNA fingerprinting)
　　Isoenzyme analysis
　　Mycoplasma tests (e.g. Hoechst stain and culture)
　　Absence of other adventitious agents (e.g. viruses)
　　Stability
　　Idiotypic analysis (hybridomas)

Quality control tests applied to working cell banks
　　Viability
　　Sterility
　　DNA analysis (to confirm consistency with the MCB)
　　Mycoplasma tests
　　Absence of other adventitious agents (e.g. viruses)
　　Stability
　　Idiotypic analysis (hybridomas)

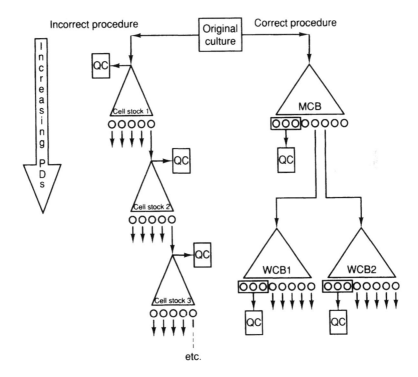

Figure 1.3.1 Cell banking procedure (QC = quality control).

Usually limits are set on the number of PDs the cells are permitted to undergo before halting the production run and recommencing with a further ampoule from the WCB. With diploid cells these parameters are ultimately delineated by the onset of senescence (i.e. failure to continue replication), and a limit for PDs is chosen somewhere short of the point at which the rate of cell doubling decreases. In the case of one human diploid cell line, MRC–5, it has now been agreed to increase the permissible upper PD limit from PD 30 to PD 40 due to improved knowledge on growth parameters (Wood & Minor 1990). In the case of heteroploid cells, the limits are defined but not as stringently controlled as with diploid cells. Furthermore, from a regulatory point of view it is more desirable to derive the whole of each product batch from a single WCB ampoule.

SCALE AND COMPOSITION OF CELL BANKS

The size of a cell bank is dependent upon the scale of operation, frequency of use and replicative capacity of the cell line in question. Typically, an MCB might contain 20–50 ampoules and a WCB 100–200 ampoules. Larger banks may be prepared for stable continuous cell lines and this is a major advantage of this type of cell line in comparison with finite cell lines of limited replicative capacity.

The number of cells per ampoule should be predetermined and validated for the ability of an ampoule of cells to regenerate a representative culture. The time taken to expand from a single ampoule to a larger culture volume will be dependent on the number of cells per ampoule, but care should be taken to avoid cryopreserving cells at excessively high densities (i.e. $>5 \times 10^6$ cells ml^{-1}), which may result in low numbers of viable cells on recovery. Liquid nitrogen storage refrigerators, even with sophisticated filling devices and alarms, are not immune to disaster, so dividing ampoule banks between facilities at different locations is an essential precaution.

EXTENDED CELL BANK

From the regulatory point of view, the characterization requirements for the cells will depend on the bulk culture process to be used. This will define the number of PDs (set within the limits discussed above) that the cells will undergo. An additional regulatory requirement is that the cells should be passaged beyond the anticipated PD level achieved at the end of each production run. This procedure should therefore establish the stability of cells well beyond the normal working limits. Therefore, after additional passaging, an extended cell bank is prepared for characterization procedures to be repeated.

THE CELL BANKING ENVIRONMENT AND PROCEDURES

Whilst the fundamental principles of cell banking are readily applied to the preparation of all animal cell stocks, the standards applied in the preparation of cell banks vary widely, depending on the intended use. Banks prepared for short-term research projects may receive only basic levels of environmental controls and record-keeping. Cells intended for involvement in product license applications and/or patent applications will need to be handled under conditions of strict and detailed documentation designed to enable traceability of all procedures and reagents. In order to ensure the absence of contaminating microorganisms in such high-grade cell banks, the air quality (in terms of freedom from airborne particles) may need to be maintained by appropriate filtered ventilation and regular testing for particles and organisms.

The major standards that can be applied to manufacturing processes are good laboratory practice (GLP) (Department of Health, 1989) and Good Manufacturing Practice (GMP) (Medicines Control Agency, 1997). Good laboratory practice provides an audited international standard that 'is concerned with the organizational processes and the conditions under which laboratory studies are planned, performed, monitored, recorded and reported' (OECD, 1982; Department of Health, 1989). Strictly speaking, GLP is not applicable to cell banking (but solely to safety testing), but application of its principles may be required as part of further procedures where the preparation of the cell products will be recorded and tested to this standard. Good manufacturing practice is to cell banking, because it is intended to ensure that 'products are consistently manufactured to a quality required by their intended use' (Sharp, 1983). Cell banking at this level not only requires the detailed documentation required under GLP, but will also involve careful environmental monitoring (as previously mentioned). single-use facilities (i.e. one cell line only per laboratory) and extensive validation of critical banking procedures. The most important validating procedures for cell banks, known as 'integrity tests', identify the capability of named operators at preparing cryopreserved banks of cells without any ampoule becoming contaminated with microorganisms. Such testing usually involves substituting a bacteriological growth medium for the cell suspensions in a procedure exactly simulating the preparation of cells for cryopreservtion. The simulated bank is then incubated at 37°C and at room temperature to expose contamination by the appearance of turbidity in the broth medium. In GMP facilities it is also important to prove that harmful material cannot persist in the environment between projects and affect subsequent banks. Thus decontamination, cleaning and fumigation procedures must be validated for their efficiency in removal of potentially deleterious agents or materials and also, indeed, the previous product.

All of the standards of working practice mentioned are regulated by official national and international bodies, which means that in order to claim compliance with any of these standards, the laboratory must be audited and accredited (GLP) or licensed (GMP) by a nationally authorized body.

FEATURES REQUIRED FOR GLP PROCEDURES

Where high-quality cell banking procedures are to be carried out, a range of laboratory features and activities must be recorded. The three standards discussed above have many of these in common, and in the case of GLP these include:

1. Job descriptions, curriculum vitae and training records for all scientific staff.
2. Maintenance of up-to-date records of studies, their nature, test systems used, completion dates and names of study directors.
3. Study plans and standard operating procedures (SOPs) with associated batch records, quality control reports and certificates of analysis.
4. Facility organization chart and site plan.
5. Details of servicing and maintenance of equipment.
6. Policies for health surveillance and training of staff.
7. List of study directors and a record of the procedure adopted for their appointment.

In order to ensure that cell banking procedures are performed in the prescribed way, with all reagents traceable to source materials, each with a certificate of analysis, a full set of SOPs is required that includes specimen record sheets. An example of the way in which SOPs can control cell banking procedures is given in Figure 1.3.2.

In terms of quality management of manufacturing production, a more recent development is the harmonization of several European quality standards into what is now known as BS EN ISO 9001. This standard can be used for any manufacturing or service activity and is readily applicable to cell banking and to research and development. This standard is also able to provide the background for cell culture in which GLP and GMP standards can be applied.

CONCLUSION

The seed stock principle for the preparation and quality control of MCBs and WCBs is an essential part of cell banking procedures, regardless of the intended use for the cells. The size of each bank and the level of quality control used are generally dependent on individual requirements and facilities. However, where a cell line or its products is to be involved in procedures under the jurisdiction of regulatory authorities, the cell banking procedures, characterization and quality control must address specific requirements for procedures and documentation. Finally, the intended use of cells or their product must determine the level of quality assurance required, and the person responsible for exploitation of the cells or product must identify the appropriate quality standard and select an agency accredited to that standard.

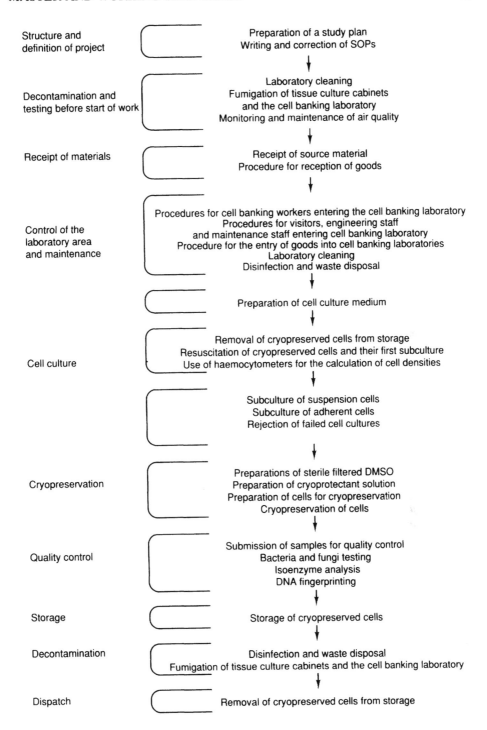

Figure 1.3.2 Flow diagram of cell banking and associated SOPs.

REFERENCES

Center for Biologics Evaluation and Research (1993) *Points to Consider in the Characterization of Cell Lines Used to Produce Biologicals*. Food and Drugs Administration, Bethesda, USA.

Department of Health Guidelines on Good Laboratory Practice: The United Kingdom Compliance Programme (1989) Department of Health, London.

Hayflick L (1989) History of cell substrates for human biologials. *Developments in Biological Standardization* 70: 11–26.

Medicines Control Agency (MCA) (1997) Rules and guidance for pharmaceutical manufacturers and distributors. The Stationary Office, PO Box 276, London, UK.

Montagnon BJ (1989) Polio and rabies vaccines produced in continuous cell lines: reality for Vero cell line. *Developments in Biological Standardization* 70: 27–47.

OECD (1982) *Good Laboratory Practice in the Testing of Chemicals – Final Report of the OECD Expert Group on Good Laboratory Practice*. Organization for Economic Cooperation and Development, Paris.

Wood DJ & Minor PD (1990) Meeting report: use of human diploid cells in production. *Biologicals* 18: 143–146.

World Health Organization (1987) *Acceptability of Cell Substrates for Production of Biologicals*, Technical Report Series 747. WHO, Geneva.

1.4 IDENTITY TESTING – AN OVERVIEW

The consequences of using a cell line whose identity is not that claimed by the provider are clearly very serious. Failure readily to identify an occurrence of cell line cross-contamination or switching during tissue culture procedures could completely invalidate a body of research or completely abort a production process. Cell line authentication is therefore an essential requirement for all cell culture laboratories and should be carried out at both the earliest passages of cultures and at regular intervals thereafter (Hay, 1988). The occurrence of cross-contamination is not merely anecdotal; documented cases have been widely reported (Nelson-Rees et al., 1981; van Helden et al., 1988). Some earlier reports indicated that levels of cross-contamination may exceed 30% of cultures tested (Halton et al., 1983). Preferential inclusion of suspect cultures in these reports means that this figure probably represented an overestimation of the problem. Nevertheless, experience at the ECACC in the authentication of cell stocks from a wide range of laboratories indicates that cross-contamination is a neglected problem. The classic example is that of HeLa contamination (Nelson-Rees et al., 1981). Early studies used conventional cytogenetic analysis in association with isoenzyme analysis to verify the species of origin. This is particularly easy in the case of the HeLa cell line because it has characteristic cytogenetic markers, and for isoenzyme analysis type B rather than the more usual type A glucose-6-phosphate dehydrogenase is expressed. A large number of isoenzyme tests are usually required for specific identification. A summary of the standard authentication techniques is given in Table 1.4.1, but this list is by no means exhaustive.

CYTOGENETIC ANALYSIS

Microscopic examination of the chromosomal content of a cell line provides a direct method of confirming the species of origin and allows the detection of gross aberrations in chromosome number and/or morphology. Cytogenetic analysis is very useful for specific identification of cell lines with unique chromosome markers and has proved useful to differentiate cell lines that were apparently identical by isoenzyme analysis. In one study of 47 cell lines reported by O'Brien et al. (1980), two cell lines could not be differentiated by eight separate enzyme tests but were readily distinguished by karyology. However, it should be borne in mind that very careful interpretation in the light of considerable experience would be required to differentiate cell lines of normal karyotype beyond the level of species. In addition,

---

Cell and Tissue Culture: Laboratory Procedures in Biotechnology, edited by A. Doyle and J.B. Griffiths.
© 1998 John Wiley & Sons Ltd.

Table 1.4.1 Comparison of key features of authentication techniques

	Cytogenetic analysis	Isoenzyme analysis	DNA fingerprinting/profiling
Determination of species	✓[a]	✓	
Rapid identification of a cell line			✓
Detection of cell line variation	✓		✓
Recognized by regulatory authorities	✓	✓	?[b]

[a] ✓ = useful applications.
[b] ? = under consideration.

many species have not been studied extensively using karyology and therefore the prevalence of chromosome markers in natural populations, and hence in cell lines, may be unknown. Cytogenetic analysis is time consuming and, to some extent, subjective. In most routine and research laboratories handling a wide range of cell lines, full characterization and regular monitoring of identity using the above technique would be impracticable. However, developments in flow cytometric analysis of chromosomes may enable routine chromosome analysis in the future.

ISOENZYME ANALYSIS

Isoenzyme analysis is useful for the speciation of cell lines and for the detection of contamination of one cell line with another. The method utilizes the property of isoenzymes having similar substrate specificity but different molecular structures. This affects their electrophoretic mobility. Thus each species will have a characteristic mobility pattern of isoenzymes. While the species of origin of a cell line can usually be determined with only two isoenzyme tests (lactate dehydrogenase and glucose-6-phosphate dehydrogenase), specific identification of a cell line would require a larger battery of tests (Halton et al., 1983). A system in which eight isoenzyme activities were investigated offers a useful compromise (O'Brien et al., 1980). This procedure retained the advantage of rapid testing while also giving a useful level of specificity for identification purposes. Certain proteins characterized in isoenzyme analysis may undergo post-translational modification, which varies with the cell type. This offers additional characteristics for identification that can also act as markers for tissue of origin or developmental stage (Wright et al., 1981). However, in analysis of enzyme systems with multi-allelic products, low-level expression of some components may lead to unreliable detection and hence mis-identification. In conclusion, the use of a wide range of isoenzyme tests for accurate identification of cell lines requires a detailed knowledge of each isoenzyme system in each species to be studied.

DNA FINGERPRINTING AND DNA PROFILING

Multilocus DNA probes such as those of Jeffreys et al. (1985), oligonucleotides homologous to very simple sequence repeats (Ali et al., 1986) and the DNA sequence of the M13 protein III gene (Ryskov et al., 1988) interact with a widely dispersed range of loci throughout the genomes of many organisms and produce unique fingerprints in Southern blot analysis of cell line DNA. An alternative approach is to analyse single loci bearing variable numbers of tandem repeats (VNTRs), which produce simple DNA profiles (as opposed to fingerprints) comprising one or two bands. The key advantages of this technique in identification are, firstly, the apparent simplicity of scoring such profiles and, secondly, that methods based on the polymerase chain reaction are available (Tautz, 1989; Jeffreys et al., 1991). Probes for single loci are used under highly specific conditions, and, consequently, each one may be useful in only a single species, unlike multilocus fingerprinting. In visualizing many loci, DNA fingerprinting techniques enable recognition of changes in a cell line that might go unnoticed when using single-locus methods. Multilocus DNA fingerprinting can therefore identify cell line variation (Thacker et al., 1988) and cross-contamination (van Helden et al., 1988), and can be used to confirm stability and consistent quality in cell stocks from diverse species. Furthermore, the development of a library of fingerprints from original documented material may help in the validation of suspect cell banks by comparison against the definitive version.

The standard techniques for cell authentication (cytogenetics and isoenzyme analysis) have been invaluable to cell biologists wishing to establish cell identity and to exclude cross-contamination. However, the availability of an ever-increasing number and diversity of cell lines indicates the need for rapid and simple techniques that also enable highly specific cell line identification. In a routine quality control setting where diverse cell lines are involved, current cytogenetic methods are impractical, whereas simplified isoenzyme procedures can offer at least species identification. DNA fingerprinting and profiling are important and timely additions to the cell biologists' battery of authentication techniques. In particular, multilocus DNA fingerprinting in a single test provides specific cell line identification for many species, combined with a screen for cross-contaminated or switched cultures. All the authentication techniques described should be viewed as complementary and each one will have areas in which it is especially useful. However, molecular genetic techniques are proving extremely powerful and sensitive for routine checks on the consistency of cell cultures.

REFERENCES

Ali S, Muller CR & Epplen JT (1986) DNA fingerprinting by oligonucleotide probes specific for simple repeats. *Human Genetics* 74: 239–243.

Halton DM, Peterson WD & Hukku B (1983) Cell culture quality control by rapid isoenzymatic characterisation. *In Vitro* 19: 16–24.

Hay RJ (1988) The seed stock concept and quality control for cell lines, *Analytical Biochemistry* 171: 225–237.

Jeffreys AJ, Wilson V & Thein SL (1985) Individual specific 'fingerprints' of human DNA. *Nature* 316: 76–79.

Jeffreys AJ, McLeod A, Tamaki K, Neil DL & Monckton DG (1991) Minisatellite repeat coding as a digital approach to DNA typing. *Nature* 354: 204–210.

Nelson-Rees WA, Daniels DW & Flandermeyer RR (1981) Cross-contamination of cells in culture. *Science* 212: 446–452.

O'Brien SJ, Shannon JE & Gail MH (1980) A melocular approach to the identification and individualization of human and animal cells in culture: isoenzyme and alloenzyme genetic signatures. *In Vitro* 16: 119–135.

Ryskov AP, Jincharadze AG, Prosnyak AG, Ivanov PL & Limborskaya (1988) M13 phage as a universal marker for DNA fingerprinting of animals, plants and microorganisms. *FEBS Letters* 233: 388–392.

Tautz D (1989) Hypervariability of simple sequences as a general source of polymorphic DNA markers. *Nucleic Acids Research* 17: 6469–6471.

Thacker J, Webb MBT & Debenham PG (1988) Fingerprinting cell lines: use of human hypervariable DNA probes to characterize mammalian cell cultures. *Somatic Cell and Molecular Genetics* 14: 519–525.

van Helden PD, Wild IJF, Albrecht CF, Theron E, Thornley AL & Hoal-van Helden EG (1988) Cross-contamination of human oesophageal squamous carcinoma cell lines detected by DNA fingerprinting analysis. *Cancer Research* 48: 5660–5662.

Wright WC, Daniels WP & Fogh J (1981) Distinction of seventy one cultured human tumour cell lines by polymorphic enzyme analysis. *Journal of the National Cancer Institute* 66: 239–247.

1.5 DNA FINGERPRINTING

The importance of DNA fingerprinting in the quality control of cell lines is increasingly recognized due to the advantages that this technique has in comparison with the standard methods of isoenzyme analysis and cytogenics. Described here are two widely applicable multilocus fingerprinting methods using M13 phage DNA (Vassart et al., 1987) and the Alec Jeffreys probes 33.15 and 33.6 (Jeffreys et al., 1985a,b). The Jeffreys method has the advantage of a centralized quality control system already established by Zeneca Cellmark Diagnostics (Abingdon, Oxon., UK). The methodology used can be summarized in the following stages:

1. Extraction of genomic DNA from cells.
2. Restriction digests of genomic DNA.
3. Electrophoretic separation and fixation of genomic DNA fragments.
4. Probe radiolabelling.
5. Hybridization and stringency washing.
6. Autoradiography.

Much of the methodology for DNA preparation, and Southern blot analysis used for DNA fingerprinting, is published elsewhere (Honma et al., 1993) and the reader should refer to this for further details (particularly for steps 1–3 above) and to the procedures supplied with the Jeffreys multilocus probes).

PRELIMINARY PROCEDURE: PROBE PREPARATION

Probes 33.15 and 33.6

Materials and equipment

- Probe/primer mixture (Zeneca Cellmark Diagnostics, Abingdon, UK)
- Labelling buffer (as recommended by probe manufacturer)
- $\alpha[^{32}P]dGTP$
- 1 U μl^{-1} Klenow enzyme (Boehringer, Mannheim)

The Jeffreys probes 33.6 and 33.15 are available for research purposes only, in a high-quality form, with primers added ready for labelling with ^{32}P by random primer extension. A recommended primer extension method of labelling is provided by the manufacturer, which involves a preparation time of less than 2 h.

M13mp8 DNA probe

Materials and equipment

- 30 ng of M13mp8 DNA
- Multiprime labelling kit (Amersham International)

M13 phage DNA, in the form of SS M13mp8 DNA can be labelled by random primer extension using the Multiprime kit (Amersham International) according to the manufacturer's instructions.

Probe purification

Unincorporated nucleotides can be removed by NACS prepac (Life Technologies Ltd, Paisley, Scotland) or NICK columns (Pharmacia) according to the manufacturer's instructions. The reaction mixture is reboiled and chilled prior to addition to the hybridization solution.

PROCEDURE: HYBRIDIZATION

Reagents and solutions

- 33.6 and 33.15 multilocus method: the reagents are described in procedures supplied by the probe manufacturer (Zeneca Cellmark Diagnostics)
- M13 multilocus method. Prehybridization and hybridization solution: 0.263 M disodium hydrogen phosphate, 1mM EDTA (pH 8.0), 7% sodium dodecylsulphate (SDS) and 1% bovine serum albumin (Fraction V, Sigma)
- Stringency wash solution: ×2 standard saline citrate (SSC)/0.1% SDS

Materials and equipment

- Prehybridization and hybridization solutions
- One aliquot of freshly labelled probe, boiled and chilled
- Stringency wash solution
- Shaking water-bath
- X-ray film (Genetic Research Instrumentation Ltd, Dunmow, Essex, UK)
- Autoradiography cassette with intensifying screen (Genetic Research Instrumentation Ltd)
- Hand-held Geiger-Muller radiation monitor

A variety of hybridization conditions have been described for different fingerprinting probes (Jeffreys *et al.*, 1985a, b; Chen *et al.*, 1990; Vassart *et al.*, 1987). The method of Church and Gilbert (1984) is particularly useful due to the simplicity of the hybridization solution.

1. Prehybridize the Southern membranes for a few hours or overnight in a sealed polythene box containing sufficient prehybridization solution to allow free

movement of the membranes. Prehybridization can be carried out at 55°C (for M13) or 62°C (Jeffreys probes) in a shaking water-bath.
2. Replace the prehybridization solution with hybridization solution containing approximately 10^6 counts min^{-1} ml^{-1} of purified, labelled DNA probe. Allow hybridization to proceed at 55°C (for M13) or 62°C (Jeffreys probes) overnight with shaking.
3. Carry out stringency washes at 55°C (for M13) or 62°C (Jeffreys probes) with shaking. At least one change of wash solution should be made and washing continued until areas of the membranes without any fixed DNA give a background reading on a hand-held Geiger-Muller radiation monitor (Mininstruments, Burnham-on-Crouch, Essex, UK).
4. Wrap each membrane individually in film (e.g. 'Saranwrap', Genetic Research Instrumentation Ltd).
5. Fix the membranes in autoradiography cassettes and place two X-ray films over them. Keep the sealed cassettes at −80°C (the period of exposure required is determined by developing top film after 12 h of exposure).
6. Label the final autoradiographs and store in a dry, dust-free envelope.

Note: Hybridized nylon membranes should not be allowed to dry out and can be dehybridized, for repeat probing, by two 30-min washes at 45°C, firstly in 0.4 M NaOH and secondly in 0.2 M Tris (pH 7.5) plus 0.1% SDS and × 0.1 SSC (as recommended by the manufacturer of Hybond N membrane. Amersham International).

DISCUSSION

Background information

In the quality control of cell lines, isoenzyme analysis and cytogenetic analysis have, until recently, been the standard accepted techniques. However, to ensure specific identification of a cell line, both techniques involve extensive and time-consuming investigations. Attempts to simplify such investigations inevitably limit their ability to differentiate cell lines. DNA fingerprinting offers a unique way of identifying cell lines and their cross-contamination by a single test. This technique identifies a cell line by visualizing the structure of the repetitive component of genomic DNA, which is extremely variable between individuals.

A range of fingerprinting methods are available but they generally fall into one of two groups: single-locus and multilocus methods.

Single-locus methods identify polymorphism at specific genetic loci (Honma *et al.*, 1993). These may be restricted in their degree of polymorphism and are often applicable to only a narrow range of species or strains of a particular animal. However, in multilocus methods the molecular probes are used at lower stringencies and detect polymorphism at a number of related loci. In this way more information is given, increasing the potential for differentiating cell lines whilst using a single probe test that can be useful over a wide species range.

Methods using DNA sequences from the M13 phage are used very successfully in diverse organisms. Oligonucleotide probes for the simple repeats of micro-satellite DNA are also extensively used.

The multilocus method discovered by Alec Jeffreys has proved useful in a wide range of animal and plant species and is currently one of the most widely used fingerprinting methods in population studies. This method, using probes 33.6 and 33.15, is the only fingerprinting method for which a formal international quality control system is operated (Zeneca Cellmark Diagnostics), and offers a unique opportunity for the standardization of cell line quality control procedures in widely distributed laboratories.

Critical parameters

The condition of cell cultures used for fingerprinting, and care in the preparation of genomic DNA, are critical features in achieving readily interpreted and reproducible results. It is advisable to ensure that culture samples to be compared are standardized in terms of culture conditions, growth state and method of sampling. When preparing DNA, strict adherence to the established procedure is important, to maintain a reliable fingerprinting system. Samples of cells or DNA kept as controls for long-term quality assurance procedures are best stored at $-80°C$ or in liquid nitrogen. Inappropriately stored samples may suffer from degradation of DNA, which, even when not particularly marked, will cause smearing of fingerprint bands, will increase the background signal in each lane and can make detailed interpretation very difficult. The quality of prepared DNA should be checked by mini-gel analysis of a small sample to identify a distinct, undegraded, high-molecular-weight band of genomic DNA. Quantitation of DNA is particularly critical for multilocus fingerprinting, and fluorimetric assay (e.g. using a TK100 fluorimeter, Hoeffer, Newcastle, Staffordshire, UK) provides a high level of accuracy and sensitivity and does not suffer interference from RNA. Restriction digests with HinfI and HaeIII should be analysed by mini-gel electrophoresis to ensure the absence of visible quantities of high-molecular-weight DNA representing partial digestion.

Troubleshooting

Care taken in the preparation of the gel, in electrophoresis and during all subsequent stages of gel handling to avoid distortion, should be repaid by straight and parallel lanes of equal molecular weight migration. Uniform migration of consecutive gels is important to aid comparisons of fingerprints between different membranes. The use of a standard DNA sample on each side of the gel, in addition to molecular weight markers, will aid interpretation considerably.

Occasional samples may show a shift of all bands, due to a slight change in the migration rate of all DNA fragments. This effect is usually easily recognized if the sample was run in parallel with other related samples (i.e. showing a high degree of band sharing). The problem can also be overcome by the use of an internal molecular weight control in each lane (e.g. lambda HindIII markers,

Sigma), which can be detected after fingerprint hybridization by rehybridizing with labelled lambda phage DNA.

Reproducibility of electrophoresis and Southern blotting is essential and can be maintained by careful quality control tests of the pH and conductivity of key reagents (e.g. electrophoresis buffer and SSC). The quality of fingerprints will be affected by poor Southern transfer of DNA. Therefore, transfer should always be checked by restaining blotted gels in electrophoresis buffer plus ethidium bromide (0.5 mg l^{-1}) for 30 min. Inspection of the restained gel on a UV-transilluminator should reveal no signs of residual DNA molecular weight markers.

Another very important contribution to the achievement of clearly interpretable fingerprints is the use of high-quality preparations of labelled probe. The probe DNA should have a high degree of purity. Separation of labelled probe from unincorporated radioactive nucleotides is particularly important to obtain good-quality results with a low background signal. Quality control of the probe's purity and level of incorporation of labelled nucleotide (e.g. by comparison of total and precipitable radioactivity in probe preparations) is vital to ensure that probe preparations used are of an acceptable standard. Ideally, probes should always be used on the day of radiolabelling to avoid degradation. Using probes produced commercially under conditions of quality assurance (e.g. probes 33.6 and 33.15) means that many of the problems encountered in producing high-quality labelled probe can be avoided.

Time considerations

In order to produce fingerprints of uniform intensity, autoradiography exposure times may have to be varied slightly to compensate for small variations between hybridized Southern membranes. The time required to achieve optimal fingerprint exposure is best predicted by the use of two X-ray films. Development of the first film after 12 h permits a fairly accurate assessment of the total exposure time required. In addition the radioactivity on each membrane can be monitored to indicate exposure time. These variables may lead to the appearance of faint additional bands in the fingerprints from hybridized membranes given longer exposure times. This, along with the possibility of minor variations in fragment migration between gels, demonstrates the importance of comparing related samples on a single membrane.

Preparation of DNA and restriction digests normally takes 2–3 days, with Southern blot procedures requiring a further overnight step (unless vacuum blotting or electroblotting techniques are used). Probe labelling, hybridization and setting up autoradiographs are completed within 2 days and autoradiographic exposure should then take between 1 and 5 days, depending on the method used. Most incubation times may not be critical, but once a reliable procedure has been developed, casual deviation from the established procedures should be strictly avoided in order to provide reproducible fingerprint quality.

Literature review

The technique of DNA fingerprinting was first established by Jeffreys *et al.* (1985a, b), who discovered mini-satellite probes with the capability to produce unique human DNA fingerprints. The alleles identified were found to be distributed among offspring in a Mendelian fashion. This technique is now widely used in population studies of various animal and plant species. Multilocus fingerprinting using gene sequences from the M13 phage genome was developed by Vassart et al. (1987) and has been used in diverse organisms and in cell line authentication (Devor et al., 1988). The Jeffreys probes have been used in cell culture technology to identify cross-contamination and genetic changes in cell lines from humans and other species (Thacker et al., 1988; van Helden et al., 1988; Armour et al., 1989). Probes 33.15 and 33.6 have been used in the analysis of closely related cell lines (Gilbert et al., 1990) and these probes have also been useful in the quality control of cell stocks for a wide species range (Stacey et al., 1991). Techniques for analysing microsatellite DNA using oligonucleotides (Tautz, 1989) and PCR (Stacey et al., 1997) may also be of use in the analysis of cell line DNA and have been used in a wide variety of species.

REFERENCES

Armour JA, Tael I, Thein SL, Fey MF & Jeffreys AJ (1989) Analysis of somatic mutations at human minisatellite loci in tumours and cell lines. *Genomics* 4: 328–334.

Chen P. Hayward NK, Kidson C & Ellem KAO (1990) Conditions for generating well-resolved human DNA fingerprints using M13 phage DNA. *Nucleic Acids Research* 18: 1065.

Church GM & Gilbert W (1984) Genomic sequencing. *Proceedings of the National Academy of Sciences of the USA* 81: 1991.

Devor EJ, Ivanovich AK, Hickok JM & Todd RD (1988) A rapid method of confirming cell line identity: DNA fingerprinting and minisatellite probe from M13 bacteriophage. *Biotechniques* 6: 200–202.

Gilbert DA, Reid YA, Mitchell GH, Pee D, White C, Hay RJ & O'Brien SJ (1990) Applications of DNA fingerprints for cell line individualisation, *American Journal of Human Genetics* 47: 499–514.

Honma M, Stacey G & Mizusawa H (1993) DNA profiling with polymorphic DNA markers. In: *Cell and Tissue Culture: Laboratory Procedures*, pp. 9A: 5.1–5.13. Wiley, Chichester.

Jeffreys AJ, Wilson V & Thein SL (1985a) Hypervariable minisatellite regions in human DNA. *Nature* 314: 67–73.

Jeffreys AJ, Wilson V & Thein SL (1985b) Individual specific DNA fingerprints of human DNA. *Nature* 316: 76–79.

Stacey GN, Bolton BJ & Doyle A (1991) DNA fingerprinting in the quality control of cell banks. In: Burke T, Dolf G, Jeffreys AJ & Wolff R (eds) *DNA Fingerprinting Approaches and Applications,* Proceedings of the First International Conference on DNA Fingerprinting, pp. 361–370. Birckhauser, Berlin.

Stacey GN, Hoelzl H, Stephenson JR & Doyle A (1997) Authentication of animal cell cultures by direct visualisation of DNA, Aldolase gene PCR and isoenzyme analysis. *Biologicals* 25: 75–83.

Tautz D (1989) Hypervariability of simple sequences as a general source for polymorphic DNA marker. *Nucleic Acids Research* 17: 6463–6471.

Thacker JM, Webb MJK & Debenham PG (1988) Fingerprinting cell lines: use of human hypervariable DNA probes to characterise mammalian cell cultures. *Somatic Cellular and Molecular Genetics* 14: 519–525.

van Helden PD, Wiid IJ, Albrecht CF, Theoron E, Thornley AL & Hoal-van Helden EG (1988) Cross-contamination of human oesophageal squamous carcinoma cell lines detected by DNA fingerprinting analysis. *Cancer Research* 48: 5660–5662.

Vassart G, Georges M, Monsieur R, Brocas H, Lecarre AS & Christophe D (1987) A sequence in M13 phage detects hypervariable minisatellites in human and animal DNA. *Science* 235: 683–684.

1.6 DETECTION OF MYCOPLASMA

Mycoplasma is a generic term given to organisms of the order Mycoplasmatales that can infect cell cultures. Those that belong to the families Mycoplasmataceae (*Mycoplasma*) and Acholeplasmataceae (*Acholeplasma*) are of particular interest.

The first observation of mycoplasma infection of cell cultures was by Robinson *et al.* (1956). The incidence of such infection has since been found to vary from laboratory to laboratory. At present 12% of cell lines received by the ECACC are infected but this may be an uncharacteristically low figure because many lines are screened for mycoplasma prior to deposition.

Mycoplasmas differ from other prokaryotes by their lack of a cell wall. They are unable to produce even precursors of bacterial cell wall polymers, unlike L-forms of bacteria that can do so under the right environmental conditions. Their size is another distinguishing feature; they are the smallest self-replicating prokaryotes, with coccoid forms of only 0.3 μm diameter capable of reproduction. Their genome size is approximately one-sixth that of *Escherichia coli*.

The importance of mycoplasma detection in cell cultures should not be underestimated. The concentration of mycoplasmas in the supernatant can be typically in the region of 10^6–10^8 mycoplasmas ml^{-1}. Additionally, mycoplasmas will cytadsorb to the host cells. They do not necessarily manifest themselves in the manner of most bacterial or fungal contaminants, e.g. pH change or culture turbidity. It is important therefore to adopt an active routine detection procedure. Mycoplasmas have been shown to elicit various effects, including the following:

- Induction of chromosome aberrations (Aula & Nichols, 1967)
- Induction of morphological alterations, including cytopathology (Butler & Leach, 1964)
- Interference in the rate of growth of cells (McGarrity *et al.*, 1980)
- Influence of nucleic acid (Levine *et al.*, 1968) and amino acid (Stanbridge *et al.*, 1971) metabolism
- Induction of membrane alteration (Wise *et al.*, 1978) and even cell transformation (MacPherson & Russel, 1966)

Mycoplasma contamination is usually caused by any of five common species. The organisms and their natural hosts are *M. hyorhinis* (pig), *M. arginini* (cow), *M. orale* (man), *A. laidlawii* (cow), and *M. fermentans* (man).

Acholeplasma alone has no sterol requirement.

A range of assay techniques are available for the detection of mycoplasma contamination. These include staining, culture, DNA probes and co-cultivation. To remove the risk of false positives and false negatives, two methods at least should be employed. The ECACC recommends the use of enrichment broth and agar culture and Hoechst 33258 DNA staining.

PROCEDURE: DNA STAIN

 This stain is heat and light sensitive. The toxic properties of Hoechst 33258 are unknown, and therefore gloves should be worn when handling the powder or solution.

Reagents and solutions

Use analytical grade (Analar) reagents and distilled water.

Carnoy's fixative

For each preparation:

- Methanol, 3ml
- Acetic acid (glacial), 1 ml
- Prepare fresh as required

Hoechst stain stock solution (100 ml)

bisbenzimide-Hoechst 33258:

- Add 10 mg of Hoechst 33258 to 100 ml of water
- Filter-sterilize using a 0.2 μm filter
- Wrap the container in aluminium foil and store in the dark at 4°C

Hoechst stain working solution (50 ml)

Prepare fresh each time as necessary by adding 50 μl of stock solution to 50 ml of water.

Mountant

Take 22.2 ml of citric acid (0.1 M) and 27.8 ml of disodium phosphate (0.2 M) and:

- Autoclave
- Mix with 50 ml of glycerol
- Adjust pH to 5.5
- Filter-sterilize and store at 4°C

Agar media

Mycoplasma agar base (80 ml) (Oxoid, Basingstoke, Hants, UK): add 2.8 g of agar base to 80 ml of water, mix and autoclave at 15 psi for 15 min. Prepare fresh as necessary.

Pig serum (10 ml)

Dispense in universal containers in 10 ml aliquots. Heat to 56°C for 45 min and store at −30°C.

Yeast extract (10 ml) (Oxoid, Basingstoke, Hants, UK)

Add 7 g to 100 ml of water, mix and autoclave at 15 psi for 15 min. Dispense in universal containers in 10 ml aliquots and store at 4°C.

Agar preparation

Prepared medium must be used within 10 days.

1. Autoclave agar at 15 psi for 15 min. Cool to 50°C and mix with the other constituents (which have been warmed to 50°C).
2. Dispense 8 ml per 5 cm diameter Petri dish. Store at 4°C in sealed plastic bags.

Broth media

Mycoplasma broth base (70 ml) (Oxoid, Basingstoke, Hants, UK): Add 2 g of broth base to 70 ml of water, mix and autoclave at 15 psi for 15 min.

Horse serum (Tissue Culture Services, Botolph Claydon, Buckingham, UK)

Dispense in universal containers in 20 ml aliquots, do not heat inactivate and store at −30°C.

Yeast extract (Oxoid, Basingstoke, Hants, UK)

Prepare as for agar media.

Broth preparation

Mix constituents, dispense in glass vials in 1.8 ml aliquots and store at 40°C. Complete medium may be stored without deterioration for several weeks.

Materials and equipment

- Carnoy's fixative
- Methanol, 3 ml

- Acetic acid
- Hoechst stain stock solution
- Hoechst stain working solution (prepare fresh as required)
- Mountant
- Fluorescence microscope equipped with epifluorescence (e.g. Zeiss) and 340–380 nm excitation filter and 430 nm suppression filter

Note: Both methods used for the detection of mycoplasma have the following general principles in common. The cells to be tested should, before testing, complete at least two passages in antibiotic-free media. Infection may be hidden by the presence of antibiotics. Cell cultures tested from frozen ampoules should undergo at least two passages because cryoprotectants may also mask infection.

1. Using a routine method of subculture, harvest adherent cells, e.g. trypsin, and resuspend in the original culture medium to a concentration of about 5×10^5 cells ml^{-1}.
2. Test the suspension lines direct from the culture at about 5×10^5 cells ml^{-1}. *Note:* Experience of working with any particular cell line should remove the absolute necessity for an accurate cell count. An adequate number of cells should be added to dishes so that a semi-confluent spread of cells on the coverslip is obtained at the time of observation (at 1 day and 3 days post-incubation). Cell overgrowth would make interpretation difficult. If it is thought that after 3 days' growth the media would be exhausted, the cells should be resuspended in a mixture of their original medium and fresh culture media.
3. Add 2–3-ml test cells to each of two tissue culture dishes containing glass coverslips.
4. Incubate at 36 ±1°C in a humidified 5% CO_2/95% air atmosphere for 12–24 h.
5. Remove one dish and leave the remaining dish for a further 48 h.
6. Before fixing the cells, examine them under an inverted microscope for bacterial and fungal contamination at ×100 magnification.
7. Fix the cells by adding approximately 2 ml of Carnoy's fixative dropwise to the edge of the dish and leave for 3 min at room temperature. It is particularly important for suspension cultures to add fixative in this way because it avoids cells being swept to one side of the culture dish.
8. Decant the fixative to a waste bottle (avoiding fumes) for careful disposal, add a further 2 ml aliquot of fixative to the dish and leave for 3 min at room temperature.
9. Decant the fixative to a waste bottle.
10. Allow the coverslip to air-dry. Invert the lid of the dish and use forceps to rest the coverslip against the lid for about 30 min.
11. Add 2 ml of Hoechst stain (wearing gloves) to the coverslip and leave for 5 min at room temperature. Shield the coverslip from direct light at this point.
12. Decant the stain to a waste bottle.
13. Add one drop of mountant to a labelled slide. Place the coverslip cell-side down, on a slide.
14. Under UV epifluorescence examine the slide at ×100 magnification with oil immersion.

ALTERNATIVE PROCEDURE: USE OF INDICATOR CELL LINES

An alternative to the described procedure is the use of an indicator cell line onto which the test cells are inoculatd. The advantages are: standardization of the system; increased surface area of cytoplasm to reveal mycoplasm; and mycoplasma screening of serum and other cell culture reagents that may be inoculated onto the indicator line. Positive and negative controls may be included in each assay. However, it may prove impracticable for many laboratories to grow mycoplasma cultures to use as positive controls. The disadvantages of using an indicator line include the time and effort required for preparation.

A recommended indicator cell line is the Vero African green monkey kidney cell line, which has a high cytoplasm/nucleus ratio. The indicator cells are added at a concentration of approximately 1×10^4 cells ml^{-1} to sterile coverslips in tissue culture dishes 10–24 h before inoculation. The test sample should be added at a concentration of approximately 5×10^5 cells ml^{-1} to give a semi-confluent monolayer at the time of observation (at 1 and 3 days post-inoculation of the sample). Slides are prepared in the same way as those prepared by the direct method.

SUPPLEMENTARY PROCEDURE: MICROBIOLOGICAL CULTURE

1. Using a sterile swab, harvest adherent cells. Resuspend them in the culture medium to a concentration of about 5×10^5 cells ml^{-1}.
2. Test the suspension cells direct at about 5×10^5 cells ml^{-1}.
3. Inoculate agar plate with 0.1 ml of the test sample.
4. Inoculate broth with 0.2 ml of the test sample.
5. Incubate test plate under anaerobic conditions (Gaspak, BBL Microbiology Systems, Cockeysville. MD, USA) at 36 ±1°C for 21 days.
6. At approximately 7 and 14 days post-inoculation, subculture test broth onto agar plates and incubate anaerobically at 36 ±1°C.
7. After 7, 14 and 21 days' incubation, agar plates are examined under ×40 or ×100 magnification using an inverted microscope.

Culture of mycoplasma

1. Each new batch of media ingredients should be subject to quality control before agar or broth preparation. It is especially important to show that new batches of pig and horse serum can support the growth of a representative sample of species found infecting cell cultures, e.g. any two of *M. orale, M. hominis, M. fermentans, M. arginini, M, hyorhinis* or *A. laidlawii*. The National Collection of Type Cultures (Colindale, London, UK) or the American Type Culture Collection (Manassas, Virginia, USA) may supply type strains, or wild-type strains may be used. Stock positive control cultures may be kept frozen at –70°C in mycoplasma broth.

2. Quality control of each batch of complete broth media should be performed before use by adding at least two different strains of positive control mycoplasmas. A non-inoculated broth should be incubated as a negative control.
3. Mycoplasma agar medium should be shown to be able to support mycoplasma growth. The test can be performed at the time of the test sample inoculation. A non-inoculated plate should also be included as a negative control.
4. It is necessary to be able to distinguish 'pseudocolonies' and cell aggregates from mycoplasma colonies on agar. Pseudocolonies are caused by crystal formation and may possibly increase in size. They are distinguished from genuine mycoplasma colonies by using Dienes stain, which does not stain pseudocolonies but stains mycoplasma colonies blue. Most fungal and bacterial colonies also appear colourless. Cell aggregates, however, do not increase in size and thus are more easily distinguished. An indicator of genuine mycoplasma colonies is their typical 'fried egg' appearance on agar. This, however, is not always apparent in primary isolates. By using a sterile bacteriological loop, cells may be disrupted, leaving the agar surface free of aggregates, but mycoplasma colonies, because of the nature of their growth, will leave a central core embedded in the agar.

SUPPLEMENTARY PROCEDURE: ELIMINATION OF CONTAMINATION

Removal of broken cells debris before Hoechst staining

Cell debris in a culture often gives a microscopic image similar to that of mycoplasma contamination after the cells are stained with Hoechst 33258. To avoid possible 'false-positive' results of mycoplasma contamination, samples (culture supernatant of a test cell line) should be filtered aseptically using a membrane filter with 0.8-μm pore size to remove cell debris.

Contaminating mycoplasmas in animal cell culture usually grow up to a concentration of 10^5–10^8 ml^{-1}. Therefore, although 90%, at most, of mycoplasmas will be trapped on the filter (Ohno & Takeuchi, 1990), the mycoplasmas that pass through the membrane are detectable by direct Hoechst staining on a slide-glass or by staining after propagation in a host cell culture.

DISCUSSION

DNA Stain

The fluorochrome dye Hoechst 33258 binds specifically to DNA. Cultures infected with mycoplasma are seen under fluorescence microscopy as fluorescing nuclei with extranuclear mycoplasmal DNA (Plate 1.6.1a, b) whereas uninfected cell cultures contain fluorescing nuclei against a negative background (Plate 1.6.1b, c).

Mycoplasmas may appear as: filamentous forms, some of which may branch, indicating a culture in logarithmic growth: or as cocci, which is typical of an aged

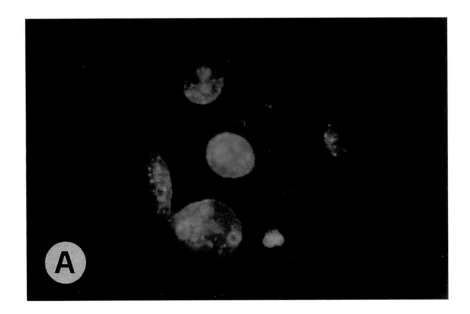

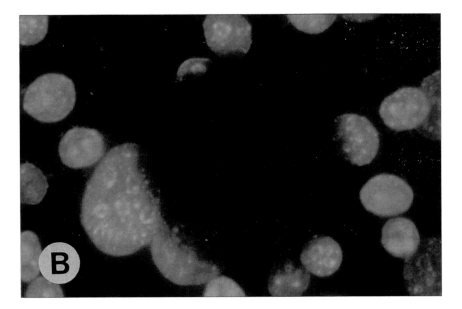

Plate 1.6.1 Mycoplasma-infected cell cultures stained with Hoechst 33258 and mounted in plain buffered glycerol (A, B) or in the presence of 0.01% 1,4-p-phenylenediamine (PPD) (C, D). The field diaphragm of the microscope was partially closed and photographs of the central area of each field (A, plain; B, 0.01% PPD) were taken immediately. The central areas delimited by the partially closed diaphragm were exposed to ultraviolet light for 2 min, and then the whole fields were immediately photographed after opening the diaphragm (B, plain; D, 0.01% PPD). In the central area of the culture mounted in plain buffered glycerol (B, central area) fluorescence is virtually abolished, whereas the staining intensity of the culture mounted in the presence of 0.01% PPD is unaffected (D). Each series of photographs was taken with the same exposure time, using a ×100/NA 1.3 oil-immersion fluorite objective.

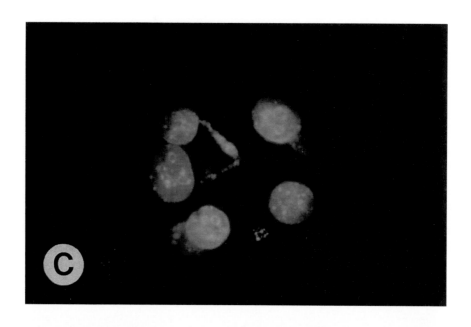

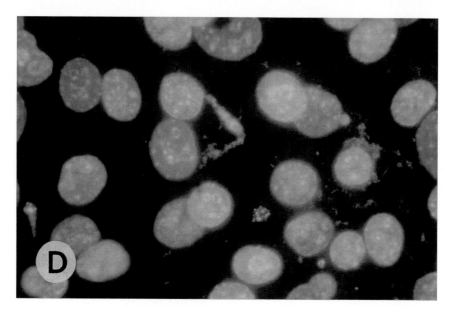

mycoplasma culture. Some slide preparations may contain extranuclear fluorescence produced by disintegrating nuclei and these should not be confused with mycoplasmas. Normally, fluorescence of this type is not uniform in size and is too large to be mycoplasma. Contaminating fungi or bacteria will also stain using this technique but will appear much brighter and larger than mycoplsmas.

The main advantages of the staining method are that results are obtained speedily and the, as yet, non-cultivable *M. hyorhinis* strains that have been detected in cell cultures can be observed.

Both positive and negative slides may be kept in the dark for several weeks without deterioration.

REFERENCES

American Type Culture Collection, 10801 University Boulevard, Manassas, Virginia, 20110 2209, USA.

Aula P & Nichols WW (1967) The cytogenetic effects of mycoplasma in human leucocyte cultures. *Journal of Cell Physiology* 70: 281–290.

Butler M & Leach RH (1964) A mycoplasma which induces acidity and cytopathic effect in tissue culture. *Journal of General Microbiology* 34: 285–294.

European Collection of Cell Cultures, (CANR), Porton Down, Salisbury, Wiltshire, SP4 0JG, UK.

Levine EM, Thomas L. McGregor D, Hayflick L & Eagle M (1968) Altered nucleic acid metabolism in human cell cultures infected with mycoplasma. *Proceedings of the National Academy of Sciences of the USA* 60: 583–589.

MacPherson I & Russel W (1966) Transformations in hamster cells mediated by mycoplasmas. *Nature* 210: 1343–1345.

McGarrity GJ, Phillips D & Vaidya A (1980) Mycoplasmal infection of lymphocyte cultures: Infection with *M. salivarium*, *In Vitro* 16: 346–356.

National Collection of Type Cultures, Central Public Health Laboratory, 61 Colindale Avenue, Colindale, London NW9 5HT, UK.

Ohno T & Takeuchi M (1990) Test for mycoplasma contamination. Standardized protocols for quality control of animal cell lines. Report from JTCA Cell Bank Committee, *Tissue Culture Research Communications* 9: Suppl. 9–11.

Robinson LB. Wichelhausen RB & Roizman B (1956) Contamination of human cell cultures by pleuro-pneumonia-like organisms. *Science* 124: 1147–1148.

Stanbridge EJ, Hayflick L & Perkins FT (1971) Modification of amino acid concentrations induced by mycoplasmas in cell culture medium. *Nature (London) New Biology* 232: 242–244.

Wise KS. Cassell GH & Action RT (1978) Selective association of murine T lymphoblastoid cell surface alloantigens with mycoplasma hyorhinis. *Proceedings of the National Academy of Sciences of the USA* 75: 4479–4483.

1.7 MYCOPLASMA DETECTION METHODS USING PCR

Five species are considered to be responsible for more than 95% of cell culture contaminations: *M. arginini, M. fermentans, M. orale, M. hyorhinis* and *A. laidlawii*.

Many different techniques, such as bacteriological culture, DNA staining using fluorochrome and immunological or biochemical methods, are available to detect mycoplasma contamination (see section 1.6). However, none seem to be fully efficient, so a combination of different methods is often necessary. Molecular tools such as hybridization using rDNA gene probes or polymerase chain reaction (PCR) have been developed over the past few years. Several studies using 16S rDNA-based PCR concluded that PCR seems to be a very convenient method for routine detection of cell culture contaminations (Spaepen, 1992; Teyssou, 1993; van Kuppeveld, 1994).

Materials and equipment

- Programmable heating block
- Gel electrophoresis unit
- Gel staining tray
- UV transilluminator

 UV light is damaging to eyes. Always wear protective glasses.

A photograph of the gel can be taken using a bellows-type camera equipped with a Polaroid MP4 system using 3000 iso-Polaroid 667 film. An orange filter is necessary to obtain a good film image. The time of exposure required is 0.5–1 s with a fully opened diaphragm.

Reagents and solutions

For PCR procedure

- Taq DNA polymerase. Store the polymerase at −20°C (as recommended by the manufacturer) and place it on ice before use. The ×10 Taq buffer is supplied with the enzyme by the manufacturer
- Nucleotides (dNTP): dATP, dTTP, dCTP, dGTP
- Primers (see Table 1.7.1)
- Controls. Several controls must be used: distilled water and uninfected cells as negative controls; cells infected with *Mycoplasma* spp. and/or *A. laidlawii* as positive controls

---

Cell and Tissue Culture: Laboratory Procedures in Biotechnology, edited by A. Doyle and J.B. Griffiths.
© 1998 John Wiley & Sons Ltd.

Table 1.7.1 Specific primers for PCR

Primer name	Sequence	Position[a]
molli1	5'-TACGGGAGGAGCAGTA-3'	343-359
molli2a	5'-TCAAGATAAAGTCATTTCCT-3'	463-482
molli2b	5'-TACCGTCAATTTTTAATTTTT-3'	451-471

[a]Position numbers are relative to the *Escherichia coli* 16S rDNA nucleotide sequence. The forward primer molli1 and the reverse primer molli2a are used to detect *Mycoplasma* species. The forward primer molli1 and the reverse primer molli2b are used to detect *Acholeplasma* species. Each set of primers are used separately.

For analysis of amplified samples

- TAE buffer, pH 7.5–7.8 (Sambrook *et al.*, 1989): (×50 stock solution, per litre), (242 g of Tris base, 57.1 ml of glacial acetic acid and 100 ml of 0.5 M EDTA). This buffer can be stored at room temperature. A ×1 solution can be obtained by preparing 20 ml of the ×50 stock solution in 980 ml of water.
- Acrylamide and bisacrylamide
  Acrylamide is a potent neurotoxin. Wear gloves and a mask when weighing powdered acrylamide and bisacrylamide. For maximum safety and convenience, we recommend the use of a commercially available ready-made stock solution of ×2 acrylamide–bisacrylamide mix (Bioprobe, 93100 Montreuil sous Bois, France).
- Ammonium persulphate
- TEMED (tetramethylethylenediamine)
- Ethidium bromide solution: 10 mg ml^{-1} stock solution made up of 100 mg of ethidium bromide and 10 ml of water

  Ethidium bromide is both an irritant and a mutagen! Please handle carefully and do not discard it into the environment.
- ×10 loading buffer: 0.25% bromophenol blue and 50% glycerol in water

PROCEDURE: AMPLIFICATION

Amplifications are performed in a final volume of 50 μl.

Reaction mix

1. Prepare the reaction mixture within each tube: 2.5 U of Taq polymerase; 200 μmol l^{-1} of each dNTP; 1 μmol l^{-1} of each primer (either molli1 + molli2a or molli1 + molli2b).
2. Add 5 μl of ×10 Taq buffer and adjust the final volume to 40 μl using water.
3. Vortex briefly.
4. Add 50 μl of mineral oil to each tube.
5. Place the tubes in ice.

Samples to be tested

1. Into each tube, transfer directly 10 µl of controls or cell suspensions to be tested under the mineral oil (no pretreatment of cell suspension is required).
2. Place the tubes in the programmable heating block at 95°C for 5 min.

Amplification steps

1. Perform 35 amplification cycles. Each cycle consists of three steps:
   - denaturation at 95°C for 1 min,
   - annealing at 55°C for 1 min,
   - extension at 72°C for 2 min.
2. Perform a final extension step at 72°C for 10 min.

SUPPLEMENTARY PROCEDURE: ANALYSIS OF AMPLIFIED SAMPLES

1. Remove the tubes from the programmable heating block. If the samples are not analysed at once, they can be stored at –20°C until they are analysed.
2. Prepare the 8% polyacrylamide gel: acrylamide/bisacrylamide (25:1), ×50 TAE, 10% ammonium persulphate, TEMED and water.
3. Prepare the samples by adding 2µl of ×10 loading buffer to 8 µl of the amplified product.
4. Distribute 3–5 µl of the loading sample onto the gel.
5. Do not forget to include the appropriate molecular weight marker (1 kb ladder or marker V from Boehringer, Mannheim).
6. Perform the electrophoresis (3 h of 18 W) with ×1 TAE buffer.
7. When separation is achieved, stain the gel by placing it in a solution of ethidium bromide (0.5 mg ml^{-1}) for 15 min.
8. Place the gel on the UV transilluminator and photograph it, as shown in Figure 1.7.1; with the primer set molli1 + molli2a a characteristic 142 bp PCR product is obtained, whereas a 128 bp specific DNA fragment is observed with the primer set molli1 + molli2b.

Agarose gel can be used instead of the acrylamide gel, for which a horizontal electrophoresis unit is required. A 1.5% (w/v) gel is prepared by adding the powdered agarose to ×1 TAE buffer. The electrophoresis is performed in ×1 TAE buffer of a voltage of 1-5 V cm^{-1} (measured as the distance between the electrodes).

Caution! Contamination may occur easily and invalidate the results. Avoiding false positive reactions needs the use of several precautions (Kwok & Higuchi, 1989):

- Physically isolate PCR preparations and products. Each step of PCR must be performed in separate rooms. All reagents must be prepared, aliquoted and stored in a free PCR-amplified product area

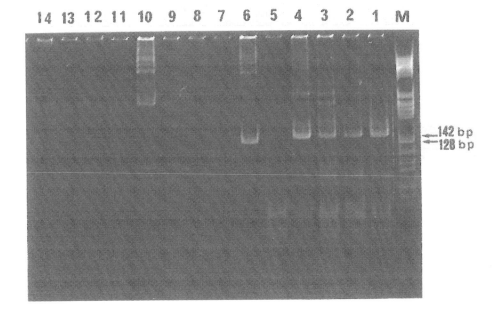

Figure 1.7.1 Detection of mycoplasma by PCR in experimentally infected Vero cells. Amplifications performed with the primer set molli1 + molli2a: (lane 1) Vero cells infected with *M. arginini*; (lane 2) Vero cells infected with *M. fermentans*; (lane 3) Vero cells infected with *M. hyorhinis*; (lane 4) Vero cells infected with *M. orale*; (lane 5) Vero cells infected with *A. laidlawii*; (lane 11) mollicute-free Vero cells; (lane 13) negative control (distilled water). Amplifications performed with the primer set molli1 + molli2b: (lane 6) Vero cells infected with *A. laidlawii*; (lane 7) Vero cells infected with *M. arginini*; (lane 8) Vero cells infected with *M. orale*; (lane 9) Vero cells infected with *M. fermentans*; (lane 10) Vero cells infected with *M. hyorhinis*; (lane 12) mollicute-free Vero cells; (lane 14) negative control.

- Aliquot reagents
- Use disposable gloves
- Avoid splashes. Use caps that do not require excessive force to remove. If splashing occurs, change gloves
- Use positive displacement pipettes
- Premix reagents as often as possible
- Add DNA last and cap each tube before proceeding to the next tube

DISCUSSION

The PCR method that we have described is able to detect cell cultures containing contaminating mycoplasma species with great sensitivity (1.10 cfu/10 µl) and seems to be more sensitive than the other techniques available. Also, PCR analysis can be performed from lyophilized cell cultures, which facilitates the transport of samples. Moreover, sample preparation is very simple and does not require any DNA extraction, so the results are obtained in 1 day.

However, several disadvantages can be reported. This PCR method requires two different PCR systems: molli1 + molli2a to detect all the *Mycoplasma* and *Spiroplasma* spp. and molli1 + molli2b to detect *Acholeplasma* spp. The set of primers molli1 + molli2a amplifies two phylogenetically closely related species – *Clostridium ramosum* and *C. inocuum* – but these species are rarely found as cell culture contaminants. Moreover, this method does not allow bacterial identification of the detected species or antibiotic susceptibility testing.

Is there a better PCR technique? Over the past few years different authors have described other 16S rDNA-based PCR methods. Spaepen *et al.* (1992) used a nested PCR system with great sensitivity, but the use of a second amplified cycle dramatically increased the risk of DNA carryover contaminations. van Kuppeveld *et al.* (1994) reported a single PCR system that seems to be very suitable to detect cell culture contamination but it requires a DNA extraction stage, which is very time consuming. Moreover, a new marked PCR method is available (Stratagene, CA). The primers used make it possible rapidly to (4–5 h) test eukaryotic cells for mycoplasma infection but this method seems to be less sensitive than our PCR technique.

Finally, all these PCR methods are convenient tools for routine detection of Mollicutes cell culture contaminants and can be used as a substitute for the classical methods.

REFERENCES

Kwok SK & Higuchi R (1989) Avoiding false positives with PCR. *Nature, London* 339: 237–238.

Sambrook J, Fritsh EF & Maniatis P (1989) *Molecular Cloning: a Laboratory Manual*, 2nd edn. Cold Spring Harbor Laboratory Press, Cold Spring Harbor, NY.

Spaepen M, Angulo AF, Marynen P & Cassiman JJ (1992) Detection of bacterial and mycoplasma contamination in cell cultures by polymerase chain reaction. *FEBS Microbiology Letters* 99: 89–94.

Teyssou R, Poutiers F, Saillard C, Grau O, Laigret F, Bove JM & Bebear C (1993) Detection of Mollicutes contamination in cell cultures by 16S rDNA amplification. *Molecular and Cellular Probes* 7: 209–216.

van Kuppeveld FJM, Johansson KE, Galama JMD, Kissing J, Bolske G, van der Logt JTM & Matchers WJG (1994) Detection of mycoplasma contamination in cell cultures by a mycoplasma group-specific PCR. *Applied and Environmental Microbiology* 60: 149–152.

1.8 BACTERIA AND FUNGI

For any reasonable quality control programme involving cell lines it is absolutely necessary to include routine culture tests for bacteria and fungi. Most often the presence of such contaminants will be obvious even to the novice. This is especially true when the cultures are set up in the absence of all antibiotics – a practice that is strongly recommended. However, while overt microbial contamination may be more common, many organisms will grow slowly in the usual cell cultures, sometimes obscured by animal cell debris, and remain unnoticed. The steps outlined below will reveal all of the most usual bacterial or fungal organisms that might be expected to thrive in conventional cell culture systems.

PROCEDURE: DETECTION OF BACTERIA AND FUNGI IN CELL CULTURES

Materials and equipment

- Dehydrated media for bacterial/fungal cultivation
- Bacto Sabouraud dextrose broth (Difco)
- Bacto thioglycollate medium (Difco)
- Trypticase soy broth powder (Baltimore Biological Laboratory)
- Bacto brain heart infusion agar (Difco)
- Bacto blood agar base (Difco)
- YM broth base (Difco)
- Nutrient broth base (Difco)
- Bacto yeast extract (Difco)
- Reference microbial cultures for positive controls
- Hotplates and stirring apparatus
- GasPak Anaerobic System (Baltimore Biological Laboratory)

To examine cell cultures or suspect media for bacterial or fungal contaminants, proceed as follows:

1. Using an inverted microscope, equipped with phase contrast optics if possible, examine cell culture vessels individually. Scrutiny should be especially rigorous in cases in which large-scale production is involved. Check each culture first using low power. The suppliers listed provide the specific media required but other suitable vendors exist. Batches of media should be tested for optimal growth promotion before use in cell culture quality control.
2. After moving the cultures to a suitable isolated area, remove aliquots of fluid from cultures that are suspect and retain these for further examination. Alternatively, autoclave and discard all such cultures.

Cell and Tissue Culture: Laboratory Procedures in Biotechnology, edited by A. Doyle and J.B. Griffiths.
© 1998 John Wiley & Sons Ltd.

3. Prepare wet mounts using drops of the test fluids and observe under high power.
4. Prepare smears, heat-fix and stain by any conventional method (e.g. Wright's stain), and then examine under oil immersion.
5. Consult Cour *et al.* (1979) and Freshney (1987) for photomicrographs of representative contaminants and further details.

Microscopic examination is only sufficient for detection of gross contaminations and even some of these cannot be readily detected by simple observation. Therefore, an extensive series of culture tests is also required to provide reasonable assurance that a cell line stock or medium is free of fungi and bacteria.

To perform these tests on standard cell cultures or stocks of frozen cells:

1. Pool and mix the contents of about 5% of the ampoules from each freeze-lot prepared using narrow-width pipettes. Aliquots from cultures to be tested should generally include some of the monolayer or cell suspension. It is commonly recommended that antibiotics be omitted from media used to cultivate and preserve stock cell populations. If antibiotics are used, the pooled suspension should be centrifuged at 2000 g for 20 min and the pellet resuspended in antibiotic-free medium. A series of three such washes with antibiotic-free medium prior to testing will reduce the concentration of antibiotics that could obscure contamination.
2. From each pool, inoculate each of the test media listed in Table 1.8.1 with a minimum of 0.3 ml of the test cell suspension and incubate under the conditions indicated. Include positive and negative controls comprising a suitable range of bacteria and fungi that might be anticipated. A recommended grouping consists of *Pseudomonas aeruginosa*, *Micrococcus salivarius*, *Escherichia coli*, *Bacteroides distasonis*, *Penicillium notatum*, *Aspergillus niger* and *Candida albicans*.
3. Observe as suggested for 14–21 days before concluding that the test is negative. Contamination is indicated if colonies appear on solid media or if any of the liquid media become turbid.

Table 1.8.1 Regimen for detecting bacterial or fungal contamination in cell cultures

Test medium	Temperature (°C)	Aerobic state	Observation time (days)
Blood agar			
Fresh, defibrinated	37	Aerobic	14
with rabbit blood (5%)	37	Anaerobic	14
Thioglycollate broth	37, 26	Aerobic	14
Trypticase soy broth	37, 26	Aerobic	14
Brain heart infusion broth	37, 26	Aerobic	14
Sabouraud broth	37, 26	Aerobic	21
YM broth	37, 26	Aerobic	21
Nutrient broth with 2% yeast	37, 26	Aerobic	21

For further details, see Cour et al. (1979).

DISCUSSION

While this procedure will permit detection of most common bacterial and fungal organisms that grow in cell cultures, it has been noted that at least one, very fastidious, bacterial strain initially escaped observation. This was present in nine different cultures from a single clinical laboratory in the USA submitted for testing and expansion under a government contract. The organism grew extremely slowly but could be detected after 3 weeks' incubation with cell cultures that had no antibiotics and had no fluid changes. Samples so developed were inoculated to sheep blood agar plates and New York City broth (ATCC medium 1685). The organism could be observed during a subsequent 6-week incubation period at 37°C.

The Bacteriology Department at the ATCC determined the appropriate culture conditions for this microorganism, and tentatively identified it as a *Corynebacterium*. Antibiotic sensitivity tests revealed bacteriostasis with some compounds but no bactericidal antibiotics have yet been found.

This incident emphasizes the critical importance of diligent testing of cell cultures for contaminant microorganisms. By combining procedures such as those described here with procedures included elsewhere in this volume (e.g. fluorescent or nucleic acid probes for mycoplasma and viruses) one can be more certain that clean cell cultures are available for experimentation.

REFERENCES

Cour I, Maxwell G & Hay RJ (1979) Tests for bacterial and fungal contaminants in cell cultures as applied at the ATCC. *Tissue Culture Association Manual* 5: 1157–1160.

Freshney RI (1987) *Culture of Animal Cells – A Manual of Basic Techniques*, 2nd edn. Alan R Liss, New York.

1.9 ELIMINATION OF CONTAMINATION

In the event of cell cultures becoming contaminated with bacteria, fungi or mycoplasmas, the best course of action is to discard the culture, check cell culture reagents for contamination, thoroughly disinfect all safety cabinets and work surfaces and resuscitate a fresh culture from previously frozen stock. In the case of contamination with a spore-forming organism, and where such facilities exist, room fumigation may also be advisable.

However, in the case of irreplaceable stocks this course of action may not be possible and antibiotic treatment may be necessary to eliminate the contamination.

PROCEDURE: ERADICATION

Materials

- Antibiotic of choice (see Tables 1.9.1–1.9.3)
- Appropriate cell culture growth medium

1. Culture cells in the presence of the chosen antibiotic for 10–14 days. Each passage should be performed at the highest dilution of cells at which growth occurs. If the contaminant is still detectable after this time, it is unlikely that the antibiotic used will prove successful, and another antibiotic should be tried.
2. During the course of antibiotic treatment and also 5–7 days after treatment in the case of bacteria and fungi, and 25–30 days after treatment for mycoplasma-contaminated lines, re-test the culture by an appropriate method. Regular routine testing should, however, occur after this period. This is particularly important in the case of mycoplasmas, where low levels of infection may persist after antibiotic treatment.

DISCUSSION

The antibiotic of choice used to eliminate any contaminant is dependent upon the sensitivity of the organism in question. If facilities are available to identify the organism and perform antibiotic sensitivity assays, an appropriate antibiotic may be chosen for elimination and used at minimum concentration.

In the case of bacteria and fungi a Gram stain would give an indication of which antibiotic to use (see Tables 1.9.1 and 1.9.2).

Table 1.9.1 Antibiotics active against bacteria

Antibiotic[a]	Working concentration (mg l^{-1})	Gram-positive bacteria	Gram-negative bacteria
Ampicillin	100	✓	✓
Cephalothin	100	✓	✓
Gentamicin	50	✓	✓
Kanamycin	100	✓	✓
Neomycin	50	✓	✓
Penicillin V	100	✓	
Polymyxin B	50		✓
Streptomycin	100	✓	✓
Tetracycline	10	✓	✓

[a] Available from Sigma.

Table 1.9.2 Antibiotics active against fungi

Amphotericin B (Sigma)	2.5 mg l^{-1}
Ketaconazole (Sigma)	10 mg l^{-1}
Nystatin (Sigma)	50 mg l^{-1}

Table 1.9.3 Antibiotics active against mycoplasmas

Ciprofloxacin (Bayer)	20 mg l^{-1}
MRA[a] (ICN–Flow)	0.5 mg l^{-1}

[a] MRA = Mycoplasma Removal Agent.

Unlike bacterial or fungal contamination, mycoplasma infection is not always detectable in a cell culture by the usual microscopic methods.

Various antibiotics have been investigated for their ability to eradicate mycoplasma from cell lines. Historically the antibiotics of choice were either minocycline or tiamulin. However, many cell culture mycoplasmas are now resistant to these antibiotics. Today the most effective are the quinalone antibiotics ciprofloxacin (Mowles, 1988) and MRA (see Table 1.9.3).

It is extremely important that after antibiotic treatment the cells are maintained in antibiotic-free media. Lack of evidence of mycoplasma infection does not necessarily indicate that the culture is free of such infection, because the level of infection may be below the limit of detection. For this reason it is suggested that antibiotic-free growth be continued to allow any residual infection to reach levels that are detectable. If, after this period, no mycoplasmas are detected, the line may be considered to be mycoplasma-free.

Other methods of mycoplasma elimination

For most laboratories, antibiotic treatment is the most convenient method of eradication.

Other methods are available that vary in their degree of success. These include passage in athymic nude mice (Van Diggelen et al., 1977), growth of cells in rabbit or guinea pig serum (Nair, 1985) and the use of nucleic acid analogues (Marcus et al., 1980)

REFERENCES

Marcus M, Lavi U, Nattenberg A, Rottem S & Markowitz D (1980) Selective killing of mycoplasma from contaminated mammalian cells in cell cultures. *Nature (London)* 285: 659–699.

Mowles JM (1988) The use of Ciprofloxacin for the elimination of mycoplasmas from naturally infected cell lines. *Cytotechnology* 1: 355–358.

Nair CN (1985) Elimination of mycoplasma contaminants from cell cultures with animal serum. *Proceedings of the Society for Experimental Biology and Medicine* 179: 254–258.

Van Diggelen OP, Shin S & Phillips DM (1977) Reduction in cellular tumourgenicity after mycoplasma infection and elimination of mycoplasma from infected cultures by passage in nude mice. *Cancer Research* 37: 2680–2687.

CHAPTER 2

CELL QUANTIFICATION

2.1 OVERVIEW

To measure performance, to achieve reproducibility or to make comparative studies, a means of quantifying the cell population is needed. Classically, direct counts of cell numbers using a microscopic counting chamber (haemocytometer), usually in conjunction with a vital stain (e.g. Trypan blue) to distinguish viable and non-viable cells, is used. However, all vital stains are subjective and cannot give absolute values, and cell numbers take no account of differences in cell size/mass. The method is simple, quick and cheap, and requires only a small fraction of the total cells from a cell suspension.

Automation of cell counting (total cells) is possible with electronic counters, especially for non-clumping single suspension cells. A method that is widely used for total cell numbers is counting cell nuclei after dissolving the cytoplasm. This is particularly useful for large clumps of cells, where cells are inaccessible (e.g. in matrices) or where cells are difficult to trypsinize off substrates (e.g. microcarriers). If determination of viability is the prime consideration, then the most accurate method is to count the cells that have the ability to divide by the mitotic index method.

If cell mass, rather than number, is the important factor then a cell constituent has to be measured. The options are dry weight (large numbers required for accuracy), protein (probably the best parameter), DNA, protein nitrogen or lipids (Freshney, 1994).

It may not be possible to perform a direct cell count, e.g. during a monolayer culture or for cells immobilized in matrices, so an indirect measurement has to be carried out. Examples are glucose or oxygen consumption rates, which can be converted to cell numbers from predetermined standard curves. Other metabolites include lactic and pyruvic acid and carbon dioxide. Again there are reservations about such methods because these metabolic rates are not constant throughout the growth cycle of a specific cell and may also be influenced by changes in the culture that are the subject of the investigation. However, they do provide a good indication of the viable cell mass in a culture, if not an absolute value, and are widely used to give growth yields or specific utilization rates. The advantage is, of course, that the method is non-destructive. A useful parameter being increasingly used is lactate dehydrogenase activity released into the medium. This is released by dying/dead cells and therefore gives a quantitative measurement of the non-viable population. This method can be modified to measure viable cell mass by a reverse titration procedure, i.e. by controlled lysis of the cells and measuring the increase in enzyme activity.

These biochemical measurements are best used over a time-course in culture so that successive readings will show a definite trend in the culture dynamics, e.g.

stationary, growing or dying. The rule is not to rely on one method alone, but to combine two or more. Mention should be made of the use of radioisotopes to measure rates of synthesis (e.g. protein, DNA and RNA synthesis) (Freshney, 1994). However, these are usually kept for specific situations rather than general use.

There are many specific methods available that are suitable for a certain set of conditions or for a particular experimental aim (Patterson, 1979; Griffiths, 1985). These include radiochromium labelling (radiochromium binds to viable protein and is released on cell damage/death), cellular ATP, intracellular volume (by ratios of radiochemicals), specific stains (e.g. MTT (3-(4,5-dimethylthiazol-2-yl)-2,5-diphenyltetrazolium bromide), triphenyltetrazolium chloride, neutral red), redox potential, turbidity, calorimetry (Griffiths, 1985) and biomass monitors.

In conclusion, there is no ideal method and the nature of the study often determines the parameter to be measured. It is best not to rely on one method alone because a combination of several methods will give a far more accurate picture (e.g. cell counts for numbers and cell protein for mass will show growth even if no division is occurring). Also account must be taken of the fact that all viability determinations will stress the cell (e.g. dilution, mixing, time factor, trypsinization, etc.). However, because most determinations are to show relative changes, or specific metabolic activities, the inaccuracies should be remembered rather than be a concern, so that the results are not over-interpreted. On the basis that the standard methods such as haemocytometer counting, plating efficiency, electronic counting, etc. are well documented in all cell culture text books (e.g. *Culture of Animal Cells: a Manual of Basic Techniques* (Freshney, 1994), as well as the *Cell and Tissue Culture: Laboratory Procedures* manual (Doyle *et al.*, 1993), only the more specific techniques have been selected for this chapter.

REFERENCES

Doyle A, Griffiths JB & Newall D (eds) (1993) *Cell and Tissue Culture: Laboratory Procedures*. Wiley, Chichester.

Freshney RI (1994) *Culture of Animal Cells: a Manual of Basic Techniques*. Wiley–Liss, New York.

Griffiths JB (1985) Cell biology; experimental aspects. In: Spier RE & Griffiths JB (eds) *Animal Cell Biotechnology*, vol. 1, pp. 49–58. Academic Press, Orlando.

Patterson MK (1979) Measurement of growth and viability of cells in culture. In: Jakoby WB & Paston IH (eds) *Cell Culture*, vol. 11, pp. 141–151. Academic Press, New York.

2.2 HAEMOCYTOMETER CELL COUNTS AND VIABILITY STUDIES

In order to ensure that cell cultures have reached the optimum level of growth before routine subculture or freezing, it is helpful to obtain an accurate cell count and a measure of the percentage viability of the cell population.

The most common routine method for cell counting that is efficient and accurate is with the use of a haemocytometer. There are several types on the market, of which the Improved Neubauer has proved most popular.

A thick, flat counting chamber coverslip rests on the counting chamber at a distance of 0.1 mm above the base of the slide. The base of the slide has rulings accurately engraved on it, comprising 1-mm squares, some of which are further divided into smaller squares.

When cell suspensions are allowed to fill the chamber, they can be observed under a microscope and the cells counted in a chosen number of ruled squares. From these counts, the cell count per millilitre of suspension can be calculated. Hybridoma cells and others that grow in suspension may be counted directly. Cell lines that are attached will need to be removed from the tissue culture flask by trypsinization. Because accuracy of counting requires a minimum of approximately 10^5 cells ml^{-1} it may be necessary to resuspend the cells in a smaller volume of medium.

To ensure that a cell culture is growing exponentially it is useful to know the percentage viability and percentage of dead cells and hence the stage of growth of the cells. This can be estimated by their appearance under the microscope, because live healthy cells are usually round, refractile and relatively small in comparison to dead cells, which can appear larger, crenated and non-refractile when in suspension. The use of viability stains such as Trypan blue ensures a more quantitative analysis of the condition of the culture. Trypan blue is a stain that will only enter across the membranes of dead/non-viable cells.

When a cell suspension is diluted with Trypan blue, viable cells stay small, round and refractile. Non-viable cells become swollen, larger and dark blue. Both the total count of cells per millilitre and percentage of viable cells can be determined.

Cell and Tissue Culture: Laboratory Procedures in Biotechnology, edited by A. Doyle and J.B. Griffiths.
© 1998 John Wiley & Sons Ltd.

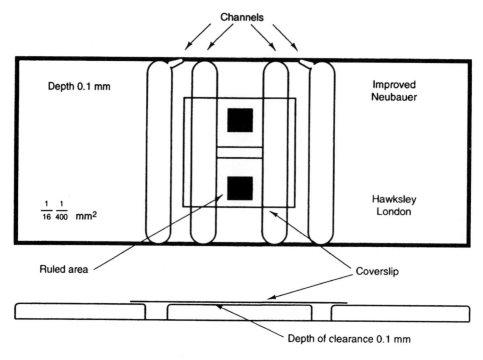

Figure 2.2.1 Improved Neubauer haemocytometer with coverslip.

PROCEDURE: HAEMOCYTOMETER CELL COUNT

Materials and equipment

- 0.4 g of Trypan blue in 100 ml of physiological saline; pass through a 0.22-µm filter to remove any debris
- Haemocytometer with coverslip – improved Neubauer British Standard for Haemocytometer Counting Chambers, BS 748:1963 (Figure 2.2.1)
- Hand-held counter
- Microscope – low power, ×40 to ×100 magnification

Trypan blue is harmful if ingested or inhaled. It is irritating to the eyes, harmful by skin contact and has been found to cause cancer in laboratory animals. Appropriate precautions should be taken when handling Trypan blue and the use of an extraction hood and gloves is advised.

1. Thoroughly clean the haemocytometer and coverslip and wipe both with 70% alcohol before use.
2. Moisten the edges of the coverslip or breathe on the chamber to provide moisture before placing the coverslip centrally over the counting area and across the grooves.
3. Gently move the coverslip back and forth over the chamber until Newton's rings (rainbow-like interference patterns) appear, indicating that the coverslip

Table 2.2.1 Examples of suitable dilutions

Cells	Trypan blue	Dilution
0.1 ml	0.1 ml	2-fold
0.1 ml	0.3 ml	4-fold
0.1 ml	0.9 ml	10-fold

is in the correct position to allow accurate counting, i.e. the depth of the counting chamber is now 0.1 mm.

4. Mix the cell suspension gently and add an aliquot to the Trypan blue solution (see Table 2.2.1). The dilution will depend on the cell concentration and may need to be adjusted to achieve the appropriate range of cells to be counted (see Step 6). Draw a sample into a Pasteur pipette after mixing thoroughly and allow the tip of the pipette to rest at the junction between the counting chamber and the coverslip. Draw the cell suspension in to fill the chamber. No pressure is required because the fluid will be drawn into the chamber by capillarity. Both halves of the chamber should be filled to allow for counting in duplicate.
5. Using a light microscope at low power, focus on the counting chamber.
6. Count the number of cells (stained and unstained separately) in 1-mm² areas (see Figure 2.2.2) until at least 200 unstained cells have been counted. As a rule, the cells in the left-hand and top grid markings should be included in a square, and those in the right-hand and bottom markings excluded (see Figure 2.2.3).
7. Count the viable and non-viable cells in both halves of the chamber.
8. Calculations:
   - Total number of viable cells = $A \times B \times C \times 10^4$
   - Total dead cell count = $A \times B \times D \times 10^4$
   - Total cell count = viable cell count + dead cell count
   - % Viability = $\dfrac{\text{Viable cell count}}{\text{Total cell count}} \times 100\%$

   where A = volume of cells, B = dilution factor in Trypan blue (Table 2.2.1), C = mean number of unstained cells (i.e. unstained count divided by the number of areas counted, D = mean number of dead/stained cells and 10^4 is the conversion factor for 0.1 mm³ to ml.

DISCUSSION

Troubleshooting

The counting chamber and coverslip should be scrupulously clean and without scratches. The clarity of the rulings on the chamber is vital and metallized glass improves both clarity and visibility.

Errors can be caused during sampling, dilution, mixing and filling the chamber and by inaccurate counting.

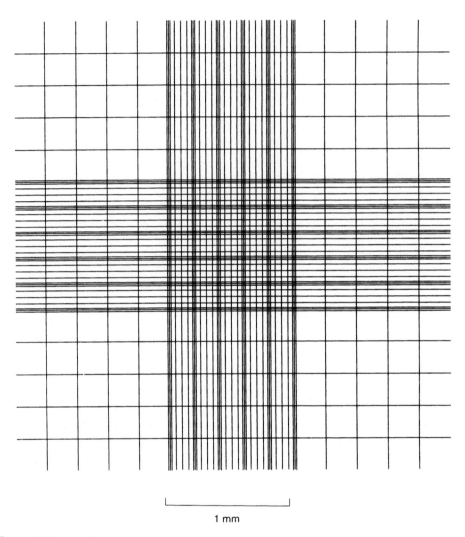

1 mm

Figure 2.2.2 Magnified view of the cell counting chamber grid. The central 1-mm square area is divided into 25 smaller squares, each 1/25 mm². These are enclosed by triple ruled lines and are further subdivided into 16 squares, each 1/400 mm², to ensure accurate counting of very small cells, e.g. erythrocytes.

The following can cause inaccuracy of the haemocytometer method by affecting the volume of the chamber:

- Overflowing the chamber and allowing sample to run into the channels/moat
- Incompletely filling the chamber
- Air bubbles or debris in the chamber

There is an inherent error in the method due to random distribution of the cells

HAEMOCYTOMETER CELL COUNTS AND VIABILITY STUDIES

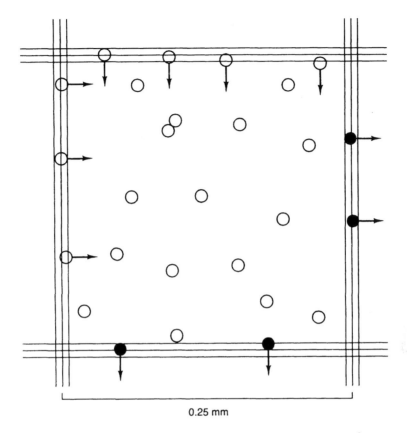

Figure 2.2.3 Diagram indicating which cells are to be counted: ○, count; ●, count on next grid.

in the chamber. The error is reduced by increasing the number of cells to be counted and duplicating the counts.

Time considerations

It is essential to examine the cells within 5 min of immersing them in Trypan blue.

2.3 MTT ASSAY

The MTT Assay (Mosmann, 1983) is a sensitive, quantitative and reliable colorimetric assay that measures viability, proliferation and activation of cells. The assay is based on the capacity of mitochondrial dehydrogenase enzymes in living cells to convert the yellow water-soluble substrate 3-(4,5-dimethylthiazol-2-yl)-2,5-diphenyl tetrazolium bromide (MTT) into a dark blue formazan product that is insoluble in water. The amount of formazan produced is directly proportional to the cell number in a range of cell lines (Mosmann, 1983; Gerlier & Thomasset 1986; Grailer *et al.*, 1988; Al-Rubeai & Spier, 1989). The results are consistent with those obtained from [^{3}H]thymidine uptake assays. The MTT assay is more useful in the detection of cells that are not dividing but are still active. It can, therefore, be used to distinguish between proliferation and cell activation (Gerlier & Thomasset, 1986). The technique permits the processing of a large number of samples with a high degree of precision using a multiwell scanning spectrophotometer (micro-ELISA reader).

PROCEDURE: MTT ASSAY – SUSPENSION OR MONOLAYER CELLS

Materials and equipment

- MTT stain
- HCl
- Propan-2-ol
- 96-Well microtitre plate
- Micro-ELISA reader

 Gloves and eye/face protection should be worn throughout the procedure because MTT is a mutagenic and toxic agent. Reference should be made to hazard codes and instructions for protection.

1. Prepare an MTT stock solution of 5 mg ml^{-1} (Sigma, St Louis) in phosphate-buffered saline (PBS), pH 7.5, and filter through a 0.22-μ filter to sterilize and remove the small amount of insoluble residue.
2. To 100 μl of cell suspension or cell monolayer in each microtitre well, add 10 μl of MTT (5 mg ml^{-1}).
3. Incubate in a humidified incubator at 37°C for 3 h.
4. Add 100 μl 0.04 M HCl in propan-2-ol to each well and mix thoroughly to dissolve insoluble blue formazan crystals.

5. Read plate on a micro-ELISA reader using a test wavelength of 570 nm and a reference wavelength of 630 nm.

Plates should be read within 5 min of adding acidified propan-2-ol.

The MTT assay can also be used to quantify cell activation by determining the maximal velocity (V) of the reaction (Gerlier & Thomasset, 1986). Alternatively, in order to compare two states of activation of a given cell, the MTT formazan produced by the same number of viable cells can be measured.

A linear relationship between cell number and MTT formazan exists up to 10^5 cells per well for a 2.5-h incubation with MTT (see Figure 2.3.1). For experiments involving the use of high cell density or for obtaining absolute cell number it is necessary that a standard curve be constructed by making a series of cell-doubling dilutions with 100 µl well^{-1}, using medium from the same culture.

ALTERNATIVE PROCEDURE: MTT ASSAY – IMMOBILIZED CELLS (AL-RUBEAI ET AL., 1990)

1. Prepare a solution of MTT in PBS at a concentration of 1 mg ml^{-1}. Add 1 ml of MTT solution to 0.5 ml of immobilized matrices (e.g. beads).
2. Incubate at 37°C for 3 h to allow MTT to diffuse throughout the matrices and react with the cells.
3. Sonicate for 15 s, using a sonic probe to release formazan products from the matrix.
4. Centrifuge at 180 g for 2 min.

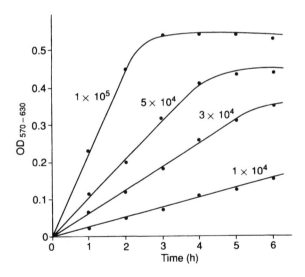

Figure 2.3.1 Relationship between hybridoma viable cell number and amount of formazan as affected by incubation time. Reproduced from Al-Rubeai (1977) by permission of John Wiley & Sons Ltd.

5. Measure optical absorbance at 570 nm. It may be convenient to dilute suspension 10 times with PBS prior to reading.

REFERENCES

Al-Rubeai M (1993) The MTT assay for cell viability. In: Doyle A & Griffiths JB (eds) *Mammalian Cell Culture: Essential Techniques*, pp. 174. Wiley, Chichester.

Al-Rubeai M & Spier R (1989) Use of the MTT assay for the study of hybridoma cells in homogeneous and heterogeneous cultures. In: Spier RE, Griffiths J, Stephenne J and Crooy P (eds) *Advances in Animal Cell Biology and Technology for Bioprocesses*, pp. 143–155. Butterworth, London.

Al-Rubeai M, Musgrave SG, Lambe CA, Walker AG, Evans NH & Spier RE (1990) Methods for the estimation of the number and quality of animal cells immobilized in carbohydrate gels. *Enzyme and Microbial Technology* 12: 459–463.

Gerlier D & Thomasset N (1986) Use of MTT colorimetric assay to measure cell activation. *Journal of Immunological Methods* 94: 57–63.

Grailer A, Sollinger HW & Burlingham WJ (1988) A rapid assay for measuring both colony size and cytolytic activity of limiting dilution microcultures. *Journal of Immunological Methods* 107: 111–117.

Mosmann T (1983) Rapid colorimetric assay for cellular growth and survival: application to proliferation and cytotoxicity assays. *Journal of Immunological Methods* 65: 55–63.

2.4 NEUTRAL RED (NR) ASSAY

Neutral red (3-amino-7-dimethyl-2-methylphenazine hydrochloride) is a water-soluble, weakly basic, supravital dye that accumulates in lysosomes of viable cells. The neutral red (NR) assay is an *in vitro* cell viability test that was developed and extensively studied for *in vitro* cytotoxicity determinations (Babich & Borenfreund, 1990). Earlier *in vitro* tissue culture studies using the NR dye were developed for assessments of viral cytopathogenicity (Finter, 1969) and for immunotoxicity assays (Mullbacher *et al.*, 1984). The NR assay is based on the incorporation of NR into the lysosomes of viable cells after their incubation with test agents. Cellular uptake of dye is accomplished by passive transport across the plasma membrane. Accumulation of NR within lysosomes occurs either from the binding of NR to fixed acidic charges, such as those of polysaccharides, within the lysosomal matrix, or from the trapping of the protonated form of NR within the acid milieu of the lysosomes. In damaged or dead cells NR is no longer retained in the cytoplasmic vacuoles and the plasma membrane does not act as a barrier to retain the NR within the cells (Bulychev *et al.*, 1978).

Principle of the assay

In the protocol for the NR assay, individual wells of a 96-well tissue culture microtitre plate are inoculated with medium containing cells, to achieve semi-confluence at the time of addition of test agents. This seeding density allows for the target cells to complete an additional replication cycle. As such, the NR assay has the potential to measure cytotoxic, cytostatic and proliferative effects of the test agents. After 1–3 days of incubation, depending on the growth kinetics of the specific cell type, the medium is replaced with an unamended (control) medium or with a medium amended with varied concentrations of the test agents. The cells are exposed for 24 h, or longer if the target cells are required to metabolize the parent test agents to cytotoxic intermediates. Alternatively, a hepatic S9 microsomal fraction from Arochlor-induced rats can be added as the bioactivating system. Following exposure to the test agent, the medium is replaced with medium containing NR and the plate is returned to the incubator to allow for uptake of the dye. The cells are then rapidly washed with a mild fixative. Dead cells or those with damage to the plasma or lysosomal membranes cannot retain the dye after this washing–fixation procedure. The NR is then extracted from the intact, viable cells, i.e. those surviving the exposure to the test agents, and quantitated spectrophotometrically using a scanning-well (ELISA-type) spectrophotometer. Quantitation of the extracted dye spectrophotometrically has been correlated with the number of viable cells, both by direct cell counts and by protein determination of cell populations (Boronfreund & Puerner, 1985).

The NR assay was designed specifically to meet the needs of industrial, pharmaceutical, environmental and other testing laboratories concerned with acute toxicity testing. The numerous advantages of the NR assay, such as its simplicity, speed, economy and sensitivity, are apparent. Good reproducibility of NR cytotoxicity data among different laboratories has been demonstrated.

Selection of cell type

Human and animal (mammalian or fish) cells derived from various organs of origin have been used both as early passage cultures and as immortalized cell lines. Anchorage-dependent cells grown as monolayers readily allow for a number of media changes. Because the number of lysosomes per cell differs between tissue types, the absolute amount of NR recovered from a given cell population will vary. Control and experimental cells must therefore always be of the same type. Growth rate should be considered in determining the exposure time to a given xenobiotic, particularly if a replication cycle is expected to be affected. Other experiments might preferentially deal with confluent cultures.

For studies involving metabolism-mediated effects of xenobiotics, cells that are known to contain the requisite mixed-function oxygenases should be selected. As an alternative, an S9 hepatic microsomal fraction (see Supplementary procedure: bioactivation) can be included in the test medium (Borenfreund & Puerner, 1987). Serum concentrations should be kept as low as tolerated (or eliminated as in a defined medium) to prevent binding of the test agent to serum components.

PROCEDURE: NEUTRAL RED ASSAY

Solutions

- Neutral red (4 mg ml^{-1} stock solution): dilute 1:100 into medium, incubate overnight at 37°C and centrifuge before use
- 1% CaCl$_2$/0.5% formaldehyde: mix 6.5 ml of 37% formaldehyde with 50 ml of 10% CaCl$_2$ and 445 ml of distilled water
- 1% Acetic acid/50% ethanol: mix 4.75 ml of acetic acid with 250 ml of 95% ethanol and 245 ml of distilled water

Materials and equipment

- Complete media suitable for chosen cell type
- Culture Petri dishes
- 96-Well tissue culture plates
- Inverted microscope
- ELISA-type spectrophotometer
- Microplate shaker
- Eight-channel pipette

NEUTRAL RED (NR) ASSAY

Note: Inclusion of serum is usually required to facilitate adherence to the plastic surface.

1. Resuspend cells of actively growing culture by standard procedure.
2. Count cells and accurately allocate appropriate number suspended in medium, calculated to reach about 60–70% confluence at time of addition of test agent. Cell number will vary depending on cell size and growth rate; approximate ranges are 9×10^3 to 4×10^4 cells per well.
3. Seed 0.2 ml containing desired number of cells to each well of 96-well plates and incubate at 37°C for 24 h, or longer, to achieve desired density.
4. Remove medium and add fresh medium containing graded dilutions of test agent. Incubate for desired length of time. At least eight wells should serve as controls and contain medium without xenobiotics.
5. At this point broad toxicity ranges can be determined in preliminary experiments. Examine cultures with an inverted microscope to determine the highest tolerated dose (HTD), which is about 90% cell survival, and those concentrations leading to total cell destruction. An appropriate concentration range of test agent can then be selected for a subsequent NR assay (Borenfreund & Borrero, 1984). Examine at least 4–8 wells per concentration of test agent. Keep serum concentration as low as possible during this step (or use defined medium without serum) in order to prevent or reduce adsorption of xenobiotic to serum components.
6. After incubation for the desired time interval, remove medium with test agent and incubate cells with fresh medium containing 40 μg ml^{-1} NR dye. (An aqueous stock solution of 4 mg ml^{-1}, shielded from light by foil, can be kept for several weeks.) This medium should be prepared earlier and incubated at 37°C overnight to allow for precipitation of small dye crystals. Centrifuge medium for 10 min at 1500 g before use and add 0.2 ml of decanted supernate to each well. The first two wells on each plate should receive medium without NR and serve as blanks for spectrophotometric analysis.
7. Continue incubation for 3 h to allow for incorporation of vital dye into surviving cells.
8. Remove medium by inverting the plate and wash cells. Although washes with phosphate-buffered saline (PBS) are acceptable, a rapid rinse with a mixture of 1% calcium chloride and 0.5% formaldehyde is preferred because it is more efficient in removing extraneous dye crystals and simultaneously promoting cell adhesion.
9. Extract dye into supernate with 0.2 ml of a solution of 1% acetic acid/50% ethanol. After 10 min at room temperature and rapid agitation for a few seconds on a microtitre plate shaker, scan the plate with an ELISA-type spectrophotometer equipped with a 540-nm filter.
10. Compare absorbance of dye extracts from control cells with those of experimentals, by determining the arithmetic mean for each set of concentrations of test agent. Calculate the percentage cell population viability:

$$\% \text{ Viability} = \frac{\text{Mean absorbance of experimental cells}}{\text{Mean absorbance of control cells}}$$

11. Use individual data points for each experimental concentration (presented as the arithmetic mean ± standard error of the mean) to construct concentration–response toxicity curves. Such curves are used to calculate midpoint toxicities or NR_{50} values, as determined by linear regression analysis. Use this value for comparison and ranking against other chemicals, or for relative sensitivity determination of different cell types.

SUPPLEMENTARY PROCEDURE: PROTEIN ASSAY

Materials and equipment

- All materials and equipment previously stated for the NR assay
- Coomassie blue (Bio-Rad Laboratories, Richmond, CA)

A procedure for total protein determination can be added, after completion of the NR assay, using the same 96-well microtitre plate (Babich & Borenfreund, 1987).

1. After completion of the NR assay, wash cells with PBS and then with distilled water.
2. Lyse cells. Add 50 µl of 0.1 M NaOH to each well and incubate plate at 37°C for 2 h or, preferably, for 24 h.
3. Add 200 µl of a 1:6 dilution of Coomassie blue to each well and keep at room temperature for 30 min.
4. Measure absorbance with the scanning spectrophotometer, using a 405-nm (reference) filter and a 630-nm (absorbance) filter, and calculate as Δ 630–405.
5. Use bovine serum albumin dilutions of 20–200 µg protein well^{-1} as standards.

SUPPLEMENTARY PROCEDURE: BIOACTIVATION

Materials and equipment

- All materials and equipment as previously stated for NR assay
- Hepatic S9 microsomal fraction from Arochlor-induced rats (commercially available, should be obtained with an analysis of S9 protein content; alternatively, S9 hepatic fractions can be prepared).
- Ultra freezer (–70°C) or liquid nitrogen to store S9 microsomal fractions.

To exert a cytotoxic effect, some chemicals require bioactivation by hepatic microsomal enzymes, usually provided as a liver homogenate (S9) prepared from Arochlor-induced rats (Borenfreund & Puerner, 1987; Babich et al., 1988).

1. Dilute stock of hepatic S9 microsomal fraction with 50 mM sodium phosphate buffer (pH 7.4) to 10 mg ml^{-1} S9 protein and store in 1-ml aliquots at –70°C until use.
2. Prepare, immediately before use, 1.0 ml of S9 co-factor mixture by combining 0.1 ml of 50 mM glucose 6-phosphate, 0.1 ml of 100 mM $MgCl_2$, 0.1 ml of

30 mM KCl, 0.1 ml of 20 mM CaCl$_2$, 0.3 ml of 50 mM sodium phosphate buffer (pH 7.4), 0.1 ml of 40 mM NADP (kept frozen until needed) and 0.2 ml of 10 mg ml^{-1} S9 protein.
3. Dilute S9 co-factor of mixture 1:10 into medium containing different concentrations of test agents; dilute controls 1:10 into medium without test agents.
4. Add 0.2-ml aliquots of S9 co-factor/toxicant-amended medium to individual wells of a 96-well microtitre plate already seeded with cells and incubated for 24 h.
5. Perform NR assay as described previously.

SUPPLEMENTARY PROCEDURE: UV RADIATION

Materials and equipment

- Medium without phenol red
- Standardized radiation source
- Lid of 96-well plate covered with black tape

The NR assay can be used for studies on the effect of UV radiation on different cell types and the inhibition or enhancement of such effects by different xenobiotics (Babich & Borenfreund, 1992).

1. Remove medium in which cells have been seeded to the 96-well plates.
2. Add medium without phenol red. This medium could contain various xenobiotics.
3. Expose to desired wavelength for predetermined time intervals. Shield control cells from radiation by covering wells with taped lid.
4. Replace experimental medium with normal medium for desired time or again add a xenobiotic to be studied. Incubate at 37°C.
5. Add NR medium for 3 h for assay as described.

Literature review

The NR assay was developed initially to serve as a potential alternative to the Draize rabbit ocular irritancy test (Borenfreund & Puerner, 1985). In addition to providing information on ocular irritancy, the assay was a good predictor of *in vitro* skin irritancy and acute toxicity. The NR assay has been used to evaluate the relative potencies of a spectrum of test agents, including inorganic metals, organometals, environmental pollutants, pharmaceuticals including cancer chemotherapeutics and over-the-counter drugs, surfactants, industrial chemicals, natural toxins, complex mixtures and biomaterials. Furthermore, this assay has been applied to investigations of: metabolism-mediated cytotoxicity; structure–activity relationships (SARs) for series of related chemicals; phototoxicity; temperature–toxicity interactions; synergistic and antagonistic interactions between combinations of test agents; growth promoters and stimulators; and oxidative stress (Babich & Borenfreund, 1990, 1991; Babich *et al.*, 1993, 1995, 1996a, 1996b; Sinensky *et al.*, 1995; Babich & Babich, 1997).

REFERENCES

Babich H & Babich JP (1997) Sodium lauryl sulfate and triclosan: *in vitro* cytotoxicity studies with gingival cells. *Toxicology Letters* 91: 189–196.

Babich H & Borenfreund E (1987) Structure–activity relationship (SAR) models established *in vitro* with the neutral red cytotoxicity assay. *Toxicology In Vitro* 1: 3–9.

Babich H & Borenfreund E (1990) Application of the neutral red cytotoxicity assay to *in vitro* toxicology. *Alternatives to Laboratory Animals* 18: 129–144.

Babich H & Borenfreund E (1991) Cytotoxicity and genotoxicity assays with cultured fish cells: a review. *Toxicology in Vitro* 5: 91–100.

Babich H & Borenfreund E (1992) Neutral red assay for toxicology *in vitro*. In: Watson RR (ed.) *In Vitro Methods in Toxicology*, Chapter 17, pp. 237–251. CRC Press, Boca Raton, FL.

Babich H, Sardana MK & Borenfreund E (1988) Acute cytotoxicities of polynuclear aromatic hydrocarbons *in vitro* with the human liver tumor cell line, HepG2. *Cell Biology and Toxicology* 4: 494–301.

Babich H, Stern A & Borenfreund E (1993) Eugenol cytotoxicity evaluated with continuous cell lines. *Toxicology In Vitro* 7: 105–109.

Babich H, Wurzburger BJ, Rubin YL, Sinensky MC, Borenfreund E & Blau L (1995) An *in vitro* study on the cytotoxicity of chlorhexidine digluconate to human gingival cells. *Cell Biology and Toxicology* 11: 79–88.

Babich H, Zuckerbraun HL, Wurzburger BJ, Rubin YL, Borenfreund E & Blau L (1996a) Response of the human keratinocyte cell line, RHEK-1, to benzoyl peroxide. *Toxicology* 106: 187–196.

Babich H, Zuckerbraun HL, Barber IB, Babich SB & Borenfreund E (1996b) Cytotoxicity of sanguinarine chloride to cultured human cells from oral tissue. *Pharmacology and Toxicology* 78: 397–403.

Borenfreund E & Borrero B (1984) *In vivo* cytotoxicity assays: potential alternatives to the Draize ocular irritancy test. *Cell Biology and Toxicology* 1: 55–65.

Borenfreund E & Puerner JA (1985) Toxicity determined *in vitro* by morphological alterations and neutral red absorption. *Toxicology Letters* 24: 119–124.

Borenfreund E & JA Puerner (1986) Cytotoxicity of metals, metal–metal and metal–chelator combinations assayed *in vitro*. *Toxicology* 39: 121–134.

Borenfreund E & Puerner JA (1987) Short-term quantitative *in vitro* cytotoxicity assay involving an S9 activating system. *Cancer Letters* 34: 243–248.

Bulychev A, Trouet A & Tulkens P (1978) Uptake and intracellular distribution of neutral red in cultured fibroblasts. *Experimental Cell Research* 115: 343–355.

Finter NB (1969) Dye uptake methods for measuring viral cytopathogenicity and their application to interferon assays. *Journal of General Virology* 5: 419–427.

Mullbacher A, Parish CR & Mundy JP (1984) An improved colorimetric assay for T cell cytotoxicity *in vitro*. *Journal of Immunological Methods* 68: 205–215.

Sinensky MC, Leiser AL & Babich H (1995) Oxidative stress aspects of the cytotoxicity of carbamide peroxide: *in vitro* studies. *Toxicology Letters* 75: 101–109.

2.5 LDH ASSAY

PROCEDURE: MEASUREMENT OF LDH ACTIVITY

The measurement of lactate dehydrogenate (LDH: EC. 1.1.1.27) in culture supernatants gives a quantitative value for the loss of cell viability:

$$\text{pyruvate} + \text{NADH} + \text{H}^+ \overset{\text{LDH}}{\rightleftharpoons} \text{NAD}^+ + \text{lactate}$$

The activity of LDH can be measured as the reduction of pyruvate to lactate (Vassault, 1983). The reduction is coupled to the oxidation of NADH to NAD^+, which is followed spectrophotometrically at 340 nm. The equilibrium is far on the side of NAD^+ and lactate. Because NADH has a high absorbance at 340 nm compared with NAD^+, the reaction is measured as the rate of decrease in absorbence at 340 nm.

Although colorimetric assays based on the reduction of tetrazolium salts have been used, they have been largely superseded because of non-specific side reactions in such assay systems.

Reagents and solutions

- Buffer (Tris, 81.3 mmol l^{-1}; NaCl, 203.3 mmol l^{-1}; pH 7.2): dissolve 4.92 g of Tris and 5.95 g of NaCl in 400 ml of water and adjust to pH 7.2 at 30°C with HCl. Make up to a final volume of 500 ml with water
- NADH solution (β-NADH, 0.17 mg ml^{-1}): dissolve 3.4 mg of NADH in 20 ml of buffer
- Pyruvate solution (9.76 mmol l^{-1}): dissolve 0.107 g of monosodium pyruvate in 90 ml of buffer. Make up to a final volume of 100 ml with buffer

Note: NADH must be free of inhibitors. Discard if the powder has become yellow.

Stability of solutions

Buffer is stable at 0–4°C provided that bacterial contamination does not occur. The NADH solution is kept at 0–4°C and must be prepared fresh daily. The pyruvate solution should be dispensed into 1.5-ml aliquots and stored at –20°C. After thawing, each aliquot should be discarded. The pyruvate solution is stable for 2 months.

Materials and equipment

- NADH solution
- Pyruvate solution
- Narrow-bandwidth spectrophotometer, fitted with a thermostatted cuvette holder capable of temperature control within ±0.1°C and a chart-recorder

Assay conditions

- Incubation temperature: 30.0°C
- Wavelength: 340 nm
- Final reaction volume: 1.07 ml
- Light path: 1.0 cm

Read absorbance at room temperature against air.

1. Centrifuge the sample of medium (3000 g, 4°C, 5 min) and take the supernatant. Maintain the supernatant at 0–4°C. Do not freeze.
2. Successively pipette into the cuvette 0.833 ml of NADH solution and 0.07 ml of supernatant. Mix thoroughly.
3. Record the background change in absorbance on the chart-recorder until a stable reading has been achieved.
4. Initiate the reaction absorbance change by the addition of 0.167 ml of pyruvate solution. Mix thoroughly.
5. Continue monitoring the reaction absorbance change with the chart-recorder. A lag phase of up to 30 s may occur. The reaction should be monitored for 5–10 min or until a linear absorbance change of ≥0.2 occurs.

Calculation

1. For the background absorbance change, choose a period where the decrease in absorbance (A) is linear with time and calculate ΔA per minute.
2. Repeat Step 1 for the reaction absorbance change.
3. Calculate as follows:

$$\text{Activity} = \left(\frac{\Delta A_{react}}{\Delta t} - \frac{\Delta A_{bkgd}}{\Delta t} \right) \times \frac{V_T}{6.22 \times d \times V_s} \quad \text{units ml}^{-1}$$

where 1 unit = 1 µmol NADH consumed min^{-1};

$\Delta A_{react}/\Delta t$ is calculated in Step 2, ΔA_{bkgd} is calculated in Step 1, V_T is the total volume of the assay mixture (ml), V_s is the volume of supernatant (ml) and d is the path length (cm).

DISCUSSION

Background information

The demand for greater process productivity has led to the development of high-cell-density entrapped-culture systems, with the resulting isolation of cells in/on structures where they are difficult to access directly. Thus methods of assessing culture viability that require sampling of the cells (review: Cook & Mitchell, 1989) are not suitable for use in such systems. Non-invasive metabolic parameters (e.g. glucose uptake) can be used to monitor changes in culture viability. However, the interpretation of such data is compromised because uptake/production rates can alter as a result of the cell switching carbon source, variation of metabolic rates with growth phase of the culture, as well as fluctuation in viable cell numbers. Analysis of the release of intracellular enzymes can be used to monitor changes in culture viability. The most commonly used enzyme in cell culture studies is LDH (EC 1.1.1.27). The assumptions are made that intracellular enzymes are only released after damage to the cell membrane and that all activity is rapidly released from damaged cells. Thus the higher the rate of release of enzyme activity, the greater the extent of cell damage/death, i.e. loss of culture viability, in that period.

The advantage of such methods over measurement of metabolite levels is that the measured parameter varies in response to changes in culture viability and not cell metabolism.

Expected results

The minimum activity detected in our laboratory for medium containing heat-inactivated serum is 20–30 mU ml^{-1}, which is equivalent to ΔA min^{-1} = 0.010 (Ciba-Corning Spectrascan). However, the minimum activity detected will depend upon the batch of serum used and whether it has been heat-inactivated. Lower minimum values should be expected from serum-free and low-protein media.

Pitfalls

The release of LDH activity can be related to the total number of dead and lysed cells, as determined by conventional dye exclusion methods. This relationship allows the growth and death kinetics within the culture system to be evaluated. However, there are various phenomena that could complicate this evaluation (Marc et al., 1991; Legrand et al., 1992; Wagner et al., 1992).

The stability of LDH can vary considerably, ranging from the loss of a few per cent per day to a half-life of 12 h, depending upon the cell type.

It is assumed that the release of LDH occurs rapidly after damage to the cell membrane. However, this assumption is not necessarily correct. The release of LDH can be complete in cells that are considered dead by dye exclusion methods. Alternatively, complete release may occur only upon cell lysis. This point is further complicated because dye exclusion methods do not measure lysed cells.

The specific release rate of LDH will be modulated by culture conditions that affect the intracellular LDH content. These include parameters such as pH, dissolved oxygen concentration, medium composition and age of the culture. Although the LDH content per cell is affected by such parameters, good agreement between LDH release and cell death has been reported for a variety of bioreactor configurations and stress conditions (Marc *et al.*, 1992; Legrand *et al.*, 1992; Wagner *et al.*, 1992).

In conclusion, as the kinetics of LDH release are affected by both cell type and cultural conditions, the relationship between cell death and LDH release should be determined for each new cell line or set of culture conditions.

Alternative application

The above commentary and methodology may have given the impression that the only use in cell culture for measurements of LDH activity is in the monitoring of changes in culture viability. An alternative application is in the comparison of cell viability between different regions within entrapped cell culture systems, where it may not be possible to determine viable cell numbers directly.

The assumption behind the use of LDH as a marker for loss of culture viability is that the enzyme is only released upon damage of intact (i.e. viable) cells. Therefore, if a mixture of viable and non-viable cells is lysed, the greater the number of viable cells, the more LDH activity will be recovered.

Racher *et al.* (1990) report the use of this approach to compare the distribution of cell viability within a fixed-bed porous-sphere bioreactor. The most accurate method of estimating cell numbers within glass spheres is by doing a nuclei count, which obviously precludes estimating the number of viable cells. Samples were taken from different positions within the bed, and the cells within the spheres were lysed on incubation in buffer containing Triton-X-100; the LDH activity of the lysate was determined as described above. Normalization of the enzyme activity on protein concentration of the lysate and cell density in the bead (estimated from the nuclei count) allowed comparison of cell viability between different parts of the bed.

REFERENCES

Cook JA & Mitchell JB (1989) Viability measurements in mammalian cell systems. *Analytical Biochemistry* 179: 1–7.

Legrand C, Bour JM, Jacob C, Capiaumont J, Martial A, Marc A, Wudtke M, Kretzmer G, Demangel C, Duval D & Hache J (1992) Lactate dehydrogenase (LDH) activity of the number of dead cells in the medium of cultured eukaryotic cells as marker. *Journal of Biotechnology* 25: 231–243.

Marc A, Wagner A, Martial A, Goergen JL, Engasser JM, Geaugey V & Pinton H (1991) Potential and pitfalls of using LDH release for the evaluation of animal cell death kinetics. In: Spier R, Griffiths JB & Meigner B (eds) *Production of Biologicals from Animal Cells in Culture*, pp. 569–575. Butterworth–Heinemann, Oxford.

Racher AJ, Looby D & Griffiths JB (1990) Studies on monoclonal antibody production by a hybridoma cell line (ClE3) immobilized in a fixed bed, porosphere system. *Journal of Biotechnology* 15: 129–146.

Vassault A (1983) Lactate dehydrogenase: UV-method with pyruvate and NADH. In: Bergmeyer HU, Bergmeyer J & Grassl M (eds) *Methods of Enzymatic Analysis. III. Enzymes 1: Oxidoreductases, Transferases*, 3rd edn, pp. 118–126. Verlag-Chemie, Weinheim.

Wagner A, Marc A, Engasser JM & Einsele A (1992) The use of lactate dehydrogenase (LDH) release kinetics for the evaluation of death and growth of mammalian cells in perfusion reactors. *Biotechnology and Bioengineering* 39: 320–325.

2.6 MINIATURIZED COLORIMETRIC METHODS FOR DETERMINING CELL NUMBER

Rapid and accurate assessment of viable cell number is an important requirement in many experimental situations involving animal cell culture, e.g. measurement of growth factor activity, serum batch testing and toxicity assays.

Colony-forming efficiency/clonogenic assays are the most reliable methods for assessing viable cell number. These methods are time-consuming, however, and become impractical when many samples have to be analysed (e.g. in toxicity assessment where several chemicals at a range of concentrations and combinations need to be analysed).

Colorimetric assays that can be miniaturized into 96-well plates and measured using an ELISA reader allow many samples to be analysed rapidly, and also reduce medium and plastics costs (Cook & Mitchell, 1989). The acid phosphatase assay, terminated by NaOH addition (AP NaOH assay), has been found to be particularly useful for both serum batch testing and toxicity assessment. The usefulness and limitations of the following colorimetric assays are described.

- Cellular acid phosphatase (AP 2-h and AP NaOH) (Connolly *et al.*, 1986; Martin & Clynes, 1991)
- Neutral red (NR) uptake (Borenfreund & Puerner, 1983; Fiennes *et al.*, 1987; Triglia *et al.*, 1991)
- Crystal violet dye elution (CVDE) (Kueng *et al.*, 1989; Scragg & Ferreira, 1991)
- Tetrazolium dye (3-(4,5-dimethylthiazol-2-yl)-2,5-diphenyl tetrazolium bromide, MTT) reduction (Mossman, 1983; Alley *et al.*, 1988; Vistica *et al.*, 1991)
- Staining with sulphorhodamine B (SRB) (Skehan *et al.*, 1990)

PRELIMINARY PROCEDURE: PRETREATMENT OF CELLS

Cells are subjected to a standard pretreatment to ensure that they are in exponential phase for experiments. Pretreat cells for 48 h prior to any experiment by plating in 75-cm^2 flasks and feeding with fresh medium 24 h before use. Determine the seeding concentration for each cell line; examples are 1×10^6 cells per flask for NRK cells and 2.5×10^6 cells per flask for Hep-2 cells.

PRELIMINARY PROCEDURE: 96-WELL CELL GROWTH OR TOXICITY ASSAYS

Materials and equipment

- Trypan blue, 0.4%
- Trypsin, 0.5%
- EDTA, 0.045%
- Haemocytometer
- Coulter counter

1. Using 8-well multichannel pipettes, plate the cell suspensions (carefully mixed) into 96-well tissue culture dishes (e.g. Costar or Greiner).
2. Incubate at 37°C in a humidified 95% air/5% CO_2 atmosphere. Total volume of growth medium per well should be 100 µl.
3. After an appropriate growth period (e.g. 4–6 days, depending on cell line, for serum batch testing), determine the endpoint (as below). For toxicity tests, incubate cells in 100 µl of medium for 24 h prior to addition of test substance (at twice the final concentration in 100 µl of medium). Incubation with toxic substance prior to endpoint measurement is usually for 6 days. For measurement of Trypan blue exclusion, trypsinize cells with 0.5% trypsin (Gibco) and 0.045% EDTA (Sigma).

Trypan blue is harmful if ingested or inhaled. It is irritating to the eyes, is harmful by skin contact and has been found to cause cancer in laboratory animals. Appropriate precautions should be taken when handling Trypan blue and the use of an extraction hood and gloves is advised if handling Trypan blue in powder form.

4. For direct cell counting, pool eight replicate wells and carry out counts using haemocytometer and Coulter counter.

PRELIMINARY PROCEDURE: TRYPAN BLUE EXCLUSION METHOD FOR CELL VIABILITY ESTIMATION

The materials and equipment are the same as those used for the above 96-well assays.

1. Mix 1.0 ml of cell suspension with 0.2 ml of Trypan blue solution (0.4%, Gibco).
2. Between 5 and 15 min later, count the coloured ('non-viable') and dye-excluding ('viable') cells using a haemocytometer.

PROCEDURE: COLORIMETRIC ASSAYS: GENERAL INTRODUCTION

All of these assays are read at 570 nm (except for the AP assay, for which the wavelength is 405 nm) on a Titertek Multiskan-plus ELISA plate reader using a 620-nm filter as reference wavelength.

It is important to remove any bubbles from the wells before absorbance readings.

Materials and equipment

- Substrate-containing buffer: 10 mM p-nitrophenyl phosphate (Sigma) (added just before use to 0.1% Triton X-100 (BDH)) in 0.1 M sodium acetate (Sigma) pH 5.5 p-Nitrophenyl phosphate substrate – this hydrolyses in solution during extended storage, so it should be prepared freshly before use)
- Neutral red solution, 0.4%
- Formol/calcium mixture (10 ml of 40% formaldehyde (BDH) and 10 ml of anhydrous calcium chloride (Sigma) in 80 ml of ultrapure water)
- Acetic acid/ethanol mixture (1% glacial acetic acid in 50% ethanol)
- Phosphate-buffered saline (PBS)
- Aqueous crystal violet, 0.25% (g 100 ml^{-1}) (prefiltered through Whatman No. 1 paper)
- Glacial acetic acid, 33%
- MTT, 5 mg ml^{-1} (the MTT solution can be stored for at least 6 weeks at 2°C, protected from light, without altered assay results when compared with fresh solutions)
- Dimethylsulphoxide (DMSO)
- TCA (trichloroacetic acid)
- Sulphorhodamine B (SRB) (0.4% in 1% acetic acid)
- Tris buffer, pH 10.5
- Haemocytometer
- Coulter counter
- ELISA reader

 All the dyes/substrates should be handled and disposed of as if they were considered mutagenic.

Acid phosphatase (AP) assay

1. At end of the cell growth period, remove medium and rinse the wells in 100 µl of PBS.
2. Add 100 µl of substrate-containing buffer to each well.
3. Incubate for 2 h in a humidified incubator in 5% CO_2/95% air at 37°C. Read plates at 405 nm, and either reincubate for a further time if increased sensitivity is required or 'stop' with addition of 50 µl well^{-1} of 1 M NaOH to cause an electrophilic shift in the p-nitrophenyl chromophore and thus develop most of the yellow colour, giving greatly increased sensitivity.

Neutral red (NR) assay

1. Centrifuge a 1:80 dilution of 0.4% neutral red (Difco) solution in culture medium (final concentration, 50 µg ml^{-1}) at 1500 g for 10 min before use to remove insoluble crystals. Prepare fresh solutions each day.
2. Following removal of medium, rinse plates in PBS (200 µl well^{-1}) before adding neutral red solution (above).
3. After 3 h of incubation (in a humidified incubator in 5% CO_2/95% air at 37°C), remove neutral red and rinse wells with 50 µl of formol/calcium mixture.
4. Add 200 µl well^{-1} of acetic acid/ethanol mixture.
5. Mix plates well by gentle tapping, before reading on the ELISA reader at 570 nm.

Crystal violet dye elution (CVDE)

1. After removal of medium, rinse 96-well plates with 100 µl well^{-1} of PBS and stain with 100 µl of 0.25% (g 100 ml^{-1}) aqueous crystal violet for 10 min.
2. Rinse plates four times in tapwater.
3. Dry the outsides of the plates with paper to help avoid water stains, and then dry the plates at 37°C. When dry, add 100 µl well^{-1} of 33% glacial acetic acid (33 ml 100 ml^{-1}) (BDH) and mix the contents of each well before reading at 570 nm.

MTT assay

 MTT is believed to be mutagenic.

1. At the end of the growth or toxicity test period, add 20 µl well^{-1} of 5 mg ml^{-1} MTT.
2. Incubate the mixture for 4 h in a humidified incubator in 5% CO_2/95% air at 37°C.
3. Carefully remove the medium by pipetting (with care not to remove formazan crystals) and add 100 µl well^{-1} of DMSO.
4. Aspirate repeatedly to give a uniform colour before reading at 570 nm.

Sulphorhodamine B (SRB) assay

1. Fix cells by layering 50 µl of 50% TCA (Reidel-de-Haen) directly on top of the incubation medium, and then incubate the plate at 2°C for 1 h.
2. Rinse wells five times with tapwater, flick to remove liquid and then allow to dry thoroughly.
3. Stain cells with 200 µl well^{-1} of SRB (0.4% in 1% acetic acid) for 30 min and rinse four times in 1% glacial acetic acid.
4. When dry, add 200 µl well^{-1} of 10 mM Tris buffer (pH 10.5) to release unbound dye. After mixing, read plates at 570 nm.

DISCUSSION

Most experiments requiring assessment of growth or toxicity involve comparison of experimental 'treated' cultures with control untreated cultures. If such comparisons are to give an accurate measurement of growth stimulation or inhibition, then control and experimental wells should still be in exponential growth phase at the time of assay endpoint determination. Also, the cell number in all wells should be in the linear position of the graph of optical density versus cell number per well. To ensure that these conditions apply, each assay method should be characterized for:

1. Minimum detectable cell number per well.
2. Sensitivity (measured as, for example, a change in absorbance per change of 10 000 cells).
3. Linear range of absorbance versus cell number curve.

It is important to note that these parameters vary with:

- Cell type
- Assay duration
- Assay conditions (e.g. volume, serum concentration, serum batch)

Their actual numerical values cannot, therefore, be considered as constants.

As a general rule, the most sensitive assays (using criteria 1 and 2 above) are AP NaOH. SRB, CVDE and NR, in that order. Both AP 2-h and MTT are the least sensitive, but they do have the greatest linear ranges, and this is an important consideration in some experiments where a very wide range of cell concentrations is expected.

Before applying any of these assays to routine use, it is worthwhile validating some typical assays against colony-formation assays. For example, in toxicity assays it has been found that the colorimetric assays (and Trypan blue exclusion) correlate well with colony-formation assays after 3 or 4 days of drug treatment, but if the assessment is done after 2 days of toxin exposure, all of the methods (and Trypan blue exclusion) seriously underestimate cell kill as determined by colony-formation assays. In this case, cells that are reproductively dead still have active cellular enzymes, e.g. for NR uptake and MTT reduction. Provided that the above constraints are recognized and the appropriate preliminary data are generated, these miniaturized colorimetric assays can be of tremendous value in reducing costs and increasing throughput in cell growth or toxicity assessment.

REFERENCES

Alley, MC, Scudiero DA, Monks A, Hursey ML, Czerwinsk MJ, Fine DL, Abbott BJ, Mayo JG, Shoemaker RH & Boyd MR (1988) Feasibility of drug screening with panels of human tumor cell lines using a microculture tetrazolium assay. *Cancer Research* 48: 589–601.

Borenfreund E & Puerner JA (1983) A simple quantitative procedure using monolayer cultures for cytotoxicity assays. *Journal of Tissue Culture Methods* 9: 7–9.

Connolly DT, Knight MB, Harakas NK, Wittwer AJ & Feder J (1986) Determination of the number of endothelial cells

in culture versus an acid phosphatase assay. *Analytical Biochemistry* 152: 136–140.
Cook JA & Mitchell JB (1989) Viability measurements in mammalian cell systems. *Analytical Biochemistry* 179: 1–7.
Fiennes A, Walton J, Winterbourne D, McGlashan D & Hermon-Taylor J (1987) Quantitative correlation of neutral red dye uptake with cell numbers in human cancer cell cultures. *Cell Biology International Reports* 11: 373–378.
Kueng W, Silber E & Eppenberger U (1989) Quantification of cells cultured on 96-well plates. *Analytical Biochemistry* 182: 16–19.
Martin A & Clynes M (1991) Acid phosphatase: endpoint for *in vitro* toxicity tests. *In Vitro Cell and Developmental Biology* 27A: 183–184.
Mossman T (1983) Rapid colorimetric assay for cellular growth and survival: application to proliferation and cytotoxicity assays. *Journal of Immunological Methods* 65: 55–63.
Scragg M & Ferreira L (1991) Evaluation of differential staining procedures for quantification of fibroblasts cultured in 96-well plates. *Analytical Biochemistry* 198: 80–85.
Skehan P, Storeng R, Scudiero D, Monks A, McMahon J, Vistica D, Warren J, Bokesch H, Kenney S & Boyd MR (1990) New colorimetric assay of anticancer drug screening. *Journal of the National Cancer Institute* 82: 1107–1113.
Triglia D, Braa S, Yonan C & Naughton G (1991) *In vitro* toxicity of various classes of test agents using the neutral red assay on a human three-dimensional physiologic skin model. *In Vitro Cell and Developmental Biology* 27A: 239–244.
Vistica DT, Skehan P, Scudiero D, Monks A, Pittman A & Boyd MR (1991) Tetrazolium-based assays for cellular viability: a critical examination of selected parameters affecting formazan production. *Cancer Research* 51: 2515–2520.

CHAPTER 3

CULTURE ENVIRONMENT

3.1 OVERVIEW

The first developments in routine cell culture were made possible by the creation of media that aimed to mimic physiological fluids (blood, lymph) and conditions for cell growth (Earle *et al.*, 1943). In addition, the discovery and use of antibiotics in cell culture aided this early work. Ironically enough, the formulations were prepared very precisely with milligram quantities of essential amino acids, vitamins, salts and carbohydrates (often based on the analysis of body tissues), to be followed by the addition of an extremely ill-defined component, serum. This latter component, although vital, does contain variable quantities of growth factors; even so, problems remain with conventional media. These include the role of the energy source influencing pH change, instability of glutamine and the variability in serum, which was also a potential source of adventitious agents. In many respects, for most routine cell cultures, these early formulations are still in use, with modifications over the years aimed at overcoming the drawbacks by adding additional carbohydrate in the form of Dulbecco's modified Eagle's medium (DMEM) (Dulbecco & Freeman, 1959), for example, or using alternative buffer systems to the conventional HCO_3/CO_2, such as the Zwitterionic buffer HEPES (Good *et al.*, 1966).

Even though there have been considerable advances in the development of serum-free media, they have not totally superseded conventional serum-based formulations as yet, mainly because of cost and the requirement for tailoring each formulation to cell type. Numerous publications on the subject have been produced (Maurer, 1986) and many serum-free media are available from commercial suppliers.

Because bovine serum (foetal, newborn, adult) remains fundamental to most tissue culture, it is still a necessity to batch test a lot prior to purchase. A reliable batch of serum is a major investment for any tissue culture laboratory. Assays based on the intended use should be in place (often there are more exacting requirements for hybridoma culture) and comparisons made between the material available from alternative suppliers. If a requirement for a particular project is the shipping of materials from one country to another, then the origin of the serum is also a consideration (e.g. US Department of Agriculture restriction on importation) (Stacey *et al.*, 1998).

Another area that has seen considerable advances is that of the physical environment for cell growth. Glassware has almost entirely disappeared from most tissue culture laboratories, to be replaced by an enormous, and expensive, range of plasticware. This has enabled the scientist to derive more sophisticated culture vessels and has led to the manufacture of three-dimensional culture systems and microcarriers. An added benefit has been the reliability and reproducibility of culture systems and, as a result, the avoidance of variability and contamination

---

Cell and Tissue Culture: Laboratory Procedures in Biotechnology, edited by A. Doyle and J.B. Griffiths.
© 1998 John Wiley & Sons Ltd.

by toxic substances. The ability to culture a much wider range of differentiated cell types in highly developed environmental conditions with a host of individual growth factors and three-dimensional matrices is leading to exciting advances in tissue engineering that have enormous implications in the future treatment of human disease, e.g. organ replacement (liver, kidney, skin), gene therapy, etc.

REFERENCES

Dulbecco R & Freeman G (1959) Plaque production by the polyoma virus. *Virology* 8: 396–397.

Earle WR, Schilling EL, Stark TH, Straus NP, Brown MF & Shelton E (1943) Production of malignancy *in vitro* in the mouse fibroblast cultures and changes seen in the living cells. *Journal of the National Cancer Institute* 4: 165–212.

Good NE, Winget GD, Winter W, Connolly TN, Izawa S & Singh RMM (1966) *Biochemistry* 5: 467.

Maurer HR (1986) Towards chemically-defined, serum-free media for mammalian cell culture. In: Freshney RI (ed.) *Animal Cell Culture: A Practical Approach*, pp. 13–31. IRL Press, Oxford.

Stacey G, Doyle A & Hambleton P (eds) (1998) *Safety in Tissue Culture*. Chapman & Hall, London.

3.2 SERUM-FREE SYSTEMS

The advantages of completely defining the cell culture environment were recognized early in the development of cell and organ culture. However, several distinct advances were necessary in order to achieve the goal of maintaining functional mammalian cells in totally defined media. The roles of essential amino acids, vitamins, minerals, salts, trace metals and other nutritional components were first demonstrated by Eagle (1955a, b) and later amplified by the work of Ham (1982) and Waymouth (1972). Once these requirements were met in more complex media, it became possible to examine the role of serum, one of the last remaining, undefined components of most cell culture media (Hayashi & Sato, 1976; Barnes & Sato, 1980).

In the last two decades there has been great progress in the development of serum-free systems for the growth of mammalian cells *in vitro* (Barnes *et al.*, 1984). Work from many different laboratories and cell culture systems has contributed to our understanding of the role of serum in supporting cell growth and function. When a minimal medium such as Eagle's minimal essential medium is used, serum can remain a source of essential nutrients and co-factors, such as vitamins and some of the amino acids. A more complex medium may provide all of these nutrients and still be inadequate to support the growth of many cells in serum-free medium.

In such cases other roles of serum have been identified. These include providing hormones and growth factors, or transport and binding proteins, which present nutrients or hormones to the cell in a preferred or non-toxic form (Perez-Infante *et al.*, 1986). Serum contains attachment factors such as fibronectin and serum-spreading factor, which promote cell attachment to the substrate. Such attachment is required for the growth of some, but not all, cells *in vitro*. In addition to providing substances necessary for cell growth, serum may also play a role in stabilizing and detoxifying the culture environment. For example, serum has a significant buffering capacity and contains specific protease inhibitors, such as α_1-antitrypsin and α_2-macroglobulin. Serum albumin, present in high levels in medium containing 10% serum, may act as a non-specific inhibitor of proteolysis as well as bind fat-soluble vitamins and steroid hormones which can be toxic in their free forms. Serum components may also bind and detoxify heavy metals and reactive organics that may be present in either the water or medium components (Mather *et al.*, 1986). There are thus many roles that serum can play in a culture. Each of these will be of greater or lesser importance, depending on the cell type being studied and the culture conditions used. Therefore, devising an optimal method of serum substitution demands some understanding of both the culture system employed and the ultimate goal of the work.

Cell and Tissue Culture: Laboratory Procedures in Biotechnology, edited by A. Doyle and J.B. Griffiths.
© 1998 John Wiley & Sons Ltd.

ELIMINATION OF SERUM

The many different rationales behind growing cells in serum-free media can be divided into three categories:

- Elimination of serum *per se*
- Understanding the role of serum, nutrients, hormones and growth factors in cell physiology and endocrinology (Barnes & Sato, 1980; Bettger & McKeehan, 1986)
- Culturing and studying cell types *in vitro* that will not grow or maintain function in the presence of serum (Li *et al.*, 1996, 1997; Loo *et al.*, 1987; Roberts *et al.*, 1990)

There are several approaches to the elimination of serum. The approach, or combination of approaches, used should be determined by the goals of the work.

It may be desirable to eliminate serum, one of the most expensive medium components, in order to decrease the cost of the medium. Alternatively, the principal consideration may be to: 1) lower the levels of protein in the medium to facilitate purification of a cell-secreted protein, or 2) remove a substance in serum that may be a major contaminant of a purified protein, or decrease protein quality by proteolysis or some other alteration of the desired cell product. In these instances the means of removing the serum is less important than the end result. However, if the goal is to identify the growth-promoting substance(s) in serum, the cells should be altered as little as possible in the process of the study. Thus the goal of many of these studies has been to devise a defined culture system in which each component is a purified known substance, and in which the cells grow at the same rate as they do in serum, with no period of adaptation. The goal may be to eliminate serum components that cause loss of a specific cell function, such as a hormone response, or inhibit cell growth (Loo *et al.*, 1987). Finally the specificity of serum-free media formulation can be used to advantage to select for the growth of one cell type from among many (Mather & Sato, 1979; Roberts *et al.*, 1990; Li *et al.*, 1996). In these two instances the best serum substitution must be arrived at empirically by measuring the response to be preserved.

It has been shown that optimization of nutrient concentration and balancing the ratios of concentrations of various nutrients in the culture medium can reduce, or in some cases eliminate, the requirement for serum and serum-derived components such as hormones and growth factors (Bettger & McKeehan, 1986). This may be important where medium cost is a major concern. The optimal nutrient balance and concentrations will depend on the cell type, cell density and culture system employed. Optimizing the nutrients is a time-consuming task, but is rewarded with a culture system that has been determined to be optimal for the cell type and conditions used at a minimal cost.

SERUM SUBSTITUTION

An approach that has been widely employed in the last decade is to replace serum with a set of hormones, growth factors, attachment proteins and transport proteins that will substitute for serum in the cell culture system being used. The optimal set of factors will again be cell-type specific and vary somewhat with the culture system and the parameter being optimized, e.g. growth versus expression of a differentiated phenotype. In general, however, it has been found that the hormone requirements of the same cell type *in vitro* are similar, even if the cells are derived from different species, animals at different stages of development or primary cultures versus established cell lines. Thus primary Leydig cell cultures derived from rat, pig and mouse, and a mouse Leydig-derived cell line, all have similar, but not identical, requirements for growth and function in serum-free medium. Similarly, the medium optimized for the dog kidney proximal tubule epithelial cell line MDCK will support the growth of primary proximal tubule cells from rat, rabbit, mouse and wolf, but is not optimal for kidney epithelial cells derived from the distal tubule. By studying these types of requirements *in vitro* a greater understanding can be gained of the role these factors may play in regulating cell function *in vivo*.

Another approach to serum substitution is to adapt the cells to grow in serum-free medium with or without the optimization of nutrients or hormones described above. This was one of the first approaches taken and can provide a powerful tool for the cell biologist. Some cell types, such as myelomas, can be adapted readily to serum-free growth by a process of sequential lowering of serum until the cells are growing at a much reduced level of serum or with no serum at all. Other cell types have not been amenable to this approach. In either case, the use of a nutritionally optimized medium and a minimal hormonal supplementation may speed this process of adaptation. The advantage of this approach is that one can decide, in advance, the desired culture conditions, e.g. protein-free medium, suspension growth, etc., and impose these conditions as selection criteria. Then, if the approach is successful, the cells obtained will grow in the desired culture environment. Knowledge of the mechanism of such adaptation is not required. The disadvantage is that such adaptation may or may not be successful and may take weeks or many months to accomplish. In addition, adapting cells to various growth conditions does little to further our understanding of cell biology.

The three approaches described above may be used individually or in combination to accomplish the goal of elimination of serum from cell culture media. Some investigators strive for a completely defined culture system in which all components are known and only highly purified nutrients, growth factors, etc. are used. Optimally, the investigator should have data that support the assumption that each component is the active factor for that cell type, rather than a contaminant (e.g. platelet-derived growth factor (PDGF) is the active component of PDGF rather than TGF-β contaminating the PDGF). In some studies this level of definition may not be necessary or possible. Thus many serum-free culture systems contain serum components (e.g. bovine serum albumin or fetuin) or other undefined additives (pituitary extract or cell-produced matrix). Similarly, an investigator may choose not to optimize each nutrient individually in the basal medium but may screen several commercially available nutrient mixtures for the best medium to use for the cells of interest. In other cases the choice of the best commercial medium and a minimal hormonal supplement (e.g. insulin) may allow for a significant (10-fold) reduction in the amount of serum required for cell growth. This may be sufficient for some purposes.

DISCUSSION

Other considerations may also have a major impact on the choice of culture system and the method of eliminating serum. The requirements for maintaining cells in a serum-free medium for one passage are frequently less stringent than the requirements for extended passage of cells without serum. Nutrient and hormone requirements for the growth of cells at very low densities differ from those of the same cells maintained at high densities (e.g. greater than 5×10^5 cells ml^{-1}). Finally, it is important to measure the parameter that one wishes to optimize in the process of deriving the optimal serum-free substitute. Conditions optimal for cell growth may not be optimal for the expression of a differentiated function such as the secretion of protein X by the cells. The optimal conditions for the secretion of protein X may, in turn, differ from those optimal for the secretion of protein Y (Perez-Infante et al., 1986), or a change in supplements may cause the cells to differentiate to a non-dividing phenotype (Li et al., 1996, 1997).

While the considerations discussed above may seem daunting to an investigator wishing to devise a serum-free medium for the growth of a specific cell type, the work of many investigators over the last three decades has made the task considerably easier. There are a wide range of basal media commercially available, many of which were specifically derived for the growth of cells in low-serum or serum-free media, for example Ham's nutrient mixtures (Ham, 1982) and F-12/DMEM (Mather & Sato, 1979). The hormone supplements reported as optimal for the growth of a cell line derived from a cell type that is physiologically or embryologically related can be used as a good starting point to optimize supplements for a newly derived cell line. The types of supplements most likely to be essential (Barnes & Sato, 1980; Barnes et al., 1984) are also known. A great many of the growth factors, attachment factors and transport proteins are now available from commercial sources, significantly enhancing the ease of screening factors to derive an optimal medium supplement. There are also a number of 'serum substitutes' on the market. These may be adequate to achieve the level of cell growth, function and control of medium composition desired for some studies, but frequently will neither be chemically defined nor as good as the optimal results that could be achieved by the methods described above.

In summary, advances in our understanding of the nutritional and hormonal requirements of cells in culture and of the role played by attachment factors and transport proteins have led to the possibility that any cell can be cultured in the absence of serum. The type of serum substitute used and the means of achieving a serum-free culture system should be determined by the ultimate purpose that the culture system is to serve. The advantages of serum-free culture are manifest both in the practical realm of providing an inexpensive and simple starting material for the purification of cell-secreted proteins, such as antibodies, and for providing a more precise definition of the *in vitro* environment in order to model more closely the endocrine, autocrine and paracrine factors that regulate cell and organ function *in vivo*. Thus serum-free culture will continue to play a growing role in the next decades, both in the development of biotechnology processes and in our understanding of basic endocrinology and physiology.

REFERENCES

Barnes D & Sato G (1980) Methods for growth of cultured cells in serum-free medium. *Analytical Biochemistry* 102: 255–269.

Barnes D, Sirbasku D & Sato GH (eds) (1984) *Cell Culture Methods for Molecular and Cell Biology*, vols 1, 2 & 3. Alan R. Liss, New York.

Bettger WJ & McKeehan WL (1986) Mechanisms of cellular nutrition. *Physiological Reviews* 66: 1–35.

Eagle H (1955a) The specific amino acid requirements of a human carcinoma cell (strain HeLa) in tissue culture. *Journal of Experimental Medicine* 102: 37–48.

Eagle H (1955b) Nutrition needs of mammalian cells in tissue culture. *Science* 122: 501–504.

Ham RG (1982) Importance of basal nutrient medium in the design of hormonally defined media. In: Sato GH, Pardee AB & Sirbasku (eds) *Growth of Cells in Hormonally Defined Media*, pp. 39–60. Cold Spring Harbor Press, Cold Spring Harbor, New York.

Hayashi I & Sato GH (1976) Replacement of serum by hormones permits growth of cells in a defined medium. *Nature (London)* 259: 132–134.

Li R, Gao WQ & Mather JP (1996) Multiple factors control the proliferation and differentiation of rat early embryonic (day 9) neuroepithelial cells. *Endocrine* 5: 205–217.

Li R, Phillips DM, Moore A & Mather JP (1997) Follicle-stimulating hormone induces terminal differentiation in a pre-differentiated rat granulosa cell line (ROG). *Endocrinology* 138: 2648–2657.

Loo DT, Fuquay JI, Rawson CL & Barnes DW (1987) Extended culture of mouse embryo cells without senescence: inhibition by serum. *Science* 236: 200–202.

Mather JP & Sato GH (1979) The use of serum-free media in primary cultures. *Experimental Cell Research* 124: 215–221.

Mather J, Kaczarowski F, Gabler R & Wilkins F (1986) Effects of water purity and addition of common water contaminants on the growth of cells in serum-free media. *Biotechniques* 4: 56–63.

Perez-Infante V, Bardin CW, Gunsalus GL, Musto NA, Rich KA & Mather JP (1986) Differential regulation of testicular transferrin and androgen binding protein secretion in primary cultures of rat Sertoli cells. *Endocrinology* 118: 383–392.

Roberts PE, Phillips DM & Mather JP (1990) A novel epithelial cell from neonatal rat lung: isolation and differentiated phenotype. *American Journal of Physiology: Lung, Cellular and Molecular Physiology* 3: 415–425.

Waymouth C (1972) Construction of tissue culture media. In: Rothblat GH & Cristofalo VJ (eds) *Growth, Nutrition and Metabolism of Cells in Culture*, vol. 1, pp. 11–47. Academic Press, New York.

3.3 ADAPTATION TO SERUM-FREE CULTURE

The goal when adapting cultured cells to serum-free conditions is to maintain the original cell phenotype and avoid selecting variant subpopulations of cells. One component of the process, therefore, will be to monitor the cell characteristics of interest, such as the concentration of a secreted product.

Serum provides many different components affecting cell survival, proliferation rate and secreted or internal product concentration.

There are many advantages to removing serum from cell culture systems. The protein concentration in serum is high, generally much higher than the concentration of genetically engineered or natural products of the cells. Serum-free growth, with little or no added protein, should facilitate and simplify downstream processing and reduce costs. In addition, serum-free medium improves the reproducibility of cell growth and product yield due to the consistent and defined medium composition. In some cases, product concentration or secretion may be enhanced (Glassy et al., 1988). Finally, with the removal of serum hormones and antibodies, well-defined culture conditions provide a constant environment for studies of cell physiology, cellular immunology, cell–cell interaction and product formation kinetics. Moreover, a serum-free medium significantly reduces the risk of viral contamination.

It is very important to keep in mind the objective when adapting cells to a serum-free environment. If secretion of large amounts of a specific protein is the final goal, then measurement of that protein during the process of serum-free adaptation is critical. Secretion of large amounts of the protein may or may not coincide with doubling time. When the focus is on collecting a protein that is cell associated, high cell number is part of the final goal as well as large amounts of the protein per cell.

The process of serum-free adaptation of cells currently grown in serum begins with selecting a serum batch that supports growth of the cells in low serum concentrations and selecting a rich nutrient medium that supports cell proliferation in low serum.

The cells can then be adapted through growth in a series of progressively lower serum concentrations. To achieve the final step to serum-free condition, one of three methods can be used: addition of commercial serum substitutes, addition of defined proteins or addition of only amino acids, trace elements and other small molecules to provide a protein-free medium. The last of these alternatives is the most difficult to achieve; however, it may ultimately prove to be the most useful.

Cell and Tissue Culture: Laboratory Procedures in Biotechnology, edited by A. Doyle and J.B. Griffiths.
© 1998 John Wiley & Sons Ltd.

PRELIMINARY PROCEDURE: METHOD FOR SELECTING SERUM

Serum is classically added to cell cultures to provide hormones, growth factors, binding and transport proteins and other supplemental nutrients. Not all lots of serum have the same potential to support cell growth because they may have lower amounts of these components. Conversely, some lots may be toxic or inhibitory due to adulteration or contamination with microbes or viruses. Such sera may inhibit or kill cells at low serum concentrations due to high endotoxin levels or other inhibitory factors. This effect has nothing to do with the ability of the cells to adapt to serum-free conditions. The first step in serum-free adaptation is, therefore, to select a serum lot that can support growth of the cell line at low serum concentrations.

Materials and equipment

- Healthy culture of cells to be adapted to serum-free conditions
- Soybean trypsin inhibitor (STI): to inhibit trypsin activity resulting from use of trypsin during passage in low-serum cultures
- 3–4 serum lots to test for support of cell growth at low serum concentrations
- 2–3 rich nutrient media to test for support of cell growth at low serum concentrations
- 60-mm Petri plates for initial screening of sera and basal media

1. Choose 3–4 lots of serum to screen.
2. Harvest the cells and, if trypsin is used, add STI at 1 mg ml^{-1} to prevent residual trypsin and STI from decreasing plating efficiency and viability. It is very important to neutralize the effects of trypsin and STI. This is done most effectively by centrifuging the cells after addition of STI to remove any residual trypsin and STI activity in the wash (McKeehan, 1977).
3. Plate cells at 1×10^5, 5×10^4, 1×10^4 and 5×10^3 cells ml^{-1} in 7.5%, 5%, 2% and 1% serum for each of the test serum lots. Use the nutrient medium the cells have been growing in and plate 5 ml in each 60-mm Petri dish. Grow the cells for 10–12 days in a humidified (>95%) 5% CO_2/95% air atmosphere. Collect all the cells (trypsinization or centrifugation) from the plates and count by haemocytometer or Coulter counter. Plot growth curves for each of the serum concentrations. Select the combination with the most cell doubling at the lowest serum concentration. This will select the serum lot with fewest inhibitory factors.
4. It is also important to monitor the yield of the product of interest to determine that the serum lot does not affect its production. However, remember that the serum will not be part of the final medium and is only a tool for developing a serum-free medium. High cell proliferation is the goal.

PRELIMINARY PROCEDURE: METHOD FOR SELECTING NUTRIENT MEDIUM

The next step is choosing a nutrient medium. Amino acids supplied by high serum concentrations may enhance cell growth significantly in deficient media. As the amount of serum is decreased, the need for a balanced nutrient medium becomes more critical.

1. Use the serum lot, serum concentration and cell density chosen in the previous experiment. This should provide the most stringent test of the nutrient balance of the medium.
2. Use the type of medium the cells have been growing in as control. Test 3–4 commercially available balanced media, such as F12/DMEM (Ham's nutrient mixture/Dulbecco's modified Eagle's medium), DMEM, IMDM (Iscove's modified Dulbecco's medium), HB-CHO (Hana Biological Chinese hamster ovary medium) or RPMI-1640 (Roswell Park Memorial Institute).
3. Use STI if trypsin is used to passage the cells and centrifuge the cells once to remove residual trypsin and STI.
4. Determine growth curves and product yields in the control and test media. Select the medium that gives the most cell doublings while still maintaining one of the highest product yields.

PRELIMINARY PROCEDURE: TYPES OF SERUM-FREE MEDIA

There are three categories of serum-free media: totally protein free, those containing complex serum substitutes and those containing defined proteins and other additives.

Complex serum substitutes

There are several commercially available serum replacement products that can be added to the nutrient medium chosen in the preceding experiment to produce a serum-free medium. Examples of these products include: HL1 (Sigma), CPSR1-5 (Sigma), Nutridoma (Boehringer, Mannheim) and NuSerum (Collaborative Research). These substitutes are complex and undefined; however, they often provide good low-protein growth conditions. This is the easiest of the serum-free growth conditions to achieve with most cell types.

Addition of defined proteins

Cell lines differ in the defined proteins they require for optimal cell growth. Usually, addition of insulin (I) and transferrin (T), along with the trace element selenium (S), is sufficient to enable most cell lines to grow in rich balanced nutrient media, such as the ones listed above. However, several categories of additives may enhance or be required for cell growth and/or production of the desired cell product (Brown, 1987; Glassy *et al.*, 1988; Tecce & Terrana, 1988; Schneider, 1989):

1. Hormones – insulin, steroids, pituitary hormones, peptide growth factors, etc.
2. Binding proteins – albumin, fetuin, peptide hormones, etc.
3. Lipid sources – fatty acids, cholesterol, high-density lipoproteins and low-density lipoproteins, etc.
4. Trace elements – selenium, magnesium, iron, zinc, etc.
5. Attachment factors – fibronectin, poly-D-lysine, collagen, serum-spreading factor, etc.

A typical profile for hybridoma cells includes the ITS mixture and the small molecules ethanolamine and β-mercaptoethanol. There are many variations of this basic profile, each specific for the particular hybridoma line and nutrient medium used.

Protein-free

Complete elimination of macromolecules is desirable for many purposes but is also the most difficult to achieve. The first such medium was MCDB 302, derived in Ham's laboratory to support clonal cell growth of Chinese hamster ovary (CHO) cells without deliberate attempts to adapt cells in protein-free medium. This medium contains solely amino acids, vitamins, organic compounds and inorganic salts to replace fully the functions of serum rather than to develop new cell lines that do not need serum. More recently, nearly protein-free media have been developed for murine hybridomas, lymphocytes and hepatoma cells. These media may contain steroid hormones and lipid precursors.

Protein-free cell growth is most likely to be achieved by starting with a balanced nutrient medium of amino acids and trace elements that is designed for the cell line. For a further discussion of modified nutrient media, see the next section.

MODIFYING THE NUTRIENT MEDIUM

Two very different approaches can be used to modify the nutrient medium chosen for a particular cell line so that it will support growth in the absence of serum. The first is to use the nutrient medium plus a limiting amount of serum as a base to which purified components are added to restore growth. This process is repeated with lower amounts of serum until a serum-free medium is derived. In this way, a large array of factors can be tested for effects on cell growth and product yield (Sato & Ross, 1979).

The second approach is to redefine the nutrient medium itself so that each component is at its optimal concentration for the cell line of interest. This is a more difficult and tedious process than the first, but it provides the balanced nutrient medium necessary for derivation of protein-free growth conditions. To derive the balanced nutrient medium, start with a low serum concentration at which there is reduced cell growth. Then systematically titrate each component in the medium to determine its optimal growth-promoting concentration. After progressing through all 20 amino acids, inorganic salts, vitamins and other components in the medium, select the optimal growth response from each of the titration curves and use this new concentration to generate a new medium formulation as the starting point. Lower the serum concentration to a growth-limiting amount and again systematically titrate each amino acid for its optimal growth-promoting concentration. Use this process in combination with adding defined molecules until the need for serum is eliminated and there is a minimum of defined additives or no defined additives in the final balanced nutrient medium formulation.

The second method is a holistic approach in which the balance of amino acids in the medium is the key (Ham & McKeehan, 1978).

PROCEDURE: METHOD FOR ADAPTING CELLS TO SERUM-FREE MEDIUM

When the serum lot and nutrient medium or balanced nutrient medium have been chosen and the goals and approach have been defined, the next step is to adapt an aliquot of healthy cells to serum-free growth:

1. Maintain the cells in the serum lot and medium determined as optimal in the previous experiments. Grow the cells under these conditions until the same doubling time previously obtained in the old serum and medium is achieved.
2. Reduce the serum concentration to 7.5% and passage the cells when stable growth kinetics at this serum concentration are attained. Then reduce serum to 5% and repeat the process. The cells should adapt easily to these serum concentrations.
3. Transfer a portion of the cells to 2% serum. At this serum level it is harder for the cells to adapt. If the cells do not grow, try transferring a new aliquot of cells from 5% to 4% or 3% serum and/or try adding defined supplements (e.g. insulin). For more information on medium supplements, refer back to 'Types of serum-free media'. At each step, when the cells have become adapted, freeze some of the cells. Freeze a larger stock of vials when the cells are adapted to 2% serum. Should problems arise later on, this is a good point to return to.
4. When cells are able to grow well in 2% serum with a repeatable seeding density ($1-3 \times 10^5$ cells ml^{-1}), transfer a portion of the cells to 1% serum.
5. Continue to reduce serum from 1% to 0.75%, 0.5% and 0.2% with defined supplements added. To facilitate adaptation, use a high seeding density (5×10^5 cells ml^{-1}) by centrifuging the cells out of the old medium and using all the cells in a smaller medium volume. When the cells are adapted to growth in the lower serum concentration at high seeding density, they must be grown at lower inoculum densities ($1-10 \times 10^4$ cells ml^{-1}) to ensure that the cells are truly adapted to the new growth conditions. The number of cell doublings in the new growth conditions, not the final cell number, is the true indication of adaptation. The cells must be able to be subcultured serially in the medium and be able to achieve the same number of doublings each time to give confidence that they are truly adapted. Freeze an aliquot of the adapted cells.

6. After the cells have adapted to 0.2% serum with defined additives, the next step is to proceed to serum-free conditions. For the first few passages, a high cell inoculum is recommended. Keep the cell density high by centrifuging the cells and resuspending them at high density in fresh medium until they are able to divide. Conditioned medium can also be used as a supplement until the cells start to stabilize. As the conditioned medium is progressively removed, additional defined factors and/or increases in the concentration of those already in use may be required. The final transition to serum-free growth may take longer than the previous steps, up to 1–2 months. The total time for the serum-free adaptation process may range from 1 to 6 months.

DISCUSSION

This discussion provides an overview of the methods for selecting a serum lot and balanced nutrient medium as starting points for adapting a cultured cell line to serum-free conditions. The ultimate stringency of serum-free medium attained (containing serum substitutes, defined additives or protein-free) depends on the needs of the researcher and the characteristics of the cell line. Serum provides many different functions for the cell and there are many different types of cells, each of which requires a medium derived to meet its particular needs. The information provided here should enable the development of a serum-free medium appropriate for each particular case.

Troubleshooting

This section includes some additional points to keep in mind when adapting cells to serum-free growth.

Initial high cell density

Cells can synthesize most of the essential amino acids, but do so in amounts too low to compensate for dilution in low-density cultures. The requirement for many amino acids is, therefore, a function of cell concentration. During adaptation of cells to lower serum or serum-free conditions it may be necessary to culture the cells at high density, $3–6 \times 10^5$ cells ml^{-1}, until they have adapted. This can be accomplished by centrifuging the cells and resuspending at higher densities.

Gradual serum withdrawal

Some cells may not require a slow weaning process (serum withdrawal), while others will only adapt if the process is very gradual. Serum decreases that cause severe effects, where the vast majority of cells die, increase the possibility of variant clones emerging. A conservative approach is, therefore, recommended. Maintain viability above 50%.

Add defined components

The most frequently used additions for serum-free growth are insulin, transferrin and selenium. Recommended starting concentrations for these are 10 mg ml^{-1}, 10 mg ml^{-1} and 5×10^{-8} M, respectively. Many other additions are described in the literature for specific cell lines.

Conditioned medium

After cells have adapted to lower serum concentrations at high cell density, the next step is to decrease cell density. One method of easing the transition to lower seeding densities is addition of 5–20% conditioned medium. This gives the cells time to synthesize the medium components they need. However, conditioned medium should be used sparingly because it also contains components (proteases, lactic acid, etc.) that can be detrimental to cell growth. Use it only when the cells cannot grow in serum-free medium and there are indications that the cells are not at the stage of adaptation to make the nutrients themselves.

Growth vessel

The growth vessel and some supplements are determined by whether the cells are grown attached or in suspension. When the cells are anchorage dependent, they can be grown in plates or flasks. As serum is reduced, it may be necessary to add attachment factors to the culture. For cells grown in suspension in spinner flasks or fermenters, non-ionic surfactants (F68) that increase viscosity may be needed to minimize shear stress caused by agitation.

REFERENCES

Barnes D (1987) Serum-free animal cell culture. *Biotechniques* 5: 534–541.

Brown BL (1987) Reducing costs upfront: two methods for adapting hybridoma cells to an inexpensive, chemically defined serum-free medium. In: Seaver SS (ed.) *Commercial Production of Monoclonal Antibodies – A Guide for Scale-up*, pp. 35–47. Marcel Dekker, New York.

Clark SA & Looby D (1989) Adaptation of hybridoma cell lines to grow and secrete monoclonal antibody in serum-free/defined medium. In: Spier RE, Griffiths JB, Stephenne JS & Crooy PJ (eds) *Advances in Animal Cell Biology and Technology for Bioprocesses*, pp. 291–297. Butterworth, London.

Glassy MC, Tharakan JP & Chau PC (1988) Serum-free media in hybridoma culture and monoclonal antibody production. *Biotechnology and Bioengineering* 32: 1015–1028.

Ham RG & McKeehan WL (1978) Nutritional requirements for clonal growth of nontransformed cells. In: Katsuka H (ed.) *Nutritional Requirements of Cultured Cells*, pp. 63–116. Japan Scientific Societies Press, Tokyo, and University Park Press, Baltimore.

McKeehan WL (1977) The effects of temperature during trypsin treatment on viability and multiplication potential of single and normal human and chicken fibroblasts. *Cell Biology International Reports* 1: 335–343.

Miyaji H, Harada N, Mizukami T, Sato S, Fujiyoshi N & Itoh S (1990) Expression of human lymphotoxin in namalwa KJM-1 cells adapted to serum-free medium. *Cytotechnology* 4: 39–43.

Romijn HJ (1988) Development and advantages of serum-free, chemically defined nutrient media for culturing of nerve tissue. *Biology of the Cell* 63: 263–268.

Sato GH & Ross R (1979) *Hormones and Cell Culture*, Books A and B. Cold Spring Harbor Laboratory, Cold Spring Harbor, NY.

Schneider YJ (1989) Optimisation of hybridoma cell growth and monoclonal antibody secretion in a chemically defined, serum- and protein-free culture medium. *Journal of Immunological Methods* 116: 65–77.

Shacter M (1989) Serum-free media for bulk culture of hybridoma cells and the preparation of monoclonal antibodies. *Tibtech* 7: 248–253.

Tecce MF & Terrana B (1988) High-yield and high-degree purification of human δ-fetoprotein produced by adaptation of the human hepatoma cell line HEP G2 in a serum-free medium. *Analytical Biochemistry* 169: 306–311.

3.4 AMINO ACID METABOLISM

Amino acids play a diverse role in cellular metabolism, acting as both a nitrogen donor and a carbon source, and are a vital constituent of all tissue cell culture media. Amino acid metabolism has been reviewed in detail (Meister, 1965; Patterson, 1972). Early investigations by Eagle distinguished 13 'essential' and 7 'non-essential' amino acids. These are outlined in Table 3.4.1. This led to the development of traditional cell culture media such as modified Eagle's medium (MEM) (Eagle, 1959), Dulbecco's modified Eagle's medium (DMEM) (Dulbecco & Freeman, 1959), RPMI 1640 (Moore *et al.*, 1967) and Ham's F-12 (Ham, 1965).

The nutritional requirements for a certain metabolite, however, may also be influenced by the cell population density. For example, serine, cystine, glutamine and asparagine have been shown to be required at low, but not at high, cell densities (Eagle & Piez, 1962). This occurs in situations when the newly synthesized metabolite is lost into the medium in amounts that exceed the biosynthetic capacity of the cell. The critical population density can be regarded as that which is able to 'condition' the medium, i.e. to build up a concentration in equilibrium with the minimum effective intracellular level, before the cells die of the specific deficiency. At high cell densities, however, the cell culture medium can require supplementation with extra amino acids, in order to prevent their depletion.

It has been recognized that certain cells have a specific requirement for an amino acid, for example serine for lymphoblastoid cells (Birch & Hopkins, 1977). This may be due either to the inability of certain cells to make an amino acid, or because they are decomposed in the medium. The concentration of amino acids

Table 3.4.1 Eagle's essential and non-essential amino acids

Essential amino acids	Non-essential amino acids
Arginine	Alanine
Cystine	Asparagine
Glutamine	Aspartic acid
Histidine	Proline
Isoleucine	Glycine
Leucine	Serine
Lysine	Glutamic acid
Methionine	
Phenylalanine	
Threonine	
Tryptophan	
Tyrosine	
Valine	

Cell and Tissue Culture: Laboratory Procedures in Biotechnology, edited by A. Doyle and J.B. Griffiths.
© 1998 John Wiley & Sons Ltd.

usually limits the maximum cell number attainable, influences cell survival and growth rate and can affect the synthesis of certain proteins. Too low a concentration of an amino acid can result in rapid exhaustion from the medium, and is thus 'limiting', whereas too high a concentration can be inhibitory.

Consequently, quantitative amino acid data are important in order to optimize tissue cell culture medium. The concentration of many amino acids can be determined and the data used to calculate the rate of utilization or assimilation of the individual amino acids. By performing such analysis over an extended culture period, the pattern of amino acid metabolism can be studied. One method of performing such an investigation is outlined, and a case study given as an example.

PROCEDURE: AMINO ACID ANALYSIS

Several techniques have been developed to analyse amino acids. The following procedure describes one of the most frequently used methods for the determination of amino acid concentration, the PITC method. Free amino acids in medium samples are derivatized with phenylisothiocyanate (PITC) to produce phenylthiocarbamyl (PTC) amino acids. After separation by reverse-phase HPLC, the PTC amino acids are detected using a UV spectrophotometer at 254 nm.

Amino acids determined

The amino acids alanine, arginine, asparagine, aspartic acid, cysteine, glutamic acid, glycine, histidine, isoleucine, leucine, lysine, methionine, phenylalanine, proline, serine, theronine, tryptophan, tyrosine and valine were analysed for the determination of amino acid concentration.

Reagents and solutions

- Mobile phase A: dissolve 76.0 g of sodium acetate trihydrate in 4000 ml of water and add 2.0 ml of triethylamine and 8 mg of EDTA; adjust the pH to 6.4 with acetic acid
- Mobile phase B (60:40 CH_3CN/H_2O): mix 1200 ml of acetonitrile with 800 ml of water
- Sample dilution buffer: mix 6 ml of acetonitrile with 94 ml of mobile phase A
- Norleucine (1.5 mmol l^{-1}): dissolve 19.7 mg of norleucine in 100 ml of water and store at 4°C
- Tryptophan (2.0 mmol l^{-1}): dissolve 40.8 mg of tryptophan in 100 ml of water and store at 4°C
- Asparagine and glutamine (1.0 mmol l^{-1}): dissolve 15.0 mg of asparagine and 14.6 mg of glutamine in 100 ml of water and store at −20°C
- Standard solution: 0.33 mmol l^{-1} of alanine, arginine, asparagine, aspartic acid, cysteine, glutamine, glutamic acid, glycine, histidine, isoleucine, leucine, lysine, methionine, phenylalanine, proline, serine, threonine, tryptophan, tyrosine or valine and 0.23 mmol l^{-1} of norleucine; mix 200 µl of 2.5 mM amino acid

standard, 233 μl of 1.5 mM norleucine, 250 μl of 2.0 mM tryptophan, 500 μl of 1.0 mM asparagine and glutamine and 317 μl of water; store at −20°C
- Derivatization reagent: for 10 samples mix 280 μl of ethanol, 40 μl of triethylamine, 40 μl of PITC and 10 μl of 1.5 mM norleucine; prepare the derivatization reagent fresh for each batch of samples and use the reagent within 2 h of preparation

Materials and equipment

- Sodium acetate trihydrate
- Water (HPLC grade)
- Triethylamine (Sequanal grade)
- EDTA (Tritriplex III)
- Acetic acid (100% glacial)
- Acetonitrile
- Ethanol (99.5%)
- Phenylisothiocyanate (for sequential analysis)
- Amino acid standard (2.5 mM) for hydrolysate analysis (Beckman Ltd)
- D,L-Norleucine
- L-Asparagine monohydrate
- L-Glutamine
- L-Tryptophan
- Two HPLC pumps (Waters Model 510, Millipore Corp., Milford, MA, USA)
- Automated gradient controller (Waters Model 680, Millipore Corp., Milford, MA, USA)
- WISP auto-injector with cooling module (Waters Model 710B, Millipore Corp., Milford, MA, USA)
- Analytical column: Pico-Tag® Free Amino Acids Analysis Column, dimethyl-octadecylsilyl-bonded amorphous silica, 3.9 × 200 mm
- UV detector (LKB 2151 Variable Wavelength Monitor, Pharmacia Biosystems, Sollentuna, Sweden)
- Column heater (Jones Chromatography, Wales, UK)
- Degassing method (Waters Eluent Stabilization System, Millipore Corp., Milford, MA, USA)
- Integrator (Analog Interface Module 406 and System Gold, Beckman Instruments, San Ramon, CA, USA)

Time considerations

The time required for running one sample by this assay is approximately 8 h (preparation of sample, 5 h; run on the HPLC: standards 3 × 45 min + sample 1 × 45 min). The auto-injector can be filled with 45 different samples; each then takes 45 min to run.

Parameters

- Limit of quantification = 50 pmol
- Linear range of standard curve = 0.025–6.0 mmol l^{-1}
- Method precision = 2.6% (relative standard deviation, RSD)
- Day-to-day variation = 4.1% (RSD)

Safety considerations

Hazardous compounds: acetic acid, glacial; acetonitrile; phenylisothiocyanate; and triethylamine. With all these compounds the use of hood, goggles and protective gloves is recommended.

Instrument conditions

The instrument conditions are outlined in Table 3.4.2.

Preparation of sample

1. Use pyrolized glass tubes (400°C for 3–4 h).
2. Mix by vortexing the medium samples for a few seconds. Add 3 µl of medium sample to each glass tube.

Preparation of standard

1. Use pyrolized glass tubes.
2. Add 15 µl of standard solution to each glass tube and evaporate in a Speed-Vac. Store at –20°C.
3. Take one of the tubes and add 3 µl of water. In this standard tube there is 5 nmol of standard solution.

Table 3.4.2. Instrument conditions[a]

Time (min)	Flow (ml min^{-1})	% A	% B	Curve no.
Initial	1.0	90	10	*
1.0	1.0	90	10	6
21.0	1.0	46	54	5
21.5	1.5	0	100	6
26.0	1.5	0	100	6
26.5	1.5	90	10	6
39.0	1.5	90	10	6
39.5	1.0	90	10	6
46.0	1.0	90	10	6
49.0	1.5	0	100	6
64.0	0.0	0	100	10

[a]Analytical column: Pico-Tag™ Free Amino Acids Analysis Column; dimethyloctadecylsilyl-bonded amorphous silica; 3.9 × 300 mm. Mobile phase composition: A = 76.0 g of sodium acetate trihydrate, 2.0 g of triethylamine, 8 mg of EDTA, 4000 ml of H_2O, adjusted to pH 6.4 with acetic acid; B = 60% CH_3CN and 40% H_2O. Temperature = 46°C. Injection volume = 40 µl. Detection at 254 nm.

Sampling

1. Prepare the derivatization reagent for the standard and medium samples.
2. Add 37 µl of derivatization reagent to each sample and standard glass tubes. Mix by vortexing all tubes for a few seconds. Allow the reaction to proceed for 15–30 min at ambient temperature (20–25°C).
3. Remove excess reagent by evaporation in a Speed-Vac.
4. Reconstitute the derivatized standard in 200 µl of sample dilution buffer. Reconstitute the derivatized media samples in 60 µl of sample dilution buffer. Allow to stand for 30 min at ambient temperature.
5. Fill the auto-injector with the standard and test samples. In the first position place 140 µl of standard solution, then 60 µl of the media samples. In the last position place 60 µl of standard solution.
6. Prepare the data collection in the system software. Make a sample table and start the integrator.
7. Start the buffer pumps with initial conditions: 90:10, mobile phase A/mobile phase B. Equilibrate the system for 30 min.
8. Program the auto-injector for a run time of 44 min: three injections from the first position, and one injection from the other position. The third injection of standard from the first position is used for calibration.
9. Start the system from the auto-injector.

Calculation

F_X = conversion factor for amino acid X
C_{stX} = concentration of amino acid X in standard solution (µmol l^{-1})
C_{stNLE} = concentration of norleucine in standard solution (µmol l^{-1})
A_{stX} = area response for amino acid X in standard solution
A_{stNLE} = area response for norleucine in standard solution
C_X = concentration of amino acid X in sample (µmol l^{-1})
C_{NLE} = concentration of norleucine in sample (µmol l^{-1})
A_X = area response for amino acid X in sample
A_{NLE} = area response for norleucine in sample

Norleucine is used as an internal standard in the standard solution used for the calibration. Norleucine is also used as an internal standard in the samples. From the calibration run you get a factor F_x for each amino acid X:

$$F_X = \frac{A_{stNLE}}{A_{stX}} \times \frac{C_{stX}}{C_{stNLE}}$$

The concentration of each amino acid X in the sample (µmol l^{-1}) is:

$$C_X = F_X = \frac{A_X}{A_{NLE}} \times C_{NLE}$$

An example of the spectrum from the HPLC is shown in Figure 3.4.1, and the resulting amino acids calculated using a standard HPLC program are shown in Table 3.4.3.

AMINO ACID METABOLISM

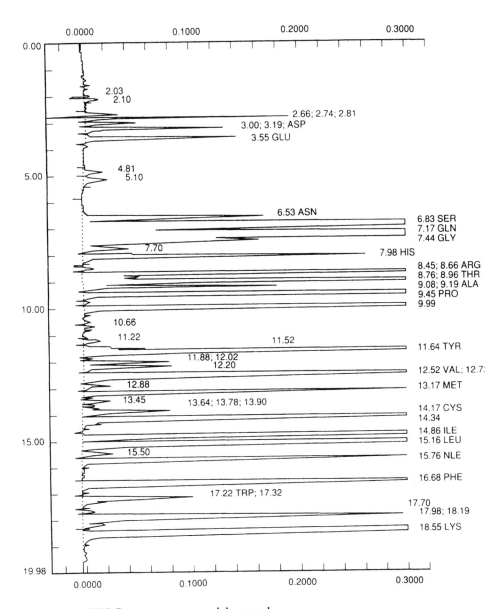

Figure 3.4.1 HPLC spectrum – a model example.

If regular amino acid samples have been taken from the cell culture over a certain time period, the rate at which the cells are utilizing or assimilating amino acids can be determined using the following formula:

Table 3.4.3 Amino acid concentrations calculated from HPLC data as shown in Figure 3.4.1.

Peak no.	Retention time	Amino acid	Concentration (nmol l^{-1})	Peak area	Peak height	Response factor
1	3.186	ASP	163.512	9.17	0.13	1.0643
2	3.549	GLU	181.510	11.53	0.14	0.9403
3	6.528	ASN	334.843	20.60	0.17	0.9705
4	6.831	SER	3913.002	232.54	1.73	1.0047
5	7.165	GLN	6854.011	323.63	2.10	1.2645
6	7.439	GLY	346.874	20.41	0.16	1.0146
7	7.981	HIS	389.688	21.79	0.26	1.0676
8	8.664	ARG	1225.510	72.37	1.10	1.0111
9	8.962	THR	1259.598	80.50	1.05	0.9342
10	9.186	ALA	216.875	12.85	0.18	1.0078
11	9.447	PRO	2111.439	138.24	2.08	0.9120
12	11.636	TYR	640.814	40.83	0.50	0.9371
13	12.525	VAL	1404.065	93.24	1.13	0.8992
14	13.167	MET	377.653	23.65	0.30	0.9536
15	14.170	CYS	407.443	42.28	0.48	0.5754
16	14.858	ILE	1426.075	92.31	1.10	0.9224
17	15.163	LEU	1581.548	97.08	1.12	0.9727
18	15.759	NLE	0.000	29.85	0.34	1.0000
19	16.685	PHE	658.524	41.22	0.45	0.9540
20	17.221	TRP	153.205	10.57	0.10	0.8654
21	18.548	LYS	1377.678	156.95	1.46	0.5241
TOTAL			25023.867	1571.60	16.11	

$$\text{Batch culture: } A.a.C. = (C_0 - C_{x-y})/E$$
$$\text{Perfusion culture: } A.a.C. = D(C_0 - C_x)/E/F$$

where $A.a.C$ = amino acid utilization/assimilation (in μmol 10^{-9} 24 h^{-1}), C_0 = amino acid concentration in incoming medium, C_x = amino acid concentration (μmol l^{-1}) on day X, C_{x-y} = amino acid concentration (μmol l^{-1}) on day X minus amino acid concentration (μmol) 24 h before (day Y), E = cell density (10^6 ml^{-1}), D = dilution rate (l 24 h^{-1}) and F = culture volume.

CASE STUDY

To emphasize the importance of controlling the amino acid concentration, an example illustrating the importance of asparagine for the production of recombinant factor VIII in DON cells will be discussed (Figure 3.4.2).

The DON cells were cultured in DMEM/F-12 on Cytodex 1 microcarriers in a 3-l perfusion bioreactor.

The flow rate was maintained at one reactor volume per day. Samples were taken daily for cell counts, factor VIII analysis and amino acid analysis. At the start of the production phase the medium was supplemented with 200 μmol of asparagine (50 μmol of asparagine was present in the unsupplemented medium).

AMINO ACID METABOLISM

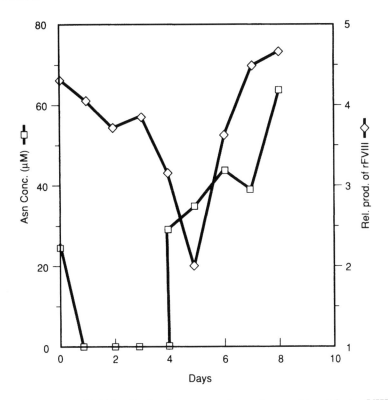

Figure 3.4.2 Culture of DON cells showing asparagine and recombinant factor VIII concentrations.

After a certain time period asparagine supplementation was stopped (day 0), as a consequence of which asparagine became exhausted from the culture medium and the production of factor VIII dropped (days 1–4). Asparagine (200 µmol) was later added to the medium (day 4), the result of which was an increase in the levels of factor VIII.

REFERENCES

Bidlingmeyer BA, Cohen SA & Tarvin TL (1984) Rapid analysis of amino acids using pre-column derivatization. *Journal of Chromatography* 336: 93–104.

Birch JR & Hopkins DW (1977) The serine and glycine requirements of cultured human lymphocyte lines. In: Acton RT & Lynn JD (eds) *Cell Culture and its Applications*, pp. 503–511. Academic Press, New York.

Dulbecco R & Freeman G (1959) Plaque production by the polyoma virus. *Virology* 8: 396–397.

Eagle H (1959) Amino acid metabolism in mammalian cell cultures. *Science* 130: 432–437.

Eagle H & Piez K (1962) The population dependent requirement by cultured mammalian cells for metabolites which they can synthesize. *Journal of Experimental Medicine* 116: 29–43.

Freshney RI (ed.) (1987) *Culture of Animal Cells – A Manual of Basic Techniques*, 2nd edn. Alan R. Liss, New York.

Ham RG (1965) Clonal growth of mammalian cells in a chemically defined synthetic medium. *Procedures of the National Academy of Sciences of the USA* 53: 288–293.

Lehninger AL (ed.) (1975) *Biochemistry – The Molecular Basis of Cell Structure and Function*, 2nd edn. Butterworth, New York.

Meister A (1965) *Biochemistry of Amino Acids*, 2nd edn. vol. 2. Academic Press, New York.

Moore GE, Gerner RE & Franklin HA (1967) Culture of normal human leucocytes. *Journal of the American Medical Association* 199: 519–524.

Patterson MK (1972) Uptake and utilisation of amino acids by cell cultures. In: Rothblat GH & Cristofalo VJ (eds) *Growth, Nutrition and Metabolism of Cells in Culture*, vol. 1, pp. 171–209. Academic Press, New York.

3.5 TISSUE CULTURE SURFACES

A tissue culture dish or flask is not an inert plastic without biological relevance, but is rather a complex physicochemical substrate having definite effects on cell morphology and growth. It is important to be aware of the characteristics of various tissue culture surfaces so that experiments can be appropriately designed and interpreted (Grinnell, 1978).

What a tissue culture surface is, how it is produced and the biological consequences of choosing one substrate over another are discussed. The most common tissue culture surface, modified polystyrene, and the effects of substrate on anchorage-dependent cells will be discussed. The differences between solid and permeable (e.g. filter membrane) substrates will also be related.

THE TREATMENT PROCESS

The tissue culture surface begins with a select grade of virgin polystyrene that has been chosen for high purity and minimal additive content. The moulded part is then subjected to conditions of either corona discharge or a gas plasma.

For plasma treatment, the moulded part is placed in a vacuum chamber under low pressure in which an electrical discharge ionizes a specific gas that is present. The resulting gas plasma generates a reaction with the polystyrene substrate, with the result that randomized chemical groups become covalently attached to the surface.

The net effect of this treatment is that the virgin polystyrene is changed from a neutral, hydrophobic surface to a derivatized, charged hydrophilic surface that allows cell spreading and attachment (Dunn, 1984). The use of different gas mixtures in the controlled environment of the plasma treatment chamber can yield a variety of different surfaces (Figure 3.5.1). For a detailed discussion on physical analysis, see Salvati *et al.* (1991).

The 'traditional' tissue culture surface is produced by exposing the surface to an oxidizing atmosphere that yields random oxygenated groups and a net-negative charge on the surface. An alternative cell culture surface containing various amino and amide functional groups produced by a mixed oxygen and nitrogen plasma (in addition to the groups produced by oxidation alone) is also commercially available (Primaria, Becton Dickinson Labware, USA). This surface has less net-negative charge, is somewhat more complex than the traditional oxygenated surface and can provide a superior cell growth substrate for certain cell types.

Cell and Tissue Culture: Laboratory Procedures in Biotechnology, edited by A. Doyle and J.B. Griffiths.
© 1998 John Wiley & Sons Ltd.

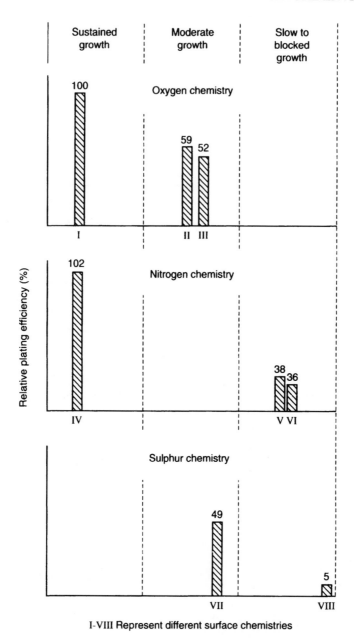

Figure 3.5.1 Growth of MRC-5 cells on experimental surfaces with varying chemistries.

STABILITY

The tissue culture surface is therefore *not* a coating, but a surface-modified polystyrene having random functional groups covalently bound to the surface. Tissue culture plasticware is subsequently sterilized to inactivate the microbial bioburden that accumulates during manufacturing. This cobalt gamma irradiation causes further randomized reactions on the tissue culture surface, but the effect of this irradiation is minimal with the routine exposures typically used for rendering the product sterile.

Tissue culture plastic can very gradually undergo further oxidation as it ages. The tissue culture surface eventually reaches a plateau of oxidation, and there is no known shelf-life restriction on tissue culture plasticware produced by plasma modification. In a practical sense, tissue culture plasticware can be stored without any known degradation.

BIOACTIVITY

Anchorage-dependent cell types must first attach and then spread on the wettable (hydrophilic) plastic substrate if cell growth and/or differentiation are to occur. It is for this reason that the moulded polystyrene flask or dish is converted to the hydrophilic tissue culture vessel. Following initial attachment and spreading, the cell/surface interaction remains important as a determinant of longer term cell morphology and growth. Figure 3.5.1 shows the dramatic difference in cell growth achieved with different surface chemistries. Furthermore, different functional groups within a basic oxygen or nitrogen chemistry show definite biological effects. With medium, cell type (MRC-5) and passage being equal, the importance of surface chemistry on cellular response is apparent.

SURFACE CHOICE AND COMPARISON

Standard tissue culture surface

Because most routine cell types were originally grown on glass, the first commercially available tissue culture surface was modelled after glass chemistry. Conventional tissue culture surfaces therefore are hydrophilic and have an oxygenated chemistry and a net-negative surface charge. This chemistry is basically the same whether the treatment process is produced by corona or plasma. This is the routine surface that is commercially available from a number of different suppliers on plastic dishes, flasks, plates and roller bottles.

Advantages

- Ready-to-use, sterile, disposable
- Stable surface/no shelf-life
- Resistant to proteolysis from cells

Disadvantage

- No inherent biological activity (no protein coating)

Biological coatings – extracellular matrices

Some primary and fastidious cell types will not attach and grow on regular tissue culture surfaces and require a protein coating to divide and become fully differentiated. In addition to polylysine and polyornithine, a variety of proteins that are derived from extracellular matrix are commercially available. Fibronectin, laminin and collagens are available as reagents and also as precoated plasticware.

Advantages:

- Coating is bioactive
- More closely resembles cell layer *in vivo*

Disadvantages:

- Definite shelf-life
- Expense of coating reagent
- Labour required for coating

Testing protocol and design

It is recommended that experimental conditions be optimized for any cell type that the laboratory is considering using on a fairly routine basis. This should include conventional tissue culture plasticware and several different biological coatings. Once the optimum substrate has been chosen, it is worthwhile performing further experiments using different serum concentrations because a change in the surface or biological matrix will often result in a change in the serum requirement. Lower serum levels in media will give lower serum background interference, as well as reducing expense due to costly serum. Serum proteins often mask the chemistry of the surface substrate. Minimal serum levels in medium should allow the chemical and biological effect of the surface to exert its influence on the cells in culture.

MICROCARRIERS

Microcarriers are spherical beads that are used to increase the effective cell growth area in culture. Due to the large increase in surface area, these microcarriers are grown in stirred suspension and are often continuously perfused with fresh medium in order to achieve very high cell densities (for a review, see Pharmacia (1982); also see Chapter 5, section 5.8).

While this is a dynamic system subject to shear force, as compared with the static system of a flask or dish, the cells growing on microcarriers are still anchorage dependent. The microcarrier bead can be made of a variety of coated and derivatized materials: glass (Solo Hill, MA, USA); dextran (Pharmacia); or modified plastic (MatTek, MA, USA). The same requirements for good surface attachment and growth still apply, and must be determined experimentally.

Advantages

- Increased surface area for cell growth
- Increased bioproduct production

Disadvantages

- Cells can be difficult to remove from beads
- Shear forces can damage fragile cells
- Requirement for perfusion or frequent medium change

POROUS MEMBRANE SYSTEMS

The previous discussions have been based on products like dishes and flasks that have a solid substrate modified for tissue culture. Cells growing on a solid substrate, however, can only receive nutrient and actively transport from the exposed (top) layer.

A system that more closely resembles the *in vivo* state is the porous membrane system. Because cells can receive nutrient from both sides of the membrane, the cell layer has an active basal and apical (or luminal) surface that allows the selective uptake and transport of nutrients and secreted products. A variety of biological functions can now be studied, such as active transport, better differentiation and co-culture of different cell types (Pitt & Gabriels, 1986; Millipore, 1990; Halleux & Schneider, 1991).

There are several commercially available products that use this principle. The device is commonly a plastic ring with a membrane attached, which can be placed into a multiwell plate or dish. Cells can be grown on one side of the membrane surface, on both sides of the membrane (transmembrane co-culture) or on both the membrane and plate well bottom for co-culture. This membrane device is commonly used in a static system (e.g. multiwell plate or culture dish) but it may also be mounted in an appropriate chamber for continuous perfusion culture. Examples of these products are the Falcon Cell Insert (Becton Dickinson Labware, NJ, USA), Millicell (Millipore, MA, USA) and Transwell (Costar, MA, USA).

There are several factors that need to be taken into consideration when using cell inserts:

1. The membrane material is important for cell attachment and growth. Common materials are polycarbonate, cellulose esters (acetate and nitrate) and PET (polyester). These materials sometimes need to be coated for cells to grow and differentiate. The different membrane materials have very different optical and mechanical properties, which must be considered when planning any experiment.
2. Pore size and density are also important in selecting a membrane. Some membranes such as polycarbonate have very high pore densities for maximum fluid transport, but are fragile and rather opaque. The pore size is also critical, depending on the experiment. For example, a 0.45-μm pore size will prevent cells from migrating through the membrane, while 3 μm can be used for transmembrane

co-culture (i.e. cytoplasmic contact) and 8–10 µm can be used for cell invasion studies. It is recommended that the manufacturer be contacted about the exact nature of the membrane being considered so that experiments can be designed appropriately.

Membrane devices are a complex system in that pore size, density and material must be taken into consideration when designing an experiment. The results can be well worth the effort, because use of a membrane system will result in a system much closer to the *in vivo* state than is possible on the solid substrate system.

Advantages

- Nutrient is accessible from both sides
- Cell transport can be polar
- More closely resembles *in vivo* compartmentalization

Disadvantages

- Currently available only in smaller sizes
- Expense of membrane devices

DISCUSSION

While tissue culture is becoming more and more a science, it is still an 'art' and, as such, the same rules of trial and error still apply: the surface needed depends on the cell type. It is best to start with regular tissue culture-treated plasticware and then to proceed to the more complex biocoatings, realizing that this is a complex factor analogous to changing media and serum. All three systems – serum, media and coating – need to be balanced in that case.

For improved differentiation and more 'physiological results' the newer membrane devices should be tried. Protein-coated membranes should prove to be a valuable tool in the future. If the result is a system close to the *in vivo* state, then the effort is well worth the time spent in experimenting with a new system. After all, that is the goal – *in vivo* simulation *in vitro*.

REFERENCES

Dunn TS (1984) The surface chemistry of tissue culture plastic (abstract). *Journal of Cell Biology* 99: 171.

Grinnell F (1978) Cellular adhesiveness and extracellular substrata. *International Review of Cytology* 43: 65–144.

Halleux C & Schneider YJ (1991) Iron absorption by intestinal epithelial cells: 1. CaCO$_2$ cells cultivated in serum-free medium on polyethylene terephtalate microporous membranes as an *in vitro* model. *In Vitro Cell Development Biology* 27A: 293–302.

Millipore (1990) *Cell Culture on Millipore Membranes: A Technical Reference Guide.* Millipore Corporation, Bedford, MA, USA.

Pharmacia (1982) *Microcarrier Cell Culture:*

Principles & Methods. Pharmacia LKB Biotechnology, Box 776, Sollentuna, Sweden S-19127.

Pitt A & Gabriels J (1986) Epithelial cell culture on microporous membranes. *American Biotechnology Laboratory* 4: 38–45.

Salvati L, Grobe G & Moulder J (1991) Surface characterization of polymer based biomaterials. *Medical Devices & Diagnostics Industry* 13: 96–102.

3.6 PLASTIC AND GLASS TISSUE CULTURE SURFACES

Cell culture first developed long before the age of readily available plastics and still in the memories of a lot of tissue culture experts are the happy hours spent in a junior capacity following a strict regime of detergent soak and wash, tapwater wash and double-distilled water rinse for the preparation of tissue culture-grade glassware. This was equally necessary for both adherent and non-adherent cell cultures because a major concern is the build-up of cellular metabolic toxins in the glass. Problems were compounded if the glassware had to be autoclaved, with the necessity of washing off bound protein that often required hot acid treatment. Most laboratories adapted existing 'off-the-shelf' glassware for this task, e.g. baby feeding bottles and Winchester bottles, before Roux flasks became the standard. These were often made of an unsuitable grade of glass and became scratched, often opaque, with use.

Fortunately, the need to use inappropriate and variable surfaces need no longer be a problem, due to the availability of high-quality disposable plasticware. This revolutionized tissue culture because a very wide range of vessels soon became available from major international suppliers competing for this lucrative business. Actually, the use of disposable plasticware is comparable to the recycling process of glassware and replacement of damaged vessels in most laboratory situations. This contribution provides basic information on what is currently available, in terms of the range of plastic and glassware, and will provide help in deciding the capacity of vessels necessary at a particular scale of culture to produce a target yield of cells. In addition, information on the major factors concerning adhesion of anchorage-dependent cells, which relates to the increasingly important topic of serum-free culture systems, is given to underline the importance of using adequate surfaces.

Adhesion

The mechanism of adhesion is a multistep process involving four phases: adsorption, contact, attachment and spreading (Grinell, 1978).

1. *Adsorption*. An important factor in adsorption to a surface is its chemical nature and, in particular, wettability. Polystyrene manufactured into vessels does not have a good chemical structure in this regard and requires treatment to produce a negatively charged surface suitable for electrostatic attachment and cell adhesion (Maroudas, 1993). The first stage of adsorption requires the presence of

protein, which may be provided by the cells themselves and/or serum. The cells produce 'micro-exudate' (Grinell, 1978) and this plays a part in cell spreading and is an important characteristic of adherent cells. The important factor is the net negative charge, and surfaces such as glass and metal that have high surface energies are thus suitable for cell attachment.
2. *Contact.* The surfaces of unattached cells possess microvilli (Springer et al., 1976), which are involved in the next stage.
3. *Attachment.* Once contact is achieved via cytoplasmic micro-extensions, these additional structures disappear from the cell surface. They are thought to be essential in overcoming electrostatic forces between the negatively charged cells and the negative charge on the attachment surface.
4. *Spreading.* Once the cell is attached, filopodia extend from the cell and are involved in the first stage of spreading. The spaces between filopodia are then filled with cytoplasm to give lamellipodia (Springer et al., 1976). The latter stage of spreading involves cytoskeletal components, including microfilaments and microtubules.

Ultimately the presence of a biochemical component on the cell – LETS (large external transformation-sensitive) glycoprotein – determines whether a cell is anchorage dependent or not (Hynes & Bye, 1974). The glycoprotein has a molecular weight of 2–2.5×10^5. Other factors influencing attachment are energy metabolism, Ca^{2+} and Mg^{2+} concentration, sulphydryl groups and incubation temperature.

Serum factors

Serum has many functions in cell culture and one key element is its role in cell attachment, spreading and growth.

It is a source of adhesion and spreading factor (ASF), fibronectin and spreading factor, all of which vary in content between different sera and batches of the same sera. If serum-free medium is used, the loss of these factors has to be compensated for by their addition; alternatively, the surfaces can be modified (Maroudas, 1993).

Review of plasticware and glassware available

Plasticware

The standards of 25 cm^2, 75 cm^2, 175 cm^2 and upwards can be variable and each supplier has to be assessed for growth capacity for the particular cell type of interest. Increasing from this scale can be achieved through the use of cell factories of roller bottles. However, this may require considerable advance preparation and such methods are very labour-intensive. The time required to trypsinize more than 10 roller bottles in a single operation must not be underestimated and, if cells are to be pooled for later manipulations, the toxic effect of long-term exposure to enzyme must be considered.

Glassware

Flasks are available in a range of sizes from 5 cm² to 200 cm². It is recommended that aluminium borosilicate glass is used because it releases significantly fewer alkaline substances than soda-lime glass. It is often the case that new bottles show poor cell adherence but after use on several occasions, with thorough washing and sterilization, cell attachment improves. Alternatively, boil glassware in weak acid. Often, attachment decreases after constant use but the surface can be re-activated with a wash in 1 mM magnesium acetate.

For suspension cultures, spinner flasks (Techne, Wheaton, Bellco, Cellon, etc.) and bottles are available, which provide sufficient capacity for most small-scale preparation of cells. Sizes range from 25 to 10 000 ml, and the stirring rate can be adapted for each cell type.

Scale-up potential

A diagrammatic representation of scale-up procedures with anchorage-dependent cells is shown in Figure 3.6.1. For most practical purposes in research laboratories the maximum number of cells required can be achieved with either modified multitray or modified roller bottle technology. Discussion in detail on scale-up is given later.

Poly-lysine as an attachment factor

Synthetic polymers of lysine can be produced in a range of molecular weights. Those in the range 30 000–300 000 have been found useful in providing a positive surface charge on glass or plastic (McKeehan & Ham, 1976).

Originally polymers of L-lysine were used but because this is the metabolically active form of the amino acid, some workers prefer to use poly-D-lysine. A simple procedure for coating surfaces follows.

PROCEDURE: A SIMPLE PROCEDURE FOR COATING SURFACES

Materials

- Poly-D-lysine, lyophilized and gamma-irradiated (Sigma)
- Poly-L-lysine, lyophilized and gamma-irradiated (Sigma)
- Distilled or cartridge water
- Bijoux or universal screw-top container

1. Select the required molecular weight range of poly-lysine, i.e. 30 000–70 000, 70 000–150 000 or over 300 000. The charge density of the molecules increases with molecular weight, so selection should be based on the degree of binding required.
2. Add sterile water to the lyophilized poly-lysine to give a final concentration of

PLASTIC AND GLASS TISSUE CULTURE SURFACES

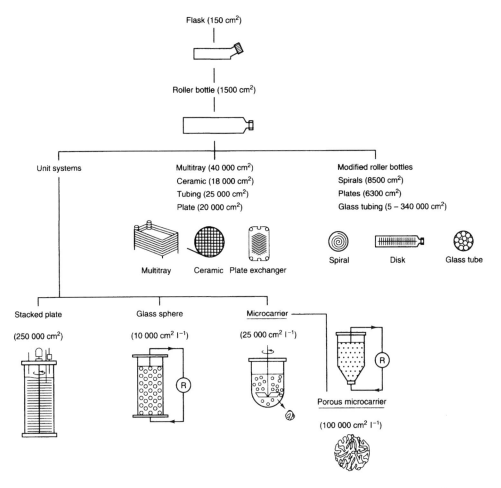

Figure 3.6.1 Diagrammatic representation of scale-up procedures with anchorage-dependent cells.

 0.1 mg ml^{-1}. Leave at room temperature until completely dissolved, which should be within a few minutes. Store the solution at 4°C.
3. Add the solution to culture flasks or dishes using 0.5–1 ml per 20–25 cm^2 surface area.
4. Cover the surface by gently tilting the vessel for a few moments. Leave for 10 min at room temperature.
5. Aspirate the solution and gently wash the surface three times with sterile distilled water.
6. Leave to dry before using. Vessels can be kept at 4°C for several weeks if not used immediately.

Solution that has been used to coat vessels can be re-used at least once. However, the charge density will diminish with each use.

REFERENCES

Grinell F (1978) Cellular adhesiveness and extracellular substrata. *International Review of Cytology* 53: 65–144.

Hynes RO & Bye JM (1974) Density and cell cycle dependence of cell surface proteins in hamster fibroblasts. *Cell* 3: 112–120.

Maroudas (1993) Physical and chemical treatments which affect cell adhesion. In: *Cell & Tissue Culture: Laboratory Procedures*, pp. 2E:2.1–2.11. Wiley, Chichester.

McKeehan WL & Ham RG (1976) Stimulation of clonal growth of normal fibroblasts with substrate coated with basic polymers. *Journal of Cell Biology* 71: 727–734.

Springer EL, Hakett AJ & Nelson-Rees WA (1976) Alteration of the cell membrane architecture during suspension and monolayer culturing. *International Journal of Cancer* 17: 407–413.

3.7 THREE-DIMENSIONAL CELL CULTURE SYSTEMS

Animal cells cultured in two-dimensional (2D) monolayers in traditional glass or plastic tissue culture flasks have been used successfully for many purposes in research and industrial production. However, such cultures may lose key phenotypic characteristics (e.g. virus susceptibility, morphology, surface markers/receptors) after repeated passage. *In vivo* the presence of three-dimensional (3D) cellular structures is critical to the correct development, function and stability of cells, tissues and organs. The characteristics that the researcher or technologist wishes to utilize are often a feature of the tissue and not individual cells, e.g. a functional bladder epithelium or crypt structures of the gut. In this section we describe some of the approaches that can be used to simulate certain features of the *in vivo* environment in an attempt to promote natural gene expression and tissue function in cultured cells. The described technologies address these features from two aspects:

1. Endogenous features (i.e. within 3D cell structures): autocrine and paracrine factors; mass transfer characteristics; establishment of appropriate cell–cell and cell–matrix contacts; cell signalling pathways; and pressure and tensile forces.
2. Exogenous features (i.e. in culture medium): hormones; nutrition, including conditioned media; pO_2; pH; pCO_2; and temperature.

Many of these specialized functions of cells are lost or expressed at low levels when they are growing in monolayers and this is due, in part, to the lack of appropriate cell–cell or cell–matrix interactions. Cultivation in a 3D system can promote or improve characteristic cell functions such as hormone secretion, production of extracellular matrix components and expression of differentiation markers.

The resulting cellular structures can be used as models for investigating development, drug metabolism, toxicity, biotransformations, pathogenesis and microorganism replication.

In industrial (e.g. recombinant protein) and medical (e.g. bioartificial organ) fields 3D cultures can be used to improve the surface area/volume ratio compared with 2D cultures, which is a useful feature where cells are used as the machinery for biological production. Such approaches promote high cell yield and increased production of cellular or recombinant proteins.

Diverse techniques for 3D culture are now available and here we will primarily consider:

- 3D multilayers or spheroids
- 3D supports, e.g. microcarriers

Cell and Tissue Culture: Laboratory Procedures in Biotechnology, edited by A. Doyle and J.B. Griffiths.
© 1998 John Wiley & Sons Ltd.

- 3D matrices, such as gels, sponges and porous microcarriers

Usually a single cell type is cultured in isolation, as in 3D systems to simulate liver function where hepatocytes are propagated to reach high organ-like densities (e.g. >10^7 cells ml^{-1}) in a 3D matrix, thereby enabling the reproduction of specific tissue-like function. Furthermore, other organ-like structures can be reproduced using co-cultures, e.g. mesenchymal and epithelial cells of intestinal origin (Goodwin et al., 1993) and foetal rat spinal cord co-cultured with human muscle cells (Mariotti et al., 1993). A further technique used to reproduce tissue-like structures is re-aggregation of primary chick cell suspensions (Funk et al., 1994). All of these approaches are very useful for the study of differentiation, cell–cell interations and tissue function, and can benefit enormously from a 3D culture approach. The applications can range from cell biology and medical research to industrial-scale production of biologicals. However, there is no single technique applicable for all purposes and in the following the key features of a number of 3D culture techniques are reviewed.

SPHEROIDS

Spheroids are suspended multicellular aggregates where cells adhere to each other instead of attaching to an artificial substrate. They are applied as models for embryogenesis of different tissues, in tumour biology as therapeutic models and also used in developmental studies. Spheroids can be generated efficiently by culture of cells on the normal tissue culture surfaces (glass and plastic) coated with agarose (Sussman & Sussman, 1961) or poly(hydroxyethyl methacrylate) (polyHEMA) (Folkman & Moscona, 1978). Short-term spheroid cultures provide useful spherical structures and in some cases spheroid culture methods significantly improve the cultivation of primary cells (Ijima et al., 1997) and the retention of differentiated characteristics.

Early research by Moscona (1961) on embryonic cells cultured as spheroids showed tissue-like structural properties and in vivo-like growth. A high differentiation capacity, which is not observed in monolayers, can be found in multicellular spheroids of hepatocytes (Tong et al., 1994), adult human glioma cells (Glimelius et al., 1988), avian foetal brain cells (Funk et al., 1994) and outer root sheath cells of human hair follicles (Limat et al., 1994). This is an especially important consideration in drug tests, where cells with a full range of natural metabolic activities are required to achieve a biochemical turnover that would be observed in tissues.

In particular, malignant human ovary and mammary carcinomas can form solid aggregates with histological similarities to the primary tumours (Schleich, 1967). In addition, the production of extracellular matrix components similar to tumours in vivo has been demonstrated by Glimelius et al. (1988) with multicellular spheroids of human glioma cells.

A series of examples reveal that cell systems like these can be used in wide ranging studies of, for example, cancer sensitivity to radiation and chemotherapy and the analysis of penetration by cytotoxic drugs in targeted tumour therapies

(Lindstrom & Carlsson, 1993; Rotmensch et al., 1994). Spheroid-like hollow bodies with a multicellular epithelial morphology have also been utilized as a model of pathogenesis of infection as a result of *Neisseria* infection. Numerous cell–cell contacts representative of cells *in vivo* have been demonstrated in spheroids from nasopharyngeal cells, e.g. junctional complexes, desmosomes, specific orientation of the cytoskeleton and cellular organelles (Boxberger et al., 1993).

MICROCARRIERS

An important feature of 3D cultures for industrial purposes is the improved surface area/volume ratio allowing high cell densities in bioreactors with relatively low costs of production and maintenance. A system often applied for 3D large-scale productions is the microcarrier bead (see Chapter 5, section 5.8). The use of porous carriers (e.g. Verax Microsphere, Cultispher G, Immobisil or Siran Porous Beads (see Chapter 5, section 5.9) has not only enabled increased unit productivity (reviewed by Looby & Griffiths, 1989; Griffiths, 1990) but allows cells to form a 3D organization within the porous microsphere.

A number of microcarrier systems have been established with primary and secondary cells from birds and mammals, permanent cell lines from fish and mammals as well as diploid human cells (Reuveny, 1985). A further possibility is the mass production of cells with the retention of differentiation potential, as can be seen with bone cells (Sautier et al., 1992) and human retinal pigment epithelial cells (Kuriyama et al., 1992). In addition to the widely used continuous cell lines for the production of biologicals, freshly harvested cells such as endothelial cells have been applied for the production of endothelium-derived relaxing factor (Bing et al., 1991).

Hepatocytes growing in microcarrier systems are used for the study of liver failure and drug metabolism. Transplantation experiments of hepatocytes grown on microcarriers showed detoxification of ammonium and reduction of bilirubin concentrations after induced acute liver failure (Nagaki et al., 1990). Furthermore, co-cultivation assays using hepatocytes and Balb/c 3T3 fibroblasts as target cells have been applied to analyse the metabolism-mediated toxicity of xenobiotics *in vitro* (Voss & Seibert, 1992).

Other examples for the successful employment of co-cultures are: the interaction between muscle and nerve cells (Shahar et al., 1985) and the co-cultivation of vascular endothelial and smooth-muscle cells using microcarrier techniques (Davies & Kerr, 1982); co-cultivation of cerebellar granule cells, cerebral cortical neurons and cortical astrocytes on collagen-coated dextran beads (Cytodex3) and the production and release of specific neurotransmitters and enzyme synthesis resembling *in vivo* interactions between neurons and astrocytes (Westergaard et al., 1991).

Microcarrier technology has provided valuable systems for industrial-scale use of animal cells. However, microcarriers have also proved to be a highly adaptable and very successful approach to studying a wide range of cell types, both in mono- and co-culture.

FILTERWELLS

In the filterwell or 'Transwell®', systems, cells are grown on a permeable membrane between two separated liquid phases (McCall et al., 1981). In this condition confluent epithelial cells can achieve their full polarized functional state without the stresses induced by 'doming' of epithelial monolayers on glass or plastic surfaces (Rabito et al., 1980). Filterwell culture has proved valuable for modelling the epithelium of the human intestine (Hidalgo et al., 1989). This technique is rapidly becoming the preferred culture system for many studies using epithelial cultures.

MATRIX SPONGES OR 3D GELS AND MATRIX SANDWICHES

Sponges made of cellulose or collagen gels can be used as matrices in Petri dishes where cells penetrate the inert material. A collagen-coating can be applied to improve cell adherence and growth. Analysis of cell characteristics in this system is readily achieved by histological analysis of sections. The influence of culture conditions on differentiation has been analysed by Bruns et al. (1994), comparing the formation of matrix substances by sheep rib perichondrium cultured on collagen sponges, fibrin gels and cellulose acetate filters. An *in vitro* model simulating normal human secretory endometrium was successfully established by Bentin-Ley et al. (1994). In this system the polarized epithelial cells were grown on 'Matrigel®', which separated the underlying collagen-embedded endometrial stromal cells. Mechanical stress-induced orientation has been studied using arterial smooth-muscle cells in 3D collagen lattices (Kanda & Matsuda, 1994). Sandwiches of collagen matrix to simulate the space of Disse in the liver have proved highly successful for the extended and improved culture of primary hepatocytes (e.g. Bader et al., 1996). Such changes in physical characteristics of culture conditions may offer valuable developments for tissue models in the future.

MICROCONTAINERS

The microcontainer technique is a recently developed method for 3D cultures (Weibezahn et al., 1994). In this patented system cells grow on the vertical walls until a multicellular layer is formed. The containers are then transferred into a vessel and perfused continuously. A controlled supply of growth media of different compositions on each side of the layer allows the culture of highly differentiating cell types and hence the establishment of tissue-like models. The full range of applications for this technique remains to be determined and it may well provide some useful *in vitro* tissue models in the future.

SIMULATED MICROGRAVITY

In order to maintain microgravity conditions achieved in space flight when cell culture experiments returned to Earth's surface, the US National Aeronautical Space Agency (NASA) laboratories developed a device called the rotating wall vessel (RWV). In this system cells are grown in a rotating body of culture medium with no direct air/medium interface. This provides for conditions of very low shear stress in which cells of different buoyant density can aggregate and proliferate together. This technique has been used for a wide range of cell cultures, including cartilage (Freed & Vunjak-Novakovic, 1993), ovarian tumour cells (Becker et al., 1993), hepatocytes (T. Battle, personal communication) and colorectal carcinoma (Goodwin et al., 1992). Co-culture systems have revealed some very useful and novel characteristics that mimic *in vivo* tissue (Goodwin et al., 1993). Thus, low shear stress technology holds great promise for the development of new *in vitro* models.

CONCLUSION

Whilst the use of standard tissue culture flasks in glass or plastic has provided much of the data on which our understanding of *in vitro* cell biology is based, a very wide range of novel culture formats are now available for the benefit of those in basic research and biotechnology. Some of these approaches, notably filterwell culture, are now fairly commonplace but others, such as microcontainers and the RWV system, remain to have their full potential identified. Recent developments in the area of 3D cell culture have been rapid. It is evident that there is still much to be learned about the way cell function is modulated in 3D cell–cell and cell–matrix interactions.

REFERENCES

Bader A, Knop E, Kern A, Boker K, Fruhauf N, Crome O, Esselman H, Pape C, Kempka G & Searing KF (1996) 3D co-culture of hepatic sinusoidal cells with primary hepatocytes – design of an organotypical node. *Exp. Cell Research*, 226(1): 223–233.

Becker JL, Prewett TL, Spaulding GF & Goodwin TJ (1993) Three-dimensional growth and differentiation of ovarian tumour cell lines in high aspect rotating wall vessel. Morphologic and embryonic considerations. *Journal of Cellular Biochemistry* 51: 283.

Bentin-Ley U, Pedersen B, Lindenberg S, Falck Larsen J, Hamberger L & Horn T (1994) Isolation and culture of human endometrial cells in a three-dimensional culture system. *Journal of Reproduction and Fertility* 101: 327–332.

Bing RJ, Binder T, Pataricza J, Kibira S & Narayan KS (1991) The use of microcarrier beads in the production of endothelium-derived relaxing factor by freshly harvested endothelial cells, *Tissue & Cell* 23: 151–159.

Boxberger HJ, Sessler MJ, Maetzel B & Meyer TF (1993) Highly polarized primary epithelial cells from human nasopharynx grown as spheroid-like vesicles. *European Journal of Cellular Biology* 62: 140–151.

Bruns J, Kersten P, Lierse W, Weiss A & Sibermann M (1994) The *in vitro* influence of different culture conditions on the potential of sheep rib perichondrium to form hyaline-like cartilage. *Virchows Archiv* 424: 169–175.

Davies PF & Kerr C (1982) Co-cultivation of vascular endothelial and smooth muscle cells using microcarrier techniques. *Experimental Cell Research* 141: 455–459.

Folkman J & Moscona A (1978) Role of cell shape in growth control. *Nature (London)* 273: 345–349.

Freed LE, Vunjak-Novakovic G & Langer R (1993) Cultivation of cell–polymer cartilage implants in bioreactors. *Journal of Cellular Biochemistry* 51: 257.

Freshney RJ (1986) *Animal Cell Culture*. IRL Press, Oxford.

Funk KA, Liu CH, Wilson BW & Higgins RJ (1994) Avian embryonic brain reaggregate culture system. *Toxicology and Applied Pharmacology* 124: 149–158.

Glimelius B, Noring B, Nedermann T & Carlsson J (1988) Extracellular matrices in multicellular spheroids of human glioma origin: increased incorporation of proteoglycans and fibronectin as compared to monolayer cultures. *Acta Pathologica et Microbiologica Scandinavica* 96: 433–444.

Goodwin TJ, Jessup JM & Wolf DA (1992) Morphological differentiation of human colorectal carcinoma in rotating wall vessels. *In Vitro Cell Development Biology* 28A: 47–60.

Goodwin TJ, Schroeder WF, Wolf DA & Moyer MP (1993) Rotating-wall vessel coculture of small intestine as a prelude to tissue modeling: aspects of simulated microgravity. *Proceedings of Society for Experimental Biology and Medicine* 202: 181–192.

Griffiths B (1990) Advances in animal cell immobilization technology. *Animal Cell Biotechnology* 4: 149–166.

Hildago IJ, Raub TJ & Borchardt RT (1989) Characterisation of the human colon carcinoma cell line (Caco-2) as a model system for intestinal epithelial permeability. *Gastroenterology* 96: 736–749.

Ijima H, Matsushita T, Nakazawa N, Koyama S, Gion T, Shirabe K, Shimada M, Takenaka K, Sugimachi K & Funatsu K (1997) Spheroid formation of primary dog hepatocytes using polyurethane foam and its application to hybrid artificial liver. In Carrondo KJT (eds) *Animal Cell Technology*, pp. 577–583.

Kanda K & Matsuda T (1994) Mechanical stress-induced orientation and ultrastructural change of smooth muscle cells cultured in three-dimensional collagen lattices. *Cell Transplantation* 3: 481–492.

Kuriyama S, Nakano T, Yoshimura N, Ohuchi T, Moritera T & Honda Y (1992) Mass cultivation of human retinal pigment epithelial cells with microcarrier. *Ophthalmologica* 205: 89–95.

Leist CH, Meyer H-P & Fiechter A (1990) Potential and problems of animal cells in suspension culture. *Journal of Biotechnology* 15: 1–46.

Limat A, Breitkreutz D, Hunziker T, Klein CE, Nozer F, Fusenig NE & Braathen LR (1994) Outer root sheath (ORS) cells organize into epidermoid cyst-like spheroids when cultured inside Matrigel: a light-microscope and immunohistological comparison between human ORS cells and interfollicular keratinocytes. *Cell and Tissue Research* 275: 169–176.

Lindstrom A & Carlsson J (1993) Penetration and binding of epidermal growth factor–dextran conjugates in spheroids of human glioma origin. *Cancer Biotherapy* 8: 145–158.

Looby D & Griffiths JB (1989) Immobilization of animal cells in fixed and fluidized porous glass sphere reactors. In: Spier RE, Griffith JB, Stephenne J & Crooy PJ (eds) *Advances in Animal Cell Biology and Technology for Bioprocesses*, pp. 336–343.

Mariotti C, Askanas V & King Engel W (1993) New organotypic model to culture the entire fetal rat spinal cord. *Journal of Neuroscience Methods* 48: 157–167.

McCall E, Povey J & Dumonde DC (1981) The culture of vascular endothelial cells on microporous membranes. *Thrombosis Research* 24: 417–431.

Moscona A (1961) Rotation-mediated histogenetic aggregation of dissociated cells. A quantifiable approach to cell interaction in vitro. *Experimental Cell Research* 22: 455–475.

Nagaki M, Kano T, Muto Y, Yamada T, Ohnishi H & Moriwaki H (1990) Effects of intraperitoneal transplantation of microcarrier-attached hepatocytes on D-galactosamine-induced acute liver failure in rats. *Gastroenterologia Japonica* 25: 78–87.

Rabito CA, Tchao R, Valentich J & Leighton J (1980) Effect of cell substratum interaction of hemicyst formation by MDCK cells. *In Vitro* 16: 461–468.

Reuveny S (1985) Microcarriers in cell culture: structure and applications. *Advances in Cell Culture* 4: 213–247.

Rotmensch J, Whitlock JL, Culbertson S, Atcher RW & Schwartz JL (1994) Comparison of sensitivities of cells to X-ray therapy, chemotherapy, and isotope therapy using a tumor spheroid model. *Gynecologic Oncology* 55: 290–293.

Sautier JM, Nefussi JR & Forest N (1992) Mineralization and bone formation on microcarrier beads with isolated rat calvaria cell population. *Calcified Tissue International* 50: 527–532.

Schleich, A (1967) Studies on aggregation of human ascites tumor cells. *European Journal of Cancer* 3: 243–246.

Shahar A, Mizrahi A, Reuveny S, Zinman T & Shainberg A (1985) Differentiation of myoblasts with nerve cells on microcarriers in culture. *Developments in Biological Standardization* 60: 263–268.

Sussman M & Sussman RR (1961) Aggregative performance. *Experimental Cell Research* 8: 91–106.

Tong JZ, Sarrazin S, Cassio D, Gauthier F & Alvarez F (1994) Application of spheroid culture to human hepatocytes and maintenance of their differentiation. *Biologie Cellulaire* 81: 77–81.

Voss JU & Seibert H (1991) Microcarrier-attached rat hepatocytes as a xenobiotic-metabolizing system in cocultures. *Cell Biology and Toxicology* 7: 387–399.

Weibezahn KF, Knedlitschek G, Dertinger H, Bier W, Schaller Th & Schubert K (1994) An *in vitro* tissue model using mechanically processed microstructures. Presented at the *41st ETCS Congress*, Verona.

Westergaard N, Sonnewald U, Peterson SB & Schousboe A (1991) Characterization of microcarrier cultures of neurons and astrocytes from cerebral cortex and cerebellum. *Neurochemistry Research* 16: 919–923.

CHAPTER 4

BIOCHEMISTRY OF CELLS IN CULTURE

4.1 OVERVIEW

Cells *in vivo* grow in a highly controlled and complex environment. On removal from their location *in situ* they are dispersed from a histological configuration, where cell contact plays an important role, into a simplified growth medium. This lacks hormonal and neuronal regulatory mechanisms and the thousands of intermediate metabolites present in body fluids. In addition the organism has highly evolved oxygenation and detoxification systems ensuring that, except in conditions of stress, an extremely uniform environment is maintained. Thus a cell in culture is in a very alien situation, the stresses of which can only be minimized by attention to environmental optimization. An important factor is the nutritional environment, which has to replace the complex body fluids and will affect all the physiological and metabolic events in the cell. Cell media were first developed with components identified in body fluids (amino acids, vitamins, etc.), plus serum, to provide the multitude of unknown metabolites. Over the years the need for serum has been reduced, or replaced, as a greater understanding of the cells' nutritional requirements has been obtained. Thus, a whole range of growth factors, hormones and inorganic ions can be added in place of serum to give a more standardized and optimized environment. This in turn not only means greater cell and product productivity, but allows more informative studies to be carried out on understanding cellular regulation of growth and metabolism. This allows greater maximization of performance in culture, whether it is to enhance secretion of monoclonal antibodies or to respond specifically to a drug in a toxicity assay. Cells in culture no longer have to be de-differentiated. Thus, the ubiquitous cell can, with the correct environment, demonstrate many differentiated characteristics.

An understanding of cell biochemistry is essential to potentiate the use of cells for all purposes. Many nutrients have been identified as essential, as have many that play a key role in cell behaviour (e.g. glutamine, glucose, oxygen). Inhibitory factors such as ammonia and lactate have also been identified. However, it must be stressed that this is only the 'tip of the iceberg' when it comes to controlling the biochemistry of the cell. The macrometabolites may have been identified, but cellular control and regulation is based on a cascade of signals from micrometabolites acting on the cell surface and in the cytoplasm. This is why studies at the molecular biology level of the role of growth factors in this cascade are essential for a full understanding of cell biochemistry.

The nutrient balance is not the only factor affecting cell biochemistry, as there are many physiochemical interactions in a culture. The need to scale up cell cultures increases the importance of understanding the effect of such factors as agitation and mixing, and of how oxygen should be supplied in the least damaging way.

Cell and Tissue Culture: Laboratory Procedures in Biotechnology, edited by A. Doyle and J.B. Griffiths.
© 1998 John Wiley & Sons Ltd.

Keeping the cells healthy and viable has always been a dominant priority. The recognition that cell death may be programmed or behavioural (apoptosis) rather than just the result of hostile environmental factors causing necrosis and lysis has opened up another means of controlling cell viability and longevity in culture. An understanding of the gene products that control these processes is gradually unfolding, leading to great expectations of improvements in our capabilities of controlling cellular functions.

It is easy to accumulate masses of metabolic data on cell performance in culture, especially at the level of oxygen, glucose and glutamine utilization. The difficult part is analysing and interpreting the data, as shown by the numerous publications giving voluminous data but totally lacking any form of subjective analysis. It is our aim to cover this complex and expanding area by looking at, for example, computer modelling of cell kinetics, in order to stimulate new approaches.

It is expected that a combination of environmental control, medium design, the control of growth factor-regulated gene expression and genetic modification of the cell will enable huge advances to be made in animal cell productivity.

4.2 QUANTITATIVE ANALYSIS OF CELL GROWTH, METABOLISM AND PRODUCT FORMATION

With the growth of the biotechnology industry there has been an ever-increasing volume of data obtained from a variety of cell types cultured under different conditions and in different culture systems. Unfortunately, the lack of standardization for data presentation makes it difficult to compare results. The frequently presented profiles of product or metabolite concentration versus time have limited value because the concentration depends directly upon the viable cell concentration. It is more useful to present results as production or consumption rates per cell (or unit mass). For example, the characteristic parameter for glucose consumption is the specific (per cell) glucose consumption rate (mmol glucose viable cell^{-1} unit time^{-1}) rather than the glucose concentration profile. The concentration of a product or nutrient is still important because it may directly affect the production or consumption rate of itself or other metabolites (Miller & Blanch, 1991). For example, the glucose consumption rate increases with glucose concentration (up to 1–5 mM). Glucose consumption is also a function of culture P_{O2} and pH. Thus it may be necessary to control parameters such as pH, P_{O2} and glucose concentration in order to examine the effects of these or other parameters (e.g. hormone levels) on cell metabolism. At a minimum it is important to report the levels of these and other parameters, such as ammonia and glutamine concentration, when presenting experimental results.

Typical commercial cell culture systems include batch or fed-batch suspension reactors and perfused immobilized-cell reactors. However, the transient nature of batch culture causes difficulties in studying the effects of external stimuli on growth, metabolism and product formation. Due to metabolite concentration gradients, and the difficulty of obtaining representative cell samples, immobilized-cell reactors are also poorly suited for the analysis of cell growth and metabolism. As a result it is desirable to use well-defined model systems. Continuous-flow suspension reactors allow metabolic parameters to be measured at steady state, after cells have adapted to new (and possibly inhibitory) conditions. Perfusion reactors (with cells immobilized on suspended or stationary supports) extend these benefits to anchorage-dependent cells, and provide model systems for cell responses *in vivo*. However, while it is instructive to study the behaviour of cells under well-defined conditions, the results obtained must be verified in the culture system selected for commercial production.

In this section methods are introduced for processing and presenting data for cell growth, metabolite consumption and product formation in a manner that facilitates

---

Cell and Tissue Culture: Laboratory Procedures in Biotechnology, edited by A. Doyle and J.B. Griffiths.
© 1998 John Wiley & Sons Ltd.

comparison between diverse reactor systems and cell types. Data analysed in this manner also facilitate application of results obtained in one reactor system to another. Equations are presented for a variety of culture configurations, such as batch, fed-batch, continuous and perfusion of immobilized cells. Specific examples are included to illustrate the methods of analysis.

ERRORS IN CALCULATIONS

Experimental errors tend to be quite large in biological systems, e.g. ± 30% in protein concentration measurements. Cell number measurements are generally no better than ± 5%. At lower viabilities (< 70% viable), accurate determination of viability and cell number is difficult, and the error in each determination may be greater than ± 10%. Errors in cell and metabolite concentration measurements lead to uncertainties in calculated parameters, such as specific growth, production and consumption rates, therefore a complex profile for these calculated parameters should not be assumed when a straight-line or simple function will suffice.

Errors for batch culture tend to be larger than those for continuous culture due to rapidly changing conditions and limited samples. When dealing with batch cultures, as many samples as possible should be taken; however, care should be taken that the culture volume is not depleted so much that the culture conditions are affected (e.g. $K_L a$ as discussed below). Continuous culture allows for more frequent samples. However, the total sample rate should not exceed the dilution rate so as to maintain a constant reactor volume.

CELL GROWTH AND DEATH RATES

The central parameter in cell culture is the viable cell concentration. The specific growth rate (μ), the specific death rate (k_d) and the fraction of viable cells (f_v) are used to characterize the proliferative state and health of a culture. These parameters are calculated from the viable and non-viable cell concentrations using equations appropriate for the type of culture vessel employed (see below). The viable cell concentration is also used to calculate the metabolic quotients (e.g. specific glucose consumption rate), as discussed in a later section. In the absence of cell concentration data, analysis is limited to ratios of metabolic rates (see below).

Free cells in suspension

Figure 4.2.1 shows a general schematic of a well-mixed suspension culture vessel. For semi-batch culture the outlet flow rate would be zero, while for batch culture both flow rates would be zero. Assuming constant density, a total mass balance around the vessel is:

$$\frac{dV}{dt} = F_i - F_o \tag{4.2.1}$$

QUANTITATIVE ANALYSIS OF CELL GROWTH

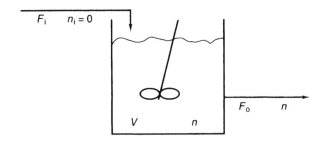

Figure 4.2.1 Schematic diagram of a continuous-flow stirred suspension vessel. For semi-batch culture $F_o = 0$ and for batch culture $F_o = F_i = 0$.

where V is the volume of medium in the vessel, F_i is the volumetric flow rate of fresh medium and F_o is the volumetric flow rate of the spent medium (including suspended cells). The balance on total cells is given by: rate of accumulation of total cells = rate at which cells are added in the feed − rate at which cells are removed in the outlet + rate of cell growth − rate of cell lysis.

The balance on total cells for a sterile feed is given by:

$$\frac{d(nV)}{dt} = 0 - nF_o - \mu_{app}nV \qquad (4.2.2)$$

where n is the total cell concentration and μ_{app} is the apparent specific growth rate, which includes cell growth and lysis.

Noting that only viable cells can divide and assuming that viable cells are not directly lysed, Equation 4.2.3 is obtained:

$$\mu_{app}n = \mu n_v - k_l n_d \qquad (4.2.3)$$

where μ is the true specific growth rate, n_v is the viable cell concentration, n_d is the dead cell concentration and k_l is the dead cell specific lysis rate. Note that cell death converts a viable cell into a dead cell, but does not directly change the total cell concentration. The viable cell balance is given by:

$$\frac{d(n_v V)}{dt} = 0 - n_v F_o - \mu_{app}n_v V \qquad (4.2.4)$$

μ_{app} is the viable cell-derived apparent specific growth rate given by:

$$\mu_{app}n_v = \mu n_v - k_d n_v \qquad (4.2.5)$$

where k_d is the specific death rate.

Continuous culture

For a constant-volume vessel $F_i = F_o = F$. Significant acid and base additions for pH control should be included in the F_i term. Reactor samples are included in the F_o term and do not affect the analysis as long as V is approximately constant. Equation 4.2.2 becomes:

$$\frac{dn}{dt} = -nD + \mu_{app}n \qquad (4.2.6)$$

where the dilution rate $D = F/V$. Equation 4.2.6 can be rearranged to obtain a general expression for μ_{app} as in Equation 4.2.7:

$$\mu_{app} = \frac{d \ln(n)}{dt} + D \qquad (4.2.7)$$

Because cell count data are often noisy, it is best to obtain the derivative by fitting $\ln(n)$ versus time with a low-order polynomial, or by drawing a smooth curve through a plot of $\ln(n)$ versus time and manually determining the slope at selected points of interest. For the special case of steady state, the derivative is zero and $\mu_{app} = D$.

Lysis of dead cells is generally not important at conditions of interest with reasonable cell viability (see below for exceptions). In that case k_l will be approximately zero and Equation 4.2.3 can be used to obtain an expression for μ as in Equation 4.2.8:

$$\mu = \mu_{app}\left(\frac{n}{n_v}\right) = \frac{\mu_{app}}{f_v} \qquad (4.2.8)$$

Note that μ diverges from μ_{app} (and hence D at steady state) as the viability decreases. This is because new cell growth must offset cell death, as well as cell removal in the outlet stream. The cell death rate can be obtained from Equations 4.2.4 and 4.2.5:

$$k_d = \mu - D - \frac{d \ln(n_v)}{dt} \qquad (4.2.9)$$

The death rate is generally low at high dilution rates and becomes more significant at low dilution rates.

Example 1

Table 4.2.1 contains cell concentration and viability data for a continuous culture experiment in which a pH step change was implemented (Miller et al., 1988). The initial steady state was established at pH 7.6. At time zero the pH was decreased to 7.1. A plot of the natural logarithm of total cell concentration versus time is shown in Figure 4.2.2a. A slope of zero is assumed for the initial (time < 0 days) and final (time > 7 days) steady states. A constant slope is taken for the exponential growth portion (1 day < time < 5.5 days). Two other slopes are also illustrated in the figure. The calculated slopes are substituted into Equation 4.2.7 with $D = 0.41$ day^{-1} to obtain μ_{app}; μ is calculated using Equation 4.2.8. The apparent and true specific growth rates are shown in Table 4.2.1 and plotted in Figure 4.2.2b. Note that μ_{app} is essentially constant during the increase in cell concentration, while μ reaches a maximum and then continually declines, as the viability increases, during this period.

QUANTITATIVE ANALYSIS OF CELL GROWTH

Table 4.2.1 Data for the growth rate example calculations illustrated in Figure 4.2.2

Culture time (days)	ln (total cells ml^{-1})	Fraction viable	Slope (day^{-1})	μ_{app} (day^{-1})	μ (day^{-1})
−0.88	14.14	0.62	0.00	0.41	0.66
−0.24	14.15	0.63	0.00	0.41	0.65
0.26	14.16	0.68	0.00	0.41	0.60
0.78	14.07	0.68	0.00	0.41	0.60
1.12	14.16	0.69	0.08	0.49	0.72
1.74	14.23	0.75	0.17	0.58	0.77
2.12	14.34	0.72	0.17	0.58	0.81
2.73	14.40	0.76	0.17	0.58	0.76
3.11	14.52	0.78	0.17	0.58	0.74
3.73	14.56	0.80	0.17	0.58	0.72
4.12	14.64	0.81	0.17	0.58	0.72
4.72	14.76	0.82	0.17	0.58	0.71
5.11	14.84	0.85	0.17	0.58	0.68
5.74	14.95	0.82	0.00	0.41	0.50
6.12	14.84	0.81	n.d.	n.d.	n.d.
6.78	14.85	0.75	−0.05	0.36	0.48
7.22	14.87	0.69	0.00	0.41	0.59
7.80	14.85	0.69	0.00	0.41	0.59
8.10	14.83	0.68	0.00	0.41	0.60
8.73	14.89	0.71	0.00	0.41	0.58
9.12	14.86	0.71	0.00	0.41	0.58
9.72	14.86	0.70	0.00	0.41	0.59

n.d. = not determined.

Batch culture

In this case $F_i = F_o = F = 0$. Reactor samples (i.e. $F_o \neq 0$) do not affect the analysis because they reduce the culture volume without altering any of the concentrations. However, acid or base additions do affect concentrations. If the volume added is significant the reactor effectively becomes a fed-batch reactor. In this case Equations 4.2.11 and 4.2.12 (below) would apply. For negligible cell lysis the expressions for μ_{app}, μ and k_d for batch culture can be obtained from Equations 4.2.7–4.2.9 by setting $D = 0$:

$$\mu_{app} = \frac{d\ln(n)}{dt}; \quad \mu = \frac{\mu_{app}}{f_v}; \quad k_d = \mu - \frac{d\ln(n_v)}{dt} \quad (4.2.10)$$

Equation 4.2.10 is valid for small k_l, which is normally the case until the culture reaches the late stationary or death phase.

Semi-batch culture

For semi-batch culture $F_o = 0$, but $F_i \neq 0$ so V is not constant. In this case it is easiest to solve Equations 4.2.2 and 4.2.4 in terms of the number of cells (nV or n_vV, respectively) in the vessel, as opposed to the cell concentrations:

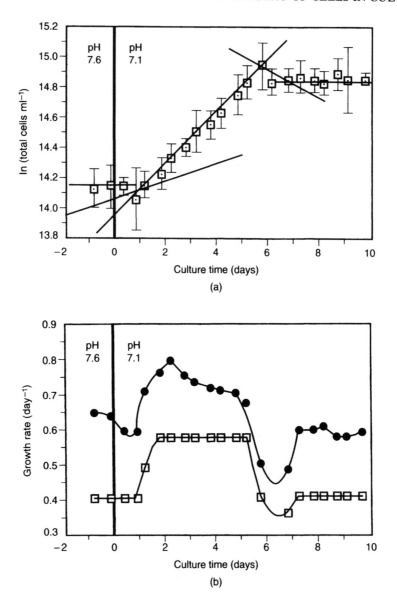

Figure 4.2.2 A pH step change continuous culture experiment. The reactor pH was decreased from 7.6 to 7.1 at time zero. The dilution rate was maintained at 0.41 day^{-1} throughout the experiment. (a) Natural logarithm of total cell concentration versus culture time. Tangent lines are shown for slope determination. Error bars represent the standard deviation based on four replicate cell counts. (b) Apparent (μ_{app}, □) and true (μ, ●) specific growth rates calculated from (a) and Equations 4.2.7 and 4.2.8. Adapted from Miller *et al.* (1988) by permission of *Biotechnology and Bioengineering*.

$$\mu_{app} = \frac{d \ln (nV)}{dt}; \quad \mu_{app} = \frac{d \ln (n_v V)}{dt} \qquad (4.2.11)$$

where $V(t) = \int F_i \, dt$. For small k_1, we again have $\mu = \mu_{app}/f_v$. From Equations 4.2.5 and 4.2.11, Equation 4.2.12 is obtained:

$$k_d = \mu - \frac{d \ln (n_v V)}{dt} \qquad (4.2.12)$$

Note that these relations can be used for periodic, as well as continuous, medium additions (i.e. for $F_i(t)$ a discrete or continuous function). Equations 4.2.11 and 4.2.12 neglect sampling ($F_o \neq 0$). In this case reactor samples do affect the culture because they reduce the volume and alter cell and nutrient concentrations. However, Equations 4.2.11 and 4.2.12 can be used as long as the samples removed represent a small fraction of the vessel volume. If large samples are removed, Equations 4.2.11 and 4.2.12 can be used for the periods between samples, with $V(t)$ corrected for the next interval.

Periodic medium replacement (semi-continuous culture)

Cells are often maintained in spinner or shake flasks by periodic replacement of spent medium (with or without replacement of suspended cells) with fresh medium. Values for μ_{app}, μ and k_d can be obtained for these systems by treating them as batch reactors during the intervals between feedings. The cell concentration at the beginning of each interval can be measured directly or calculated from the value before medium exchange and the fraction of medium replaced. Taking samples between medium exchanges will improve the accuracy of the results obtained.

Continuous culture with cell retention

The cell concentration in continuous culture is often limited by the nutrient supply. However, the cell concentration cannot be increased by simply increasing the medium flow rate because cells are washed out of the reactor at dilution rates greater than μ_{max} (the maximum specific growth rate of the cells). One way to obtain higher cell concentrations is to retain cells within the reactor and increase the medium perfusion rate. In general a purge stream is used to allow stable operation at an elevated cell concentration. A typical system is shown in Figure 4.2.3. The total cell concentration in the outlet stream is given by $n_o = \alpha_s n$. The retention ratio (fraction of cells retained in the reactor) is given by the quantity $1 - \alpha_s$. For the system shown in Figure 4.2.3, $\alpha_s = F_2/F_o$. Note that reactor samples should be included in the F_2 term for rigorous analysis. In many cases, however, the sample size is negligible compared with the perfusion rate and need not be included. Other cell retention systems, such as spin filters, settling columns and recirculation loops containing ultrafiltration membranes, can be treated in the same way as long as α_s can be measured or calculated. For constant V with $F_i = F_o = F$, the total cell balance around this system becomes:

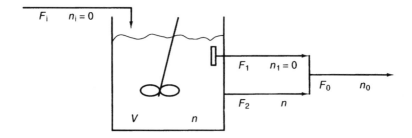

Figure 4.2.3 Schematic diagram of a continuous-flow stirred suspension vessel with cell retention. The distribution of outlet flow between F_1 (cell-free stream) and F_2 is varied to maintain the desired cell concentration in the reactor.

$$\frac{dn}{dt} = -\alpha_s nD + \mu_{app} n \quad (4.2.13)$$

Expressions for μ_{app}, μ and k_d can be obtained by replacing D with $\alpha_s D$ in Equations 4.2.7–4.2.9:

$$\mu_{app} = \frac{d \ln (n)}{dt} + \alpha_s D; \quad \mu = \frac{\mu_{app}}{f_v} \quad (4.2.14)$$

$$k_d = \mu - \alpha_s D - \frac{d \ln (n_v)}{dt}$$

At steady state $\mu_{app} = \alpha_s D < D$. It should be noted that these equations assume equal α_s values for viable and non-viable cells. This will be true for separation devices such as filters, which do not allow any cells to pass (e.g. Figure 4.2.3), but may not be true for settling columns, which generally have greater retention (i.e. smaller α_s) of viable cells. In the latter case, the investigator should measure α_s for viable and non-viable cells by taking cell counts in the reactor and at the settling column outlet. If there is no significant difference in α_s for viable and non-viable cells or if the reactor viability is very high, then the above equations may be used. Otherwise the $\alpha_s n$ term in Equation 4.2.13 must be replaced by $(\alpha_{sv} n_v + \alpha_{sd} n_d)$ where $1 - \alpha_{sv}$ and $1 - \alpha_{sd}$ represent the retention ratios for viable and non-viable cells, respectively.

Cases with significant cell lysis

Cell lysis generally becomes important when viability is low. Examples include continuous culture (with or without cell retention) at low dilution rates and the late stationary and death phases of batch culture. Cell lysis is also important in stirred reactors with very high agitation rates. In order to quantify cell growth parameters under these conditions, the lysed cells must be accounted for. One way to do this is to measure the amount of the cytosolic enzyme lactate dehydrogenase (LDH) released into the medium (see Chapter 2, section 2.5). A procedure for measuring LDH activity is described in Chapter 4, section 4.7.

Lactate dehydrogenase is a useful marker because it is released upon cell death and is stable over short periods of time (5% loss per day) (Wagner et al., 1992), so that the concentration of LDH in the medium provides an estimate of the total number of (intact plus lysed) dead cells. Equations 4.2.1–4.2.5 are still valid, but a balance needs to be included for the effective cell concentration n_e, which is equal to the total cell concentration plus the concentration of cells that have lysed (i.e. what the total cell concentration would be if no cells had lysed):

$$\frac{d(n_e V)}{dt} = 0 - n_e F_o + \mu^e_{app} n_e V \qquad (4.2.15)$$

where μ^e_{app} is the effective cell-derived apparent specific growth rate. Because new cell mass is only produced by viable cells, we have:

$$\mu^e_{app} n_e = \mu n_v \qquad (4.2.16)$$

The effective cell concentration is given by:

$$n_e = n_v + \frac{C_{LDH}}{\gamma_{LDH}} \qquad (4.2.17)$$

where C_{LDH} is the concentration of LDH in the culture medium and γ_{LDH} is the LDH content per cell; γ_{LDH} has been shown to be relatively constant in viable cells (Wagner et al., 1992). In most suspension cultures γ_{LDH} is zero in dead cells (i.e. all LDH is released upon cell death). However it has been shown that dead cells (i.e. cells that do not exclude Trypan blue) detached from microcarriers have a γ_{LDH} value approximately 50% of that for attached viable cells (Wagner et al., 1992). The dead cell LDH content should be verified for each system and γ_{LDH} can be obtained by lysing a known number of viable cells and measuring the amount of LDH released. If dead cells are found to contain LDH, then Equation 4.2.17 must be modified to give:

$$n_e = n_v + \frac{C_{LDH}}{\gamma_{LDH}} + \beta_d n_d \qquad (4.2.18)$$

where β_d is the fraction of γ_{LDH} retained by dead cells, so the amount of LDH released by dead cells is equal to $(1 - \beta_d)\gamma_{LDH}$. One way to test for the importance of cell lysis is to compare the value of C_{LDH}/γ_{LDH} with that of n_d in the culture. Significant lysis is evidenced by n_d being less than C_{LDH}/γ_{LDH} (or n_d being less than $C_{LDH}/\gamma_{LDH} + \beta_d n_d$ if dead cells retain some LDH).

In a short-term batch culture or in a continuously perfused system with low medium residence time, the degradation of LDH may not be significant and may therefore be ignored. However, if analysis is continued over several days at low viabilities or high lysis rates, LDH degradation will be significant. Therefore, the calculated effective cell concentration will be underestimated if no correction is made. The simplest modification is to increase the first day's LDH value by 5% to obtain a corrected C_{LDH} for that day. Five per cent of this value is then added to the C_{LDH} value for the next day, and the process is repeated to calculate C_{LDH} for the second day, and so on. The more rigorous approach of accounting for degradation by writing an unsteady-state mass balance on LDH and solving for an effective C_{LDH} (the concentration if no degradation occurred) at each sample point may be justified for extended experiments. An average degradation rate of 5% per day may be assumed if no data are available. However, the degradation rate may vary with culture parameters such as pH and even with cell number and viability (due to release of proteolytic enzymes). The degradation rate in a particular system can be measured by removing cells from a sample and following the LDH concentration over time.

1. *Continuous culture.* For a constant-volume vessel Equation 4.2.15 can be rearranged to give:

$$\mu_{app}^e = \frac{d \ln(n_e)}{dt} + D \qquad (4.2.19)$$

where n_e is obtained as described above. The specific growth rate is given by:

$$\mu = \mu_{app}^e \left(\frac{n_e}{n_v}\right) \qquad (4.2.20)$$

The specific death rate k_d is given by Equation 4.2.9, while k_l can be obtained from Equations 4.2.3 and 4.2.7:

$$k_l = \mu \left(\frac{n_v}{n_d}\right) - \left(D + \frac{d \ln(n)}{dt}\right)\left(\frac{n}{n_d}\right) \qquad (4.2.21)$$

2. *Batch culture.* In this case the importance of cell lysis can be estimated via the extent of deviation from a straight line in a plot of C_{LDH} versus n_d (Figure 4.2.4). As for the case without significant cell lysis, the batch parameters can be obtained by setting $D = 0$ in Equations 4.2.9 and 4.2.19–4.2.21:

$$\mu_{app}^e = \frac{d \ln(n_e)}{dt}; \quad \mu^e = \mu_{app}^e \left(\frac{n_e}{n_v}\right);$$

$$k_d = \mu - \frac{d \ln(n_v)}{dt}; \quad k_l = \mu \left(\frac{n_v}{n_d}\right) - \left(\frac{d \ln(n)}{dt}\right)\left(\frac{n}{n_d}\right) \qquad (4.2.22)$$

Cells on microcarriers

Cells on microcarriers are normally cultured in batch or perfusion reactors. In perfusion systems microcarriers are normally retained, but free cells exit with the outlet stream. The balance on total cells in a constant-volume perfusion system is given by:

$$\frac{dn}{dt} = -n_F D + \mu_{app} n \qquad (4.2.23)$$

with $n = n_A + n_F$, where n_A is the concentration of attached cells and n_F is the concentration of free cells. If attached cells are present predominantly as a monolayer on the bead, then it is generally found that all of the attached cells are

QUANTITATIVE ANALYSIS OF CELL GROWTH

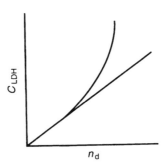

Figure 4.2.4 Schematic of medium LDH concentration (C_{LDH}) versus dead cell concentration (n_d) in batch culture. For systems with minimal cell lysis, the relationship is a straight line with slope equal to the LDH content per cell (γ_{LDH}). The curve with $C_{LDH} > n_d$ indicates significant dead cell lysis, with greater deviation from the straight line indicating more extensive lysis.

viable. In this case $n_v = n_A + f_{Fv}n_F$ (where f_{Fv} is the viable fraction of free cells) and:

$$\mu_{app}n = \mu_A n_A + \mu_F (f_{Fv}n_F) \tag{4.2.24}$$

or:

$$\mu_A = \mu_{app}\left(\frac{n}{n_A}\right) - \mu_F f_{Fv}\left(\frac{n_F}{n_A}\right) \tag{4.2.25}$$

where μ_{app} is given by Equation 4.2.23 as:

$$\mu_{app} = \frac{d\ln(n)}{dt} + D\left(\frac{n_F}{n}\right) \tag{4.2.26}$$

Because free cells are removed with the outlet medium, n_F is normally much less than n. However, the second term in the RHS of Equation 4.2.26 may be important because the increase in cell concentration often slows as cells reach confluence on the beads. Parameter μ_A can be calculated from μ_{app} if μ_F is known, and μ_F can be determined by allowing the microcarriers to settle, growing the remaining suspended cells in batch culture and analysing the data as described above. Because $n_F \ll n$, it should be possible to obtain an accurate estimate for μ_F at the conditions of interest before the nutrient supply in the medium removed from the reactor is depleted. For many cell types μ_F is very small for single cells and f_{Fv} is of the order 0.5. In such cases, the second term in Equation 4.2.25 can be neglected, especially since $n_F \ll n$.

The balance on adherent cells is given by:

$$\frac{dn_A}{dt} = \mu_{app}^A n_A = \mu_A n_A - k_R n_A \tag{4.2.27}$$

where μ_{app}^A is the apparent specific growth rate of adherent cells and k_R is the rate of cell release from the beads; μ_{app}^A can be used directly to characterize the process or it can be used with μ_A to determine k_R, as in:

$$k_R = \mu_A - \frac{d \ln(n_A)}{dt} \tag{4.2.28}$$

Although adherent cells are not removed with the outlet medium, they are removed with reactor samples. If large samples are removed for cell characterization, then the cell concentration should be adjusted for use in calculating the various specific growth rates.

For batch microcarrier culture, Equations 4.2.23–4.2.28 apply with $D = 0$. However, the contribution from free cells in Equations 4.2.25 and 4.2.26 will be greater because free cells are not removed from the system. High n_F values may lead to aggregate formation, which further complicates analysis (see below).

Cell clumps and spheroids

The situation is much more complicated for multilayers of cells on microcarriers or for cells that grow in clumps or spheroids. The specific growth rate in such systems is normally a function of cell location. For example, cells at the outside of tumour cell spheroids generally proliferate rapidly, while those further in are quiescent. For spheroids with a radius larger than about ten cell diameters, there is normally a necrotic core of dead cells. In these complex systems it is only possible to determine apparent growth rates. These can be expressed in terms of total cell numbers, viable cell numbers or spheroid volume. Another difficulty with these systems is that accurate cell counts are hindered by cell clumps that resist dissociation with trypsin. One way around this is to lyse the cells with surfactant and count the nuclei released (Lin *et al.*, 1991). However, analysis is complicated because the nuclei from recently dead cells are preserved, while those from cells dead for long periods are lost.

Immobilized-cell reactors

Cell density is rarely measured directly in immobilized-cell reactors because it is not possible to obtain a single-cell suspension suitable for counting without disturbing the reactor. Specialized techniques are available for indirect non-invasive determination of cell density, but most researchers estimate cell density via nutrient consumption rates. Methods for estimation of cell density from nutrient consumption rates and an analysis of the reliability of such estimates are discussed below.

CELL METABOLISM

The best way to characterize cell metabolic patterns is in terms of consumption rates and production rates per viable cell (collectively termed metabolic quotients). This allows much easier comparison between different cell types or culture conditions than do plots of changes in the levels of nutrients or products over time. It also shows directly how metabolic patterns change with time during batch growth

QUANTITATIVE ANALYSIS OF CELL GROWTH

or in response to changes in culture conditions for a given experiment. As discussed below, metabolic quotients can be readily calculated from metabolite and viable cell concentration data.

Metabolic quotients in stirred vessels

The equations in this section are applicable to any system in which the viable cell density can be determined. This includes all of the systems described above except for immobilized cell reactors.

Dissolved substrates and products

For stirred vessels such as those shown in Figures 4.2.1 and 4.2.3, a balance on the concentration of a dissolved metabolite M gives:

$$\frac{d(MV)}{dt} = F_i M_F - F_o M + q_M n_v V \qquad (4.2.29)$$

where M_F is the concentration in the feed stream. M is the concentration in the vessel and the metabolic quotient q_M is the net specific formation rate of M (mol M time^{-1} viable cell^{-1}). Equation 4.2.29 applies for reactors with cell retention because small metabolites are not retained along with the cells. However, the relation must be modified if product proteins are (partially) retained in the vessel.

For constant-volume vessels we obtain:

$$q_M = \frac{D(M - M_F) + (dM/dt)}{n_v} \qquad (4.2.30)$$

Note that q_M will be negative for nutrients that are consumed by the cells and positive for products. In order to obtain positive values for all of the metabolic quotients, it is customary to speak in terms of the specific substrate (nutrient) consumption rate q_S and the specific product formation rate q_P:

$$q_S = \frac{D(S_F - S) - (dS/dt)}{n_v}; \quad q_P = \frac{D(P - P_F) + (dP/dt)}{n_v} \qquad (4.2.31)$$

For batch systems, Equation 4.2.31 becomes:

$$q_S = \frac{-(dS/dt)}{n_v}; \quad q_P = \frac{(dP/dt)}{n_v} \qquad (4.2.32)$$

For semi-batch systems the metabolic quotients are as in Equation 4.2.33:

$$q_S = \frac{F_i S_F - [d(SV)/dt]}{n_v V}; \quad q_P = \frac{-F_i P_F + [d(PV)/dt]}{n_v V} \qquad (4.2.33)$$

Note that V (and possibly F_i) varies with time.

Metabolic quotients are easily determined in continuous culture. At steady state, the derivative terms in Equation 4.2.31 may be set equal to zero, thereby allowing easy calculation of the specific consumption and production rates. To calculate the metabolic quotients during transients (in continuous, batch or semi-batch culture), it is necessary to obtain dS/dt and dP/dt (or $d(SV)/dt$ and $d(PV)/dt$) for each time point. Scatter in S and P data will result in large fluctuations in q_s and q_P if the derivatives are determined by connecting points from successive samples. These fluctuations may mask trends in the data, so it is recommended that the derivatives be obtained using the slopes of smoothed curves through the S and P data points. Optionally, one could smoothe out scatter in the data by fitting the metabolite and product concentration versus time plots with low-order polynomial functions and then obtaining the derivatives analytically. Regardless of the method used to obtain the derivative, the measured values of S, P and n_v should be used for the other terms in Equations 4.2.31–4.2.33.

Example 2

Table 4.2.2 contains metabolite and cell concentration data for a continuous culture experiment with $D = 0.54$ day^{-1} in which a glucose step change was implemented (Miller *et al.*, 1989). The initial steady state was established at a feed glucose concentration of 5.2 mM, while the residual glucose concentration in the reactor was close to 0 mM. At time zero, the glucose concentration in the reactor was increased to 8.3 mM and the feed concentration was increased to 13.8 mM. The viable cell concentration increased from an initial steady-state value of $\sim 1.6 \times 10^6$ cells ml^{-1} to a final steady-state value of $\sim 2.5 \times 10^6$ cells ml^{-1} (Table 4.2.2). Figure 4.2.5 shows the glucose and lactate concentration profiles. Derivatives for the lactate concentration versus time plot were determined from the slopes of the lines shown in Figure 4.2.5b, and are shown in Table 4.2.2. Derivatives for the glucose concentration versus time plot were determined from the plots in Figure 4.2.6, and are shown in Table 4.2.2. At the initial and final steady states, the slope is taken to be zero (Figures 4.2.5a and 4.2.6c). During the first day after the step change, the slope is essentially constant, as shown in Figure 4.2.6a. The slope at the transition time of 1.44 days is determined by fitting the region near that point with a third-order polynomial (Figure 4.2.6b) and analytically obtaining the derivative. The specific glucose consumption and lactate production rates calculated from Equation 4.2.31 are presented in Figures 4.2.7a and 4.2.7b, respectively. Note that q_{glc} and q_{lac} are very sensitive to the reactor glucose concentration.

Estimation of metabolic quotients in flask cultures

Consumption and production rates are often measured by incubating cells in media and measuring the substrate or product concentration at the beginning and end of a specified period of time. Equation 4.2.32 becomes:

$$q_S = \frac{-(\Delta S/\Delta t)}{n_{Vave}}; \quad q_P = \frac{(\Delta P/\Delta t)}{n_{Vave}} \qquad (4.2.34)$$

where n_{Vave} is the average viable cell concentration. ΔS and ΔP represent the difference between the initial and final substrate and product concentrations,

Table 4.2.2 Data for the glucose step change example illustrated in Figures 4.2.5–4.2.9

Time (days)	glc (mM)	lac (mM)	n_v (cells ml^{-1} × 10^{-6})	glc slope (mM day^{-1})	lac slope (mM day^{-1})	q_{glc} (× 10^9)	q_{lac} (× 10^9)	$Y'_{lac,glc}$	q_{ATP} × 10 (P/O = 2)	q_{ATP} × 10 (P/O = 3)
-1.77	0.06	10.80	1.69	0.00	0.00	1.64	2.62	1.59	2.30	3.32
-1.15	0.11	10.56	1.67	0.00	0.00	1.65	2.57	1.56	2.36	3.42
-0.76	0.06	10.01	1.65	0.00	0.00	1.68	2.43	1.44	2.36	3.42
-0.14	0.00	11.20	1.56	0.00	0.00	1.80	2.98	1.65	2.32	3.33
0.00	8.28	10.47	n.d.	-5.93	6.67	n.d.	n.d.	n.d.	n.d.	n.d.
0.04	8.17	10.90	n.d.	-5.93	13.33	n.d.	n.d.	n.d.	n.d.	n.d.
0.09	7.89	11.31	1.63	-5.93	13.33	5.60	11.1	1.98	2.81	3.66
0.17	7.33	12.42	1.59	-5.93	13.33	5.93	11.7	1.98	2.91	3.78
0.25	6.83	14.27	1.51	-5.93	13.33	6.42	13.0	2.03	3.03	3.89
0.34	6.44	15.28	1.57	-5.93	13.33	6.31	12.9	2.04	3.09	3.99
0.51	5.33	16.90	1.75	-5.93	13.33	6.00	12.0	2.00	2.97	3.85
0.94	2.78	22.36	2.15	-5.93	13.33	5.53	11.2	2.02	2.82	3.67
1.44	0.44	29.56	2.21	-1.69	0.00	4.03	6.59	1.64	2.37	3.23
1.85	0.28	27.66	2.40	0.00	-3.30	3.04	4.26	1.40	2.14	3.00
2.27	0.33	27.22	2.64	0.00	-1.10	2.75	4.62	1.68	2.22	3.10
2.85	0.39	26.89	2.47	0.00	0.00	2.93	5.31	1.81	2.31	3.20
3.24	0.33	26.00	2.45	0.00	0.00	2.97	5.16	1.74	n.d.	n.d.
3.85	0.28	30.11	2.38	0.00	0.00	3.07	6.24	2.03	2.42	3.32
4.23	0.28	26.33	2.56	0.00	0.00	2.85	5.01	1.76	2.36	3.30
4.85	0.28	27.22	2.43	0.00	0.00	3.00	5.47	1.82	2.39	3.31
5.27	0.28	26.56	2.55	0.00	0.00	2.86	5.07	1.77	2.20	3.05
5.85	0.22	26.44	2.52	0.00	0.00	2.91	5.11	1.76	2.41	3.37

n.d. = not determined.

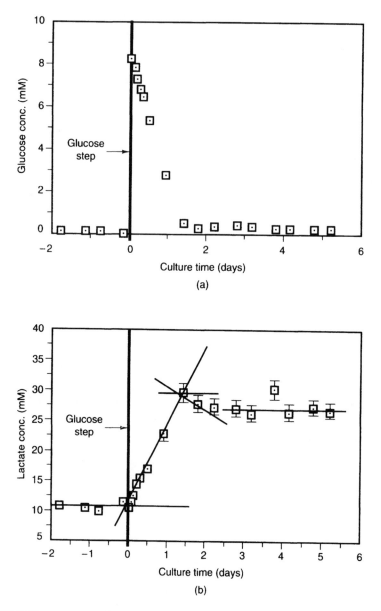

Figure 4.2.5 Glucose step change continuous culture experiment. Initial feed glucose concentration was 5.2 mM. At time zero the feed glucose concentration was increased to 13.8 mM and the reactor glucose concentration was increased to 8.3 mM. The reactor dilution rate was maintained at 0.54 day^{-1} throughout the experiment. (a) Glucose concentration (mM) as a function of culture time. The estimated error is ± 0.02 g l^{-1} based on the variability of a 2 g l^{-1} glucose standard and phosphate-buffered saline (PBS) control. (b) Lactate concentration (mM) as a function of culture time. Tangent lines are shown for slope determination. Error bars represent the standard deviation based on a standard curve. Adapted from Miller *et al.* (1989) by permission of *Biotechnology and Bioengineering*.

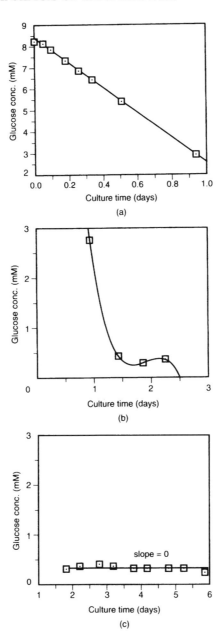

Figure 4.2.6 Plots for determination of the derivative of glucose concentration as a function of time for the glucose step change experiment described in Figure 4.2.5. The slope is zero for times less than zero (initial steady state). (a) Linear fit for the first day after the step change. Least-squares analysis gives a slope of -5.93 mM day^{-1}. (b) Polynomial fit to obtain the derivative at the transition point (1.44 days). A third-order fit gives [glucose] = $21.28 - 33.65t + 17.76t^2 - 3.08t^3$. The derivative is then $d[\text{glucose}]/dt = -33.65 + 35.52t - 9.25t^2$. For $t = 1.44$ days, the slope is -1.69 mM day^{-1}. (c) Zero slope for all points after 1.44 days (final steady state).

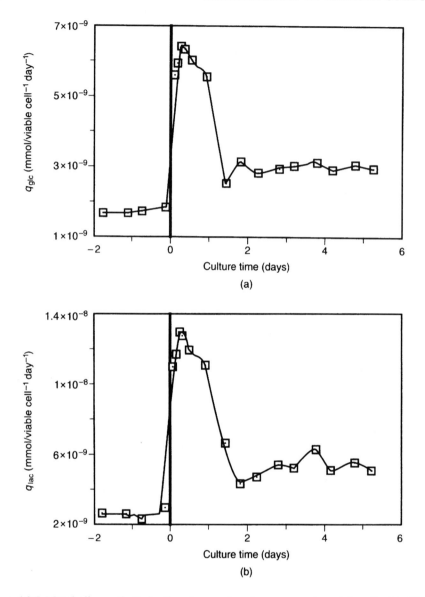

Figure 4.2.7 Metabolic quotients for the glucose step change experiment described in Figure 4.2.5. (a) Specific glucose consumption rate (mmol glucose consumed viable cell^{-1} day^{-1}) as a function of culture time. (b) Specific lactate production rate (mmol lactate produced viable cell^{-1} day^{-1}) as a function of culture time. Adapted from Miller *et al.* (1989) by permission of *Biotechnology and Bioengineering*.

respectively, and Δt represents the time between samples. It should be noted that metabolic quotients obtained over long periods reflect average values over the interval. The method used to calculate $n_{V\mathrm{ave}}$ depends upon the amount of cell growth that occurs during the time period Δt. For very long incubation times (Δt much greater than the doubling time), multiple samples should be taken. For moderate times during which significant growth occurs, one may assume a linear dependence of cell concentration on time. In this case $n_{V\text{ave}}$ is simply the arithmetic average of the initial and final cell concentrations. For non-confluent adherent cell cultures, the initial cell concentration may be estimated using plating efficiencies obtained from experiments in replicate flasks. For confluent monolayers with diminished cell growth rates, it may be acceptable to use the final cell count for $n_{V\text{ave}}$.

Oxygen

The use of oxygen in cell culture is characterized by the specific oxygen consumption rate. Cultures grown in incubators are generally oxygenated by simple diffusion. Oxygen may be provided to cells in bioreactors via perfusion with oxygenated medium or via transport from the headspace, gas bubbles and/or semipermeable tubing. In general:

$$q_{O_2} = \frac{D(C_{O_{2F}} - C_{O_2}) + K_L a[(P_{O_2}/H) - C_{O_2}] - [d(C_{O_2})/dt]}{n_v} \quad (4.2.35)$$

where C_{O_2} and $C_{O_{2F}}$ are the dissolved oxygen concentrations (mM) in the vessel and feed stream, respectively, P_{O_2} is the oxygen partial pressure (mm Hg) in the gas phase, H (~ 760 mm Hg mM^{-1} (Miller, 1987)) is the Henry's law constant for oxygen in culture medium at 37°C and $K_L a$ is the volumetric oxygen mass transfer coefficient. In most agitated vessels, the contribution from oxygen in the feed stream is negligible. This yields:

$$q_{O_2} = \frac{K_L a[(P_{O_2}/H) - C_{O_2}] - [d(C_{O_2})/dt]}{n_v} \quad (4.2.36)$$

The derivative is evaluated according to the methods described above for other metabolites. As before, point values should be used for all of the other terms. Procedures for determination of $K_L a$ are described in Chapter 4, section 4.6. Note that $K_L a$ is assumed to be constant throughout. However, large samples in batch culture could deplete the volume enough to change $K_L a$. In such circumstances, a functionality relating $K_L a$ to reactor volume would be useful.

Estimation of ATP production

The specific ATP production rate can be estimated by assuming that all ATP comes from glycolysis or oxidative phosphorylation. As a first approximation, it can be assumed that all lactate produced comes from glycolysis and that all oxygen consumed is utilized for oxidative phosphorylation. Therefore, there will be one mole of ATP produced per mole of lactate produced. Also, depending on the P/O ratio (ATP molecules formed per O atom consumed), there will be 6 moles (for P/O = 3) or 4 moles (for P/O = 2) of ATP produced per mole of O_2 consumed. Thus the specific ATP production rate can be expressed as:

$$q_{ATP} = q_{lac} + 2(P/O)q_{O_2} \quad (4.2.37)$$

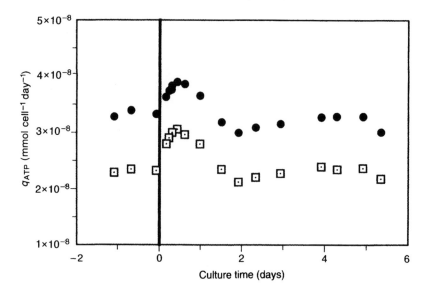

Figure 4.2.8 Specific ATP production rate as a function of culture time for P/O ratios of 3 (●) and 2 (□) for the glucose step change experiment described in Figure 4.2.5. Adapted from Miller et al. (1989) by permission of *Biotechnology and Bioengineering*.

Example 3

Data presented for the glucose step change in Example 2 can be used to estimate the specific ATP consumption rate as a function of the P/O ratio; q_{O_2} (not shown) is calculated from Equation 4.2.36, while q_{lac} (Table 4.2.2) is calculated from Equation 4.2.31 as shown in Example 2. These values are then substituted into Equation 4.2.37 to obtain the q_{ATP} values shown in Figure 4.2.8 for P/O ratios of 2 and 3.

Metabolite yield ratios

The apparent yield of lactate from glucose, $Y'_{lac,glc}$, provides an estimate of the fraction of glucose converted to lactate via glycolysis:

$$Y'_{lac,glc} = \frac{q_{lac}}{q_{glc}} \qquad (4.2.38)$$

This is an apparent yield because lactate can be produced via metabolism of other substrates such as glutamine, and because pyruvate derived from glycolysis can be converted into other compounds such as alanine. The theoretical maximum yield is 2 because no more than two molecules of lactate can be obtained from a single molecule of glucose. However, production of lactate from glutamine may result in yields greater than 2. The true yield of lactate from glucose, $Y_{lac,glc}$, would be the number of molecules of lactate directly obtained from the metabolism of glucose. True yields must be determined by metabolic labelling studies. For example, the carbon atoms in glucose could be radiolabelled so that any radioactive lactate molecules must have been produced from glucose. Nuclear magnetic resonance techniques with ^{13}C-labelled glucose can also be used.

Apparent yields are useful for characterizing and following changes in cell metabolism because changes in the apparent yield indicate changes in the use of alternative metabolic pathways. Typically, an apparent yield ratio relates the formation of a product to the consumption of a substrate. However, a yield ratio can also relate two substrates, as in the case of oxygen and glucose: $Y''_{O_2,glc}$ gives an estimate of the extent of oxidative metabolism relative to glycolysis. Other useful yield ratios include oxygen/glutamine, ammonia/glutamine and product/oxygen. Because the cell concentrations in Equation 4.2.31 cancel, there is often less scatter in apparent yield data than in the individual consumption rates.

Example 4

The apparent yield of lactate from glucose is easily calculated for the glucose step change described in Example 2. Both q_{glc} and q_{lac} are shown in Table 4.2.2 and Figure 4.2.7. From Equation 4.2.38, the apparent yield is obtained by dividing the specific lactate production rate by the specific glucose consumption rate. The resulting $Y''_{lac,glc}$ values are plotted as a function of time in Figure 4.2.9 and are listed in Table 4.2.2. Note that the fraction of glucose converted to lactate also increases at higher glucose concentrations.

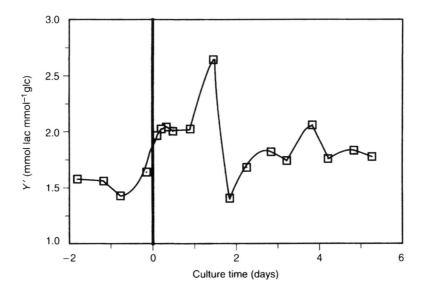

Figure 4.2.9 Apparent yield of lactate from glucose (mmol lactate produced per mmol^1 glucose consumed) as a function of culture time for the glucose step change experiment described in Figure 4.2.5. Adapted from Miller *et al.* (1989) by permission of *Biotechnology and Bioengineering*.

Calculation of apparent yields without n_v

An apparent yield is typically calculated by dividing a specific production rate by a specific consumption rate. As shown in Equations 4.2.31 and 4.2.38, the viable cell concentration terms cancel out. Therefore, the apparent yield may be calculated by evaluating the numerators in Equation 4.2.31 for product and substrate and taking their ratio. In batch culture, the analysis is even simpler. An apparent yield ratio based on Equation 4.2.32 is:

$$Y'_{P,S} = \frac{q_P}{q_S} = \frac{dP}{-dS} \qquad (4.2.39)$$

The yield is easily calculated by evaluating the derivatives of product concentration versus time and substrate concentration versus time and taking their ratio. Alternatively, product concentration is plotted against substrate concentration and the absolute value of the slope at any point is equal to the apparent yield. The important point to note here is that apparent yield ratios may be obtained even if specific consumption and production rates cannot be calculated because viable cell data are unavailable. For cases in which only endpoint data are available, an average apparent yield can be calculated. In this case, the apparent yield is simply the ratio of the change in the concentration of product to the change in the concentration of substrate:

$$Y'_{P,S} = \frac{\Delta P}{-\Delta S} \qquad (4.2.40)$$

Cell yield on substrate

Using the concept of yield ratios, we can define a quantity relating cell growth to substrate utilization:

$$Y'_{n,S} = \frac{\mu}{q_S} \qquad (4.2.41)$$

where μ is the specific growth rate and q_S is the specific consumption rate for substrate S. Yield parameter $Y'_{n,S}$ is a sensitive parameter that varies with cell growth rate and culture conditions, especially the relative amounts of different substrates, and is useful for detecting changes in substrate utilization as well as for identifying limiting nutrients in media development.

ATP maintenance energy model

The maintenance energy model (Pirt, 1975) states that energy consumption can be divided into that needed for growth and that needed for maintenance. Through the use of yield ratios, an empirical expression can be written for a specific substrate consumption rate as a function of the maintenance energy requirement and the specific production rates of all products that require the consumption of that substrate. This type of model is useful in microbial systems, for which there is frequently only one carbon source. However, such a model is not very useful for animal cell culture because multiple substrates are required and the different products can be synthesized from alternative substrates.

A more useful model is the ATP maintenance energy model. The specific utilization rate of ATP can be expressed as:

$$q_{ATP} = \frac{\mu}{Y'_{n,ATP}} = \frac{\mu}{Y_{n,ATP}} + m_{ATP} \qquad (4.2.42)$$

The advantage of Equation 4.2.42 is that all substrates provide energy via ATP so it does not matter what nutrients are used as long as q_{ATP} can be estimated (see above). In Equation 4.2.42, energy for product formation would be included in the $Y'_{n,ATP}$ term for growth-associated products and in the m_{ATP} term for non-growth-associated products. Alternatively, a specific product term could be added to Equation 4.2.42 in the form of $q_p/Y'_{P,ATP}$. Miller et al. (1989) estimated m_{ATP} and $Y'_{n,ATP}$ for hybridoma cells using Equation 4.2.42 and the information shown in Examples 2 and 3.

Immobilized-cell reactors

Estimation of cell density using metabolic quotients

It is often impossible to determine experimentally the cell concentration in immobilized-cell reactors. However, it is usually very easy to measure oxygen and metabolite concentrations in the inlet and outlet streams. From the metabolite concentration versus time data, a volumetric (per unit reactor volume rather than per cell) production rate can be determined:

$$Q_M = D(M_F - M_o) \qquad (4.2.43)$$

Here, M_o is the metabolite concentration in the outlet stream and M_F is the concentration in the feed stream to the reactor (M is often oxygen or glucose). If oxygen is also supplied via transfer from gas-filled hollow fibres, an extra term must be added to Equation 4.2.43. Terms Q_M and q_M are related by:

$$q_M = \frac{Q_M}{n_v} \qquad (4.2.44)$$

If q_M is obtained from experiments in stirred suspension or microcarrier cultures, as described above, then, it may be possible to estimate n_v from Q_M/q_M. However, care must be exercised because q_M can be altered by changes in the culture environment. For example, q_M for cells immobilized in agarose beads or hollow fibres may be different from q_M for the same cells grown in the same medium in a stirred suspension reactor (Shirai et al., 1988; Wohlpart et al., 1991). In addition, q_M may change over time due to changes in cell, nutrient and byproduct concentrations. Analysis of cell density and immobilization effects is complicated by the presence of nutrient concentration gradients. However, stirred vessels with cells immobilized in spherical beads may be used to estimate these effects if the concentrations of glucose, oxygen and other nutrients are high enough to ensure that the concentrations at the centre of the beads are in the plateau regime (i.e. metabolic quotients independent of nutrient concentrations).

Because apparent metabolic yields can be determined without knowledge of n_v, Y' values can be monitored to detect changes in metabolic patterns of cells in immobilized-cell bioreactors. For example, $Y'_{lac,glc}$ is easy to obtain and provides a good measure of the fate of the primary nutrient glucose. Relatively constant values for $Y'_{lac,glc}$, $Y'_{NH_3,glc}$, $Y'_{O_2,glc}$ and other yields provide a good indication of a stable culture environment. Under these conditions, Equation 4.2.44 may be used with confidence to estimate changes in the viable cell density.

PRODUCT FORMATION

Specific product formation rate

The characteristic parameter for product formation is the specific product formation rate, q_P. Equations 4.2.31–4.2.34 and the methods of analysis introduced for metabolic byproducts also apply to secreted protein products.

Constant q_P in batch culture

The general formula for q_P for batch growth in a stirred vessel is given by Equation 4.2.32, where P is the extracellular product concentration. Frequently, q_P is relatively constant during batch growth. One example is monoclonal antibody production by hybridoma cells (Renard et al., 1988). If it is assumed that q_P does not vary with time, then Equation 4.2.32 can be integrated to yield Equation 4.2.45:

$$P = q_P \int_0^t n_v \, dt \quad (4.2.45)$$

The integral represents the area under the viable cell curve (n_v versus time). If q_P is indeed constant during batch growth, then a plot of product concentration versus the area under the viable cell curve will yield a straight line with slope equal to q_P. This provides an easy approach for determining q_P. The integral may be evaluated by fitting the viable cells versus time curve with a polynomial and performing the integration analytically. Alternatively, the integral may be approximated by a sum of rectangles or trapezoids.

Systems with product retention

Reactors may be operated with product retention as well as cell retention. This allows for product recovery at high concentration. Retention may be accomplished by using a membrane with a pore size that is too small to allow passage of product but large enough to allow passage of metabolites and waste products. Capsules that retain cells can also retain product. In any type of product-retention system, retention may not be 100%. This does not complicate analysis as long as the retention fraction is known. For a constant-volume reactor with product retained via a membrane (similar to Figure 4.2.3), a mass balance on product gives:

$$q_P = \frac{D(\alpha_P P - P_F) + (dP/dt)}{n_v} \qquad (4.2.46)$$

where P is the concentration of product retained in the reactor and α_P is the ratio of the product concentration in the outlet stream to the product concentration in the reactor ($1 - \alpha_P$ is the retention fraction). A drop in q_P at high P levels may indicate feedback inhibition by product.

Product yield on medium components

We can define an apparent yield of product from substrate ($Y'_{P,S}$) as a measure of the efficiency of product formation that can be used to compare cells grown in different culture systems or in different media. Product P is a cellular product such as monoclonal antibody, while substrate S may be serum, glucose, glutamine, oxygen or any other important substrate. Analysis of the overall yield (from endpoint calculations) may allow for comparison of production efficiency between different reactors and culture conditions. Analysis of the yield at various time points allows for detection of changes in the mechanisms of product formation. Product yields are useful in identifying product degradation in culture (as evidenced by a decrease in $Y'_{P,S}$ for all substrates).

Immobilized-cell reactors

It is also relevant to consider the product yield on substrate in immobilized-cell reactors. The apparent yield of product from substrate, $Y'_{P,S}$, can be monitored over time to detect changes in the yield. If other yields are relatively constant, then changes in $Y'_{P,S}$ may indicate changes in the specific production rate (q_P). However, if some or all of the other calculated metabolic yields change, then it would be very difficult to attribute changes in production rate directly to changes in q_P. Inhibition of substrate consumption or growth by waste products may contribute to a decrease in product yield.

CONCLUDING REMARKS

The growth and metabolic parameters described here (see Table 4.2.3) may be used directly to compare responses by different cells in a given system or by the same cells in different culture environments. They can also be incorporated into metabolic models to provide a better understanding of the processes governing cell growth, metabolism and product formation.

Table 4.2.3 Glossary of terms

Term	Full extension
C_{LDH}	Concentration of lactate dehydrogenase (LDH) in culture (IU l^{-1})
C_{O_2}	Dissolved oxygen concentration (mM)
C_{O_2F}	Dissolved oxygen concentration in reactor feed stream (mM)
D	Reactor dilution rate (= F/V) (h^{-1})
F	Constant-volume reactor volumetric flow rate ($F_i = F_o = F$) (l h^{-1})
f_{Fv}	Viable fraction of free cells in microcarrier culture (dimensionless)
F_i	Reactor inlet medium volumetric flow rate (l h^{-1})
F_o	Reactor outlet medium volumetric flow rate (l h^{-1})
f_v	Fraction of viable cells (dimensionless)
H	Henry's law constant for oxygen (mmHg mM^{-1})
k_d	Specific death rate (h^{-1})
k_l	Specific dead cell lysis rate (h^{-1})
$k_L a$	Volumetric oxygen mass transfer coefficient (h^{-1})
k_R	Rate of cell release from microcarriers (h^{-1})
M	Metabolite concentration in reactor (mM)
m_{ATP}	ATP required for cell maintenance (mmol cell^{-1} h^{-1})
M_F	Metabolite concentration in reactor feed stream (mM)
M_O	Metabolite concentration in reactor outlet stream (mM)
n	Total cell concentration (cells ml^{-1})
n_A	Concentration of attached cells in microcarrier culture (cells ml^{-1})
n_d	Dead cell concentration (cells ml^{-1})
n_e	Effective total cell concentration if no lysis occurred (cells ml^{-1})
n_F	Concentration of free (unattached) cells in microcarrier culture (cells ml^{-1})
n_o	Total cell concentration in cell-retention reactor outlet stream (cells ml^{-1})
n_v	Viable cell concentration (cells ml^{-1})
n_{Vave}	Average viable cell concentration in the interval between samples (cells ml^{-1})
P	Product concentration in reactor (mM)
P_{O_2}	Oxyen partial pressure in the gas phase (mmHg)
P/O	Molecules of ATP generated per molecule oxygen consumed (dimensionless)
P_F	Product concentration in reactor feed stream (mM)
q_{ATP}	Specific ATP production rate (mmol cell^{-1} h^{-1})
Q_M	Volumetric production rate of M (mmol l^{-1} h^{-1})
q_M	Specific formation rate of M (mmol cell^{-1} h^{-1})
q_{O_2}	Specific oxygen consumption rate (mmol cell^{-1} h^{-1})
q_P	Specific production rate of P (mmol cell^{-1} h^{-1})
q_S	Specific consumption rate of S (mmol cell^{-1} h^{-1})
S	Substrate concentration in reactor (mM)
S_F	Substrate concentration in reactor feed stream (mM)
t	Culture time (h)
V	Reactor volume (l)
$Y''_{n,ATP}$	Apparent yield of cells from ATP (cells mmol^{-1})
$Y''_{n,S}$	Apparent yield of cells from substrate S (cells mmol^{-1})
$Y''_{P,S}$	Apparent yield of product P from substrate S (mmol mmol^{-1})
$Y_{n,ATP}$	True yield of cells from ATP (cells mmol^{-1})
α_P	Ratio of the product concentration in the outlet stream to the product concentration in the reactor for a system with product retention (dimensionless)
α_s	Ratio of the cell concentration in the outlet stream to the cell concentration in the reactor for a system with cell retention (= F_2/F_o in Figure 4.2.3) (dimensionless)
α_{sd}	Ratio of the dead cell concentration in the outlet stream to the dead cell concentration in the reactor (dimensionless)

Table 4.2.3 (*Continued*)

Term	Full extension
α_{sv}	Ratio of the viable-cell concentration in the outlet stream to the viable-cell concentration in the reactor (dimensionless)
β_d	Fraction of γ_{LDH} retained by dead cells (dimensionless)
γ_{LDH}	LDH content per cell (IU cell^{-1})
μ	True specific growth rate (h^{-1})
μ_{app}	Apparent specific growth rate (h^{-1})
μ_{app}^*	Viable cell-derived apparent specific growth rate (h^{-1})
μ_A	True specific growth rate of attached cells in microcarrier culture (h^{-1})
μ_{app}^A	Apparent specific growth rate of attached cells in microcarrier culture (h^{-1})
μ_{app}^e	Apparent specific growth rate based on effective total cell concentration (h^{-1})
μ_F	True specific growth rate of free cells in microcarrier culture (h^{-1})

ACKNOWLEDGEMENTS

This work was supported in part by NSF grant BCS-9058416 and matching funds from Schering Plough Research, Eli Lilly and Company and Lederle-Praxis Biologicals.

REFERENCES

Lin AA, Nguyen T & Miller WM (1991) A rapid method for counting cell nuclei using a particle sizer/counter. *Biotechnology Techniques* 5: 153–156.

Miller WM (1987) A kinetic analysis of hybridoma growth and metabolism. *PhD Thesis*, University of California, Berkeley, CA.

Miller WM & Blanch HW (1991) Regulation of animal cell metabolism in bioreactors. In: Ho Cs & Wang DIC (eds) *Animal Cell Bioreactors*, pp. 119–161. Butterworth–Heinemann, Stoneham, MA.

Miller WM, Blanch HW & Wilke CR (1988) A kinetic analysis of hybridoma growth and metabolism in batch and continuous suspension culture: effect of nutrient concentration, dilution rate, and pH. *Biotechnology and Bioengineering* 32: 947–965.

Miller WM, Blanch HW & Wilke CR (1989) The transient responses of hybridoma cells to nutrient additions in continuous culture: 1. Glucose pulse and step changes. *Biotechnology and Bioengineering* 33: 477–486.

Pirt SJ (1975) *Principles of Microbe and Cell Cultivation*. Blackwell Scientific Publications, Cambridge.

Renard JM, Spagnoli R, Mazier C, Salles MF & Mandine E (1988) Evidence that monoclonal antibody production kinetics is related to the integral of the viable cells curve in batch systems. *Biotechnology Letters* 10: 91–96.

Shirai Y, Hashimoto K, Yamaj H & Kawahara H (1988) Oxygen uptake rate of immobilized growing hybridoma cells. *Applied Microbiology and Biotechnology* 29: 113–118.

Wagner A, Marc A, Engasser JM & Einsele A (1992) The use of lactate dehydrogenase (LDH) release kinetics for the evaluation of death and growth of mammalian cells in perfusion reactors. *Biotechnology and Bioengineering* 39: 320–326.

Wohlpart D, Gainer J & Kirwan D (1991) Oxygen uptake by entrapped hybridoma cells. *Biotechnology and Bioengineering* 37: 1050–1053.

4.3 MODELLING

A kinetic model of mammalian cells is a quantitative description of the main phenomena that have an influence on the growth, death and metabolic activities of cells. In its simplest form a model consists of a set of mathematical relationships between the different cellular rates – growth, death, nutrient uptake and metabolite production – and the composition of the culture medium. When transferred to a computer, it provides a simulation of the time variation of the different components of the culture medium. More elaborate models, potentially capable of identifying limiting metabolic steps during biosynthesis of the desired product, may also be designed to represent changes in the intracellular content, the metabolic pathways or the cell physiology as a function of culture conditions.

As one of its major interests, a model represents an efficient tool for the kinetic analysis of cellular processes. It is able to account for the main phenomena that may simultaneously control the activities of cells. As such, depending on the culture conditions, composition of the medium and whether there is batch or continuous mode of operation, it can be used first to identify the rate-limiting factors and then to characterize quantitatively their relative importance. For instance, with a model it is possible to evaluate the kinetic effect of a depletion of glucose, glutamine and other amino acids or of an accumulation of ammonia and lactate on the rates of cell growth and death.

Because of their predictive capabilities, models are also essential tools in modern biochemical engineering for the design of processes and the optimization of media and reactor operational parameters in batch or continuous operation. They can also serve in the development of software sensors to estimate on-line the time variation of the medium composition.

The construction of a kinetic model for an animal cell culture involves several steps: a kinetic analysis of the experimental results with the formulation of hypotheses on the nature of the rate-limiting steps; the choice of rate expressions describing the influence of these phenomena on the cellular processes; evaluation of parameter values; and validation of the model with different experimental results. In this section a general methodology is described for the modelling of cell cultures, and the procedure is illustrated on the kinetics of a hybridoma cell. (For a summary of terms used, see Table 4.3.3.)

BACKGROUND FOR THE MODELLING OF MAMMALIAN CELL CULTURES

A kinetic model consists of a set of mathematical expressions that relate the rates of cellular growth and metabolism to the composition of the medium. With

---

Cell and Tissue Culture: Laboratory Procedures in Biotechnology, edited by A. Doyle and J.B. Griffiths.
© 1998 John Wiley & Sons Ltd.

mammalian cells, the rates usually measured are the rates of cell growth and death, the rates of uptake of the main nutrients, glucose and glutamine, the rates of production of the main metabolites, lactate and ammonia, and the rate of secretion of proteins. Most of the medium components, depending on their concentration, may have an influence on the metabolic activities of cells. Among the most frequently observed rate-limiting effects are the depletion of glucose and glutamine and the accumulation of ammonia and lactate.

An example of a kinetic model that takes into account the effect of these four components – glucose, glutamine, lactate and ammonia – is presented below. The different rate expressions it contains have been found correctly to simulate batch or continuous cultures of several mammalian cell lines.

Specific rate of cell growth:

$$\mu = \mu_{max} \left(\frac{[Glc]}{K_{Glc} + [Glc]} \right) \left(\frac{[Gln]}{K_{Gln} + [Gln]} \right) \left(\frac{1}{1 + ([Lac]/K_{Lac})} \right) \left(\frac{1}{1 + ([NH_4]/K_{NH4})} \right) \quad (4.3.1)$$

Specific rate of cell death:

$$k_d = A_d \left\langle \left(\frac{1}{1 + ([Gln]/C_1)} \right) + \left(\frac{1}{1 + ([Glc]/C_2)} \right) + k_1[NH_4] + k_2[Lac] \right\rangle \quad (4.3.2)$$

Specific rate of glucose consumption:

$$v_{Glc} = Y_{Glc/X} \cdot \mu + m_{Glc} \quad (4.3.3)$$

Specific rate of lactate production:

$$\pi_{Lac} = Y_{Lac/X} \cdot \mu + m_{Lac} \quad (4.3.4)$$

Specific rate of glutamine consumption:

$$v_{Gln} = Y_{Gln/X} \cdot \mu + m_{Gln} \quad (4.3.5)$$

Specific rate of ammonia production:

$$\pi_{NH4} = Y_{NH4/X} \cdot \mu + m_{NH4} \quad (4.3.6)$$

Specific rate of antibody production:

$$\pi_{MAbs} = Y_{MAbs/X} \cdot \mu + m_{MAbs} \quad (4.3.7)$$

Rate of glutamine degradation:

$$r_{Gln} = k_{deg}[Gln] \quad (4.3.8)$$

Specific rate of cell growth

The specific rate of cellular growth, μ, is defined as the number of new cells produced per unit (e.g. billion) of living cells present in the culture medium per unit time (e.g. hour).

For a given medium a cell line can be characterized by a maximum specific growth rate, μ_{max}, which is the observed growth rate in the absence of any limitations by nutrients or any inhibition by metabolites. This maximal growth rate is related to the doubling time (t_d) of the cell by the relationship:

$$t_d = 0.69/\mu_{max}$$

During culture the specific growth rate usually decreases because of either depletion of essential nutrients or accumulation of inhibitory metabolites. Equation 4.3.1 represents such a variation of specific growth rate as a function of the concentration of glucose, glutamine, lactate and ammonia in the medium. In this kinetic law the maximum specific growth rate is multiplied by four terms that describe the rate-limiting effect of each of the components.

The parameters introduced – K_{Glc}, K_{Gln}, K_{Lac}, K_{NH_4} – give the range of concentrations where either the nutrient becomes limiting or the metabolite becomes inhibitory. By modulating the values of these parameters the model can account for differences in cell sensitivities towards nutrient depletion and product inhibition.

Specific rate of cell death

The specific rate of cellular death, k_d, is defined as the number of dying cells per unit (e.g. billion) of living cells present in the culture medium per unit time (e.g. hour).

As a first approximation k_d has often been considered as a constant. Yet more detailed kinetic analyses have shown that the specific rate of cell death is also affected by the chemical composition of the medium and several physicochemical parameters, such as pH, temperature and osmotic pressure. It is often lowest at the start of the culture, and then gradually increases due either to depletion of essential nutrients or accumulation of inhibitory metabolites.

A rate expression for cell death is given in Equation 4.3.2. With its four terms it expresses the possible increase in the cell death rate due to limitations in glucose and glutamine or accumulations in lactate and ammonia. In this case, the different contributions are additive in order to take into account the effect of each component: if the expression relative to one substrate becomes equal to zero, the effects of the other components remain visible in the calculation of the specific death rate. By proper adjustment of the values of the four parameters C_1, C_2, κ_1 and κ_2 it is possible to account for differences in death kinetics from one cell line to another.

Specific rate of nutrient uptake

The specific rate of nutrient uptake, v, is defined as the number of millimoles of nutrient consumed per unit (e.g. billion) of living cells present in the culture medium and per unit time (e.g. hour).

For most cell lines it has been found to increase linearly with the specific growth rate. Thus the specific rate of glucose uptake is often expressed as a function of μ by Equation 4.3.3, which contains two parameters: the non-growth-associated specific glucose consumption rate, m_{Glc} (mmol glucose 10^{-9} cells h^{-1}), and the glucose to biomass conversion yield, $Y_{Glc/X}$ (mmol glucose 10^{-9} cells). A similar expression, Equation 4.3.5, is applicable to the specific rate of glutamine uptake.

Specific rate of metabolite and protein production

The specific rate of metabolite or protein production, π, is defined as the number of millimoles or milligrams of product excreted per unit (e.g. billion) of living cells present in the culture medium and per unit time (e.g. hour).

For most cell lines it is found to increase with the specific growth rate. Thus the specific rates of lactate production can be expressed as a function of μ by Equation 4.3.4, which contains two parameters: the non-growth-associated specific lactate production rate, m_{Lac} (mmol lactate 10^{-9} cells h^{-1}), and the lactate to biomass stoichiometric yield, $Y_{Lac/X}$ (mmol lactate 10^{-9} cells). A similar expression is often applicable to the specific rate of ammonia production (Equation 4.3.6) and antibody secretion (Equation 4.3.7).

Rate of glutamine decomposition

The spontaneous decomposition in the medium of glutamine into ammonia can be represented by a first-order rate process with respect to the glutamine concentration, as shown in Equation 4.3.8.

METHOD FOR KINETIC MODEL CONSTRUCTION

Experimental investigations

As one objective of model construction is to obtain the best fit between model simulations and experimental results, appropriate kinetic data have to be obtained. Experiments can be performed in different systems:

- *Batch cultures* are the simplest to perform, either in shake flasks or small bioreactors. However, as the concentration of all the nutrients and metabolites changes simultaneously with time, it is relatively difficult to assess the precise influence of a single medium component on cell kinetics
- *Continuous cultures* require more sophisticated equipment and long-term operation extending to several weeks. But, as a major advantage, nutrient and metabolite concentrations can be maintained constant for several days, which allows a more precise analysis of the influence of medium composition on cellular activity

When investigating cell kinetics, it is very important to control precisely the physicochemical parameters of the medium, such as temperature, pH, dissolved oxygen and osmotic pressure. It is also important to define precisely the state of the inoculum, which may have a significant effect on the progress of the culture. During the culture, one measures at regular time intervals the concentrations of living and dead cells, of the major nutrients and metabolites and of excreted proteins.

Example

The procedure of kinetic data analysis and model construction is illustrated for a hybridoma culture (cell line VO 208) in a batch system. The medium used was RPMI 1640 + 5% (v/v) foetal calf serum (FCS) + 2% (v/v) minimum essential medium (MEM) amino acids + 1% non-essential amino acids and initial glucose and glutamine concentrations of 13 mM and 4.5 mM, respectively.

The batch culture was carried out in a bioreactor (B. Braun) with a working volume of 1 l. Forty-eight hours after the last Roux bottle inoculation, the cells were inoculated at about 2×10^8 viable cells l^{-1} in the bioreactor.

The pH was maintained at a value of 7 with 0.2 M NaOH solution and gaseous CO_2. The temperature was set at 37°C and the oxygen supply regulated at 50% of air saturation with gaseous air and nitrogen. The rotating speed was 50 rpm.

During the culture, the concentrations of living and dead cells (by the Trypan blue exclusion method) were measured using a haemocytometer (Chapter 2, section 2.2): glucose, lactate and glutamine by enzymic methods; ammonia with a selective electrode; and monoclonal antibodies by ELISA assay.

The time variations of these medium compounds for the given example are presented in Figure 4.3.1.

Kinetic analysis

Prior to the construction of the kinetic model, one has to perform a detailed analysis of the experimental data in order to identify the main rate-limiting effects and the relationships that may exist between the different kinetic variables. A procedure for data analysis is described and illustrated for the previously obtained experimental results.

Identify the rate-limiting nutrients

The time variation of the different concentrations in the culture medium shows the classical mammalian cell kinetics of a batch culture: a growth phase with a maximal cell concentration of 8×10^8 viable cells l^{-1}, followed, after 100 h of culture, by a death phase. Rapid death occurs when glutamine is completely consumed in the medium, which indicates that glutamine limitation is responsible for the cessation of cell growth and increase in cell death rate. The glucose level progressively decreases and reaches a residual value near 2 mM. At this level, the rate-limiting effect of glucose on cell growth and death is probably not significant.

Identify the rate-limiting metabolites

The lactate concentration increases up to 14 mM, and the ammonia concentration up to 5.5 mM. At these levels, both lactate and ammonia can inhibit the cell growth rate or increase the cell death rate.

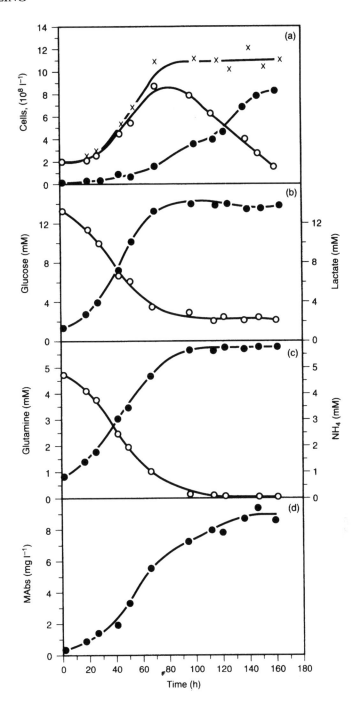

Figure 4.3.1 Evolution with time of total ((a), x), viable ((a), ○) and Trypan blue dead cells ((a), ●) and glucose ((b), ○), lactate ((b), ●), glutamine ((c), ○), ammonia ((c), ●) and monoclonal antibody ((d), ●) concentrations during VO 208 hybridoma batch culture.

Calculate the total cell production

The measured concentration of living cells is the result of two processes: the formation of new cells and the death of living cells. In order to evaluate the actual rate of new cell production, one has to calculate the total number of cells (X_t) that have been produced during the culture. As long as cell lysis can be neglected, it is given by the sum of the measured live (X_v) and dead cells (X_d):

$$X_t = X_v + X_d \tag{4.3.9}$$

The time variation of the total concentration of cells is plotted in Figure 4.3.1a.

Calculate the specific growth rate

The specific growth rate is calculated by the relationship:

$$\mu = \frac{dX_t}{X_v dt} \text{ (h}^{-1}) \tag{4.3.10}$$

Calculate the specific death rate

The specific death rate is calculated by the relationship:

$$k_d = \frac{dX_m}{X_v dt} \text{ (h}^{-1}) \tag{4.3.11}$$

Calculate the specific rates of nutrient consumption

The specific rate of glucose consumption:

$$v_{Glc} = \frac{-d[Glc]}{X_v dt} \text{ (mmol } 10^{-9} \text{ cells h}^{-1}) \tag{4.3.12}$$

The specific rate of glutamine consumption (which takes into account glutamine decomposition):

$$v_{Gln} = \frac{-d[Gln]}{X_v dt} - \frac{k_{deg}[Gln]}{X_v} \text{ (mmol } 10^{-9} \text{ cells h}^{-1}) \tag{4.3.13}$$

Calculate the specific rate of metabolites and antibody production

The specific rate of lactate production:

$$\pi_{Lac} = \frac{d[Lac]}{X_v dt} \text{ (mmol } 10^{-9} \text{ cells h}^{-1}) \tag{4.3.14}$$

The specific rate of ammonia production:

$$\pi_{NH4} = \frac{d[NH_4]}{X_v dt} - \frac{k_{deg}[Gln]}{X_v} \text{ (mmol } 10^{-9} \text{ cells h}^{-1}) \tag{4.3.15}$$

MODELLING

The specific rate of antibody production:

$$\pi_{MAbs} = \frac{d[MAbs]}{X_v dt} \quad (\text{mmol } 10^{-9} \text{ cells h}^{-1}) \quad (4.3.16)$$

Plot the different specific rates as a function of time and analyse the evolution of the metabolic activities of cells during the culture

The resulting curves in Figure 4.3.2 clearly indicate the existence of two culture periods:

- **Phase 1** lasts about 30 h and is where the different rates increase. This corresponds to the classically observed lag phase, where cells progressively adapt to their new environment.
- **Phase 2** is from 40 to 140 h, where all the specific rates progressively decrease except the cellular death rate, which continues to rise. This reduction in the specific rates of growth and metabolism is mainly due to glutamine limitation. However, the accumulation of lactate and ammonia may also have a kinetic effect.

Represent the specific rates of nutrient consumption and metabolite production as a function of the specific growth rate

All the curves obtained in Figure 4.3.3, except for antibody production, show two distinct upper and lower parts: the upper part corresponds to the initial lag phase, and the lower part to the following growth and death periods. This indicates that during the initial lag phase, at a given growth rate, cells exhibit a faster rate of nutrient uptake and metabolite production.

Analyse the lower part of the curves. Is there a linear relationship between the specific rates of cell metabolism and cell growth?

In Figure 4.3.3, the lower parts of the curves can be approximated to straight lines. This indicates that after the initial lag phase, the rate of cellular metabolism can be considered to be proportional to the rate of cellular growth.

Model development

We will only consider a kinetic model for the main culture phase after the initial lag period.

Select a rate expression for cellular growth

As suggested from the previous kinetic analysis, glutamine, lactate and ammonia have the main limiting effects on cellular growth. Thus, according to the model database in Table 4.3.1, a kinetic expression with the three rate-limiting effects is selected:

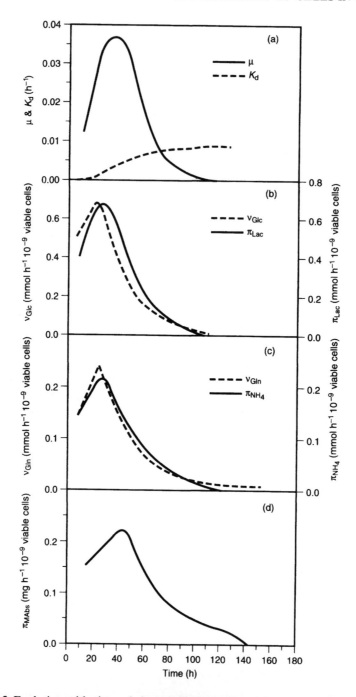

Figure 4.3.2 Evolution with time of the specific growth and death rates (a), the specific consumption rates of glucose (b) and glutamine (c) and the specific production rates of lactate (b), ammonia (c) and monoclonal antibodies (d) in VO 208 batch culture (data from Figure 4.3.1).

MODELLING

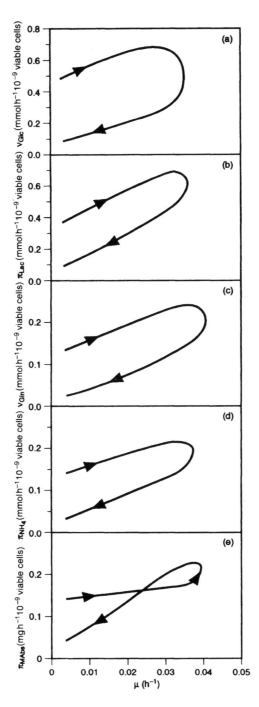

Figure 4.3.3 Relationships between specific glucose (a) and glutamine (c) consumption rates, lactate (b), ammonia (d) and monoclonal antibody (e) production rates and specific growth rate during lag and growth phases of VO 208 batch culture.

Table 4.3.1 Range of parameter values[a]

	Parameter	Order of magnitude of parameter values
Growth	μ_{max}	0.02–0.125 h^{-1}
	K_{Glc}	0.1–1 mM
	K_{Gln}	0.06–0.3 mM
	K_{Lac}	8–12 000 mM
	K_{NH_4}	1–45 mM
Death	A_d	0.008–0.08 h^{-1}
	C_1	5×10^{-6}–0.1 mM
	C_2	0.004 mM
	κ_1	0.03–0.05 mM^{-1}
	κ_2	0.01 mM^{-1}
Glucose consumption	$Y_{Glc/X}$	2–16
	m_{Glc}	0.007–0.5
Lactate production	$Y_{Lac/X}$	4–15
	m_{Lac}	0.01
Glutamine consumption	$Y_{Gln/X}$	1–3
	m_{Gln}	0.04
Ammonia production	$Y_{NH_4/X}$	0.5–2
	m_{NH_4}	0.03
Antibody production	$Y_{MAbs/X}$	4–10
	m_{MAbs}	0.04–0.06

[a] Y values are expressed in mmol 10^{-9} viable cells except for $Y_{MAb/X}$, which is expressed in mg 10^{-9} viable cells; m value units are mmol h^{-1}/10^{-9} viable cells and m_{MAbs} units are mg h^{-1}/10^{-9} viable cells.

$$\mu = \mu_{max} \left(\frac{[Gln]}{K_{Gln} + [Gln]} \right) \left(\frac{1}{1 + ([Lac]/K_{Lac})} \right) \left(\frac{1}{1 + ([NH_4]/K_{NH_4})} \right) \quad (4.3.17)$$

Select a rate expression for cellular death

Glutamine depletion as well as lactate and ammonia accumulation are also responsible for the observed increase in the specific rate of cell death. Again a three-term rate expression is selected:

$$k_d = A_d \left\{ \left(\frac{1}{1 + ([Gln]/C_1)} \right) + k_1[NH_4] + k_2[Lac] \right\} \quad (4.3.18)$$

Select the rate expressions for nutrient uptake

As linear relationships have been found for both glucose and glutamine uptake rate as a function of the specific growth rate, the kinetic expressions proposed in Equations 4.3.3 and 4.3.5 can be considered applicable:

$$v_{Glc} = Y_{Glc/X} \cdot \mu + m_{Glc} \quad (4.3.19)$$

$$v_{Glc} = Y_{Gln/X} \cdot \mu + m_{Gln} \quad (4.3.20)$$

MODELLING

Select the rate expressions for metabolite and antibody production

Linear relationships are also applicable to the specific rates of lactate, ammonia and antibody production:

$$\pi_{Lac} = Y_{Lac/X_v} \cdot \mu + m_{Lac} \qquad (4.3.21)$$

$$\pi_{NH_4} = Y_{NH_4}/Y_{NH_4/X_v} \cdot \mu + m_{NH_4} \qquad (4.3.22)$$

$$\pi_{MAbs} = Y_{MAbs/X_v} \cdot \mu + m_{MAbs} \qquad (4.3.23)$$

Write the mass balance equations for the different species

In order to simulate the time-course of a batch culture with the previously introduced rate expressions, one writes the mass balance equation for each of the considered species. They express the fact that in a batch culture the rate of variation in the concentration of a given species in the culture medium is equal to the rate of production or disappearance of the species. One thus obtains the following set of differential equations:

$$\frac{dX_v}{dt} = (\mu - k_d)X_v \qquad (4.3.24)$$

$$\frac{d[Glc]}{dt} = -v_{Glc}X_v \qquad (4.3.25)$$

$$\frac{d[Lac]}{dt} = \pi_{Lac}X_v \qquad (4.3.26)$$

$$\frac{d[Gln]}{dt} = -k_{deg}[Gln] - v_{Gln}X_v \qquad (4.3.27)$$

$$\frac{d[NH_4]}{dt} = k_{deg}[Gln] + \pi_{NH_4}X_v \qquad (4.3.28)$$

$$\frac{d[MAbs]}{dt} = \pi_{MAbs}X_v \qquad (4.3.29)$$

Integrate Equations 4.3.24–4.3.29 by a numerical method

Several standard procedures are available numerically to integrate these equations and thus calculate the time variations of the different concentrations. The Runge-Kutta method (Vetterling *et al.*, 1986) is generally advised.

Parameter identification

Finally one has to determine the values of the parameters introduced in the different rate expressions. The objective is to obtain the best agreement between the calculated time variations of the medium composition and the measured experimental results. There are two ways to evaluate the parameters: some can be calculated directly from the experimental results and others are evaluated by a curve-fitting procedure.

Evaluate the values of the parameters in the rate expressions of the specific rates of nutrient uptake and metabolite or protein productions

Each of the five linear relationships between the specific rate of cell metabolism and the specific rate of cell growth contains two parameters: a stoichiometric coefficient Y and a non-growth-associated specific rate m. From the plots shown in Figure 4.3.3 one can thus evaluate the values of ten of the parameters as the slope of the straight lines and their intercept with the y-axis. The determined values are reported in Table 4.3.2.

Evaluate the values of the other parameters by a curve-fitting procedure

The eight parameters that have been introduced in the rate expressions for cellular growth and death remain to be determined. As a first estimation one can take the lower limit of the usual range of values found for other cell lines as indicated in Table 4.3.1.

With these first values we do not obtain a correct simulation of the experimental results. Thus the values of the parameters are progressively improved by trial and error. One modifies the value of one or several of the parameters and determines if these results are an improvement in the fit between the model and the experimental results. The procedure is repeated until a satisfactory agreement is achieved. The resulting values are reported in Table 4.3.2 and the results of the final simulation are presented in Figure 4.3.4.

Table 4.3.2 Parameter values[a] for the VO 208 hybridoma cell line

	Parameter	Parameter values
Growth	μ_{max}	0.058 h^{-1}
	K_{Gln}	0.1 mM
	K_{Lac}	55 mM
	K_{NH_4}	10 mM
Death	A_d	0.008 h^{-1}
	C_1	0.1 mM
	κ_1	0.05 mM^{-1}
	κ_2	0.014 mM^{-1}
Glucose consumption	$Y_{Glc/X}$	9
	m_{Glc}	0.018
Lactate production	$Y_{Lac/X}$	13
	m_{Lac}	0.01
Glutamine consumption	$Y_{Gln/X}$	2.5
	m_{Gln}	0.037
Ammonia production	$Y_{NH_4/X}$	1.7
	m_{NH_4}	0.03
Antibody production	$Y_{MAbs/X}$	4.5
	m_{MAbs}	0.06

[a] Y values are expressed in mmol 10^{-9} viable cells except for $Y_{MAb/X}$, which is expressed in mg 10^{-9} viable cells; m value units are mmol h^{-1}/10^{-9} viable cells and m_{MAbs} units are mg h^{-1}/10^{-9} viable cells.

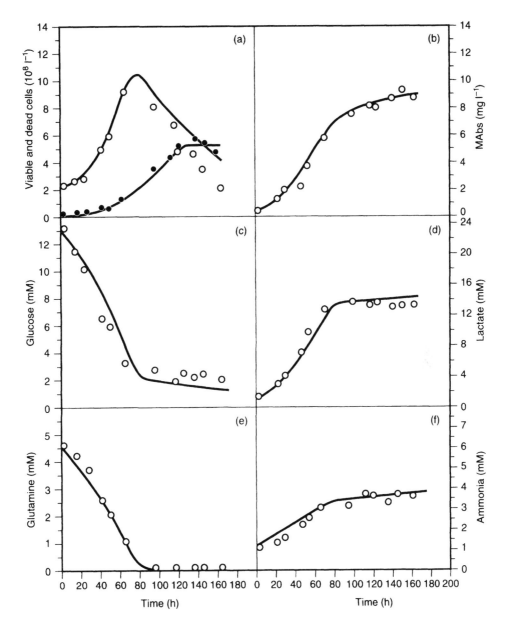

Figure 4.3.4 Experimental (○, ●) and theoretical (——) evolutions with time of viable ((a), ○) and dead ((a), ●) cell density and monoclonal antibody (b), glucose (c), lactate (d), glutamine (e) and ammonia (f) concentrations during VO 208 batch culture (RPMI containing 12.5 mM glucose, 4.5 mM Gln and 5% (v/v) FCS).

By this iterative procedure one observes that model predictions are very sensitive to the values of some of the parameters, which as a consequence have to be determined with great precision. On the contrary, for others, approximate values are sufficient because model predictions are not significantly affected by variations in these parameter values.

USE OF THE MODEL FOR THE EVALUATION OF RATE-LIMITING FACTORS

The model proposed for mammalian cell cultures provides a description of the possible influence of four of the main medium components – glucose, glutamine, lactate and ammonia – on the rates of cellular growth, death and metabolism. It contains kinetic terms that quantify the influence of each of the components either in reducing the rate of cellular growth or in increasing the rate of cell death.

During a batch or continuous culture, where the concentrations of all four species change with time, the observed reduction in growth rate or increase in death rate is generally the result of the simultaneous influence of several of these effects. From a simple kinetic analysis it is seldom possible to evaluate the relative effects of substrate depletion or metabolite accumulation on the observed kinetics. With the kinetic model as described, it becomes straightforward to characterize quantitatively the kinetic effects of the different medium components, as outlined here.

The procedure is based on the previous batch hybridoma culture, for which three growth-limiting effects – by glutamine, lactate and ammonia – were considered in the model.

Calculate the growth rate reduction factors by nutrients and metabolites

The kinetic expression for the specific rate of cellular growth (Equation 4.3.17) contains three terms that each express a possible reduction of the specific growth rate with respect to μ_{max}. These growth-rate-reducing factors are evaluated as:

$$[\text{Gln}]/(K_{\text{Gln}} + [\text{Gln}]) \text{ for glutamine}$$

$$1/(1 + [\text{Lac}]/K_{\text{Lac}}) \text{ for lactate}$$

$$1/(1 + [\text{NH}_4]/K_{\text{NH}_4}) \text{ for ammonia}$$

From the measured (or the model-calculated) time variations of the concentrations of the three different species, one estimates the values of the corresponding factors: they are equal to unity when the considered substrate or product is non-limiting and they decrease when the substrate level becomes too low or the metabolite level too high.

MODELLING 175

Calculate the factors increasing the death rate by nutrients and metabolites

The kinetic expression for the specific rate of cellular death (Equation 4.3.18) contains three terms that each express a possible increase of the specific death rate and can be evaluated as follows:

$$1/(1 + [Gln]/C_1) \text{ for glutamine}$$

$$\kappa_2 \cdot [Lac] \text{ for lactate}$$

$$\kappa_1 \cdot [NH_4] \text{ for ammonia}$$

From the measured or simulated time variations of the concentrations of the three different species, one estimates the values of the three different factors: they are equal to zero when the considered substrate or product is non-limiting and they increase when the substrate level becomes too low or the byproduct concentration too high.

Compare the kinetic effects of the nutrients and metabolites

The time variations of the different factors have been calculated and are represented in Figures 4.3.5b and 4.3.5c, together with the variation of the specific rates of cell growth and death (Figure 4.3.5a). These variations clearly indicate that during the first 70 h the slight decrease in growth rate and increase in death rate can be essentially attributed to the action of ammonia and lactate. After 80 h the sharp decrease in growth rate and increase in death rate result from the simultaneous effect of ammonia, lactate and glutamine, but glutamine depletion has the most dominant effect.

DISCUSSION

From a single batch culture it is difficult to identify precisely the kinetic effects of the four considered medium components. It would be necessary to study the kinetics of several cultures run at different initial medium concentrations of glucose and glutamine. A better quantification of the respective effects of lactate and ammonia on cell growth and death would also require additional studies on culture kinetics when adding different levels of lactate or ammonia to the initial culture medium. An alternative method would be to grow the cells in continuous mode and change separately the concentrations of the medium components in the feeding medium.

The extent of cell death is determined by measuring the cells unable to exclude the Trypan blue dye. In prolonged continuous culture or in unfavourable environments the additional phenomenon of cell lysis may also occur. The number of cells that have lysed can be determined by measuring the lactate dehydrogenase (LDH) concentration released into the medium (see Chapter 2, section 2.5). Adding the number of visible and lysed dead cells yields an evaluation of the total rate of cell death.

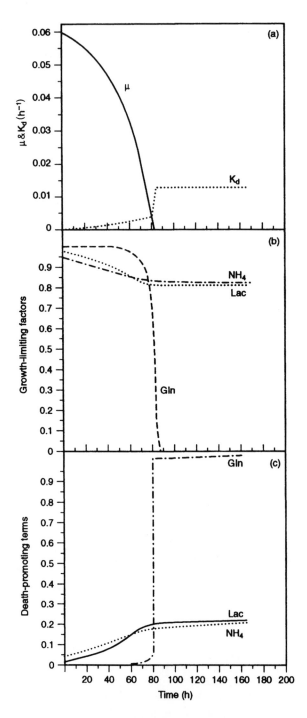

Figure 4.3.5 Modelling evolution with time of the specific growth and death rates (a) and their limiting (b) and promoting (c) factors during a batch culture.

MODELLING

According to previous kinetic studies, for many cells the production of antibody is not simply growth associated as assumed in the model presented. One may observe a specific protein production rate independent of cellular growth rate. Under these conditions a non-growth-associated type of rate expression has to be selected for protein production.

The suggested model accounts for the kinetic effects of glucose, glutamine, lactate and ammonia. Depending on the initial medium composition and on the operating conditions, additional rate-limiting effects may be observed, e.g. the influence of other amino acids, serum, dissolved oxygen, pH and osmotic pressure. The different rate expressions in the model can be modified to account for these additional effects.

A high-performance model has to be validated for different procedures: batch, fed-batch, continuous and perfused modes. Often, the extrapolation of the model to another mode needs the adjustment of different parameters.

Table 4.3.3 Nomenclature

A_d	Death constant (h^{-1})
C_1	Glutamine activation constant in specific death rate expression (mM)
C_2	Glucose activation constant in specific death rate expression (mM)
[Glc]	Glucose concentration (mM)
[Gln]	Glutamine concentration (mM)
k_d	Specific death rate (h^{-1})
k_{deg}	Glutamine degradation constant (h^{-1})
K_{Glc}	Glucose activation constant in specific growth rate expression (mM)
K_{Gln}	Glutamine activation constant in specific growth rate expression (mM)
K_{Lac}	Lactate inhibition constant in specific growth rate expression (mM)
K_{NH_4}	Ammonia inhibition constant in specific growth rate expression (mM)
[Lac]	Lactate concentration (mM)
m	Non-growth-associated specific rates (mmol or mg 10^{-9} cells h^{-1})
[MAbs]	Monoclonal antibody concentration (mg l^{-1})
(NH_4^+)	Ammonium ion concentration (mM)
r_{Gln}	Rate of glutamine degradation (mmol l^{-1} h^{-1})
X_v	Viable cell concentration (10^9 cells l^{-1})
X_m	Dead cell concentration (10^9 cells l^{-1})
Y	Growth-associated yield (mmol or mg 10^{-9} cells)
κ_1	Lactate activation constant in specific death rate expression (mM^{-1})
κ_2	Ammonia activation constant in specific death rate expression (mM^{-1})
μ	Specific growth rate (h^{-1})
μ_{max}	Maximum specific growth rate (h^{-1})
ν_{Glc}	Specific glucose consumption rate (mmol 10^{-9} cells h^{-1})
ν_{Gln}	Specific glutamine consumption rate (mmol 10^{-9} cells h^{-1})
π_{Lac}	Specific lactate production rate (mmol 10^{-9} cells h^{-1})
π_{NH_4}	Specific ammonia production rate (mmol 10^{-9} cells h^{-1})
π_{MAbs}	Specific antibody production rate (mmol 10^{-9} cells h^{-1})

REFERENCES

Vetterling TW, Teukolsky SA, Press WH & Flannery BP (1986) *Numerical Recipes: Example Book (Fortran)*. Cambridge University Press, Cambridge.

BACKGROUND READING

Batt BC & Kompala DS (1989) A structured kinetic modeling framework for the dynamics of hybridoma growth and monoclonal antibody production in continuous suspension culture. *Biotechnology and Bioengineering* 34: 515–531.

Bree MA, Dhurjati P, Geoghedan Jr RF & Robnett B (1988) Kinetic modelling of hybridoma cell growth and immunoglobulin production in a large-scale suspension culture. *Biotechnology and Bioengineering* 32: 1067–1072.

Dalili M, Sayles GD & Ollis DF (1990) Glutamine-limited batch hybridoma growth and antibody production: experiment and model. *Biotechnology and Bioengineering* 36: 74–82.

Glacken MW, Adema E & Sinskey AJ (1988) Mathematical descriptions of hybridoma culture kinetics: I. Initial metabolic rates. *Biotechnology and Bioengineering* 32: 491–506.

Goergen J-L, Marc A & Engasser J-M (1992a) Comparison of specific rates of hybridoma growth and metabolism in batch and continuous cultures. *Cytotechnology* 10: 147–155.

Goergen J-L, Marc A & Engasser J-M (1992b) Influence of medium composition on the death and lysis of hybridoma cells, in continuous culture. In: McDonald C, Griffiths JB & Spier RE (eds) *Animal Cell Technology: Developments, Processes and Products*, pp. 122–124. Butterworth, London.

Goergen JL, Marc A & Engasser JM (1993) Determination of cell lysis and death kinetics in continuous hybridoma cultures from the measurement of lactate dehydrogenase release. *Cytotechnology* 11: 189–195.

Goergen JL, Marc A & Engasser JM (1997) Kinetics and simulation of animal cell processes. In: Hauser H & Wagner R (eds) *Mammalian Cell Biotechnology in Protein Production*, pp. 345–371. Walter de Gruyter, Berlin.

Lourenço da Silva A, Marc A, Engasser JM & Goergen JL (1996) Kinetic model of hybridoma cultures for the identification of rate limiting factors and process optimisation. *Mathematics & Computers in Simulation* 1277: 1–9.

Martial A, Engasser J-M & Marc A (1990) Influence of inoculum age on hybridoma culture kinetics. *Cytotechnology* 5: 165–171.

Miller WM, Blanch HW & Wilke CR (1988) A kinetic analysis of hybridoma growth and metabolism in batch and continuous suspension culture: effect of nutrient concentration, dilution rate and pH. *Biotechnology and Bioengineering* 32: 947–965.

Taya M, Mano T & Kobayashi T (1986) Kinetic expression for human cell growth in a suspension culture system. *Journal of Fermentation Technology* 64: 347–350.

Tritsch GL & Moore GE (1962) Spontaneous decomposition of glutamine in cell culture media. *Experimental Cell Research* 3: 73–78.

4.4 CELL DEATH IN CULTURE SYSTEMS (KINETICS OF CELL DEATH)

Cell death in eukaryotic cells may be divided into two morphologically and biochemically distinct modes, those of apoptosis and necrosis. Necrosis is the classically recognized form of cell death that results from severe cellular insults and involves rapid cell swelling and lysis and is a process over which the cell has little or no control. Apoptosis, on the other hand, is the mode of cell death observed primarily under physiological conditions, and is an organized, preprogrammed response of the cell to changing environmental conditions. This form of cell death is characterized by cell shrinkage, nuclear and DNA fragmentation and breaking up of the cell into membrane-bounded vesicles, termed 'apoptotic bodies', which are subsequently ingested by neighbouring cells or macrophages *in vivo*. Apoptosis is an ATP-dependent process that in some systems also requires RNA and protein synthesis. The activation of a Ca^{2+}/Mg^{2+}-dependent, zinc-inhibitable endonuclease, which cleaves the cell's DNA into fragments of 200 base pairs or multiples thereof, is the main biochemical hallmark of apoptosis.

In cell culture systems, cell death via apoptosis is quite common and can occur under a variety of circumstances. For example, growth-factor-dependent cell lines die by apoptosis following removal of the growth-promoting agent from the culture medium (Nieto & Lopez-Rivas, 1989). Terminal differentiation of cells *in vitro* results in apoptosis (Martin *et al.*, 1990), as does normal turnover of cells with a limited lifespan, e.g. neutrophils (Savill *et al.*, 1989) or freshly isolated thymocytes (McConkey *et al.*, 1989). Altered culture medium conditions, e.g. removal of zinc from the medium (Martin *et al.*, 1991), or cultures that are allowed to overgrow to produce high cell densities, also induce apoptosis. Finally, cells may also be induced to undergo apoptosis by exposure to a wide range of cytotoxic agents (Lennon *et al.*, 1991). Under *in vivo* conditions, apoptotic cells are recognized and rapidly removed by phagocytic cells. However, in cell culture this cannot occur and instead the cells undergo secondary necrosis. Cells at this stage are Trypan blue positive. Up to this point, however, apoptotic cells maintain the ability to exclude vital dyes, and hence this method underestimates the health of a cell culture.

Reagents and solutions

Note: Lysis buffer (without proteinase K) and the TE buffer should be autoclaved to remove contaminating deoxyribonuclease activity.

- *Lysis buffer*: 10 mM EDTA, 50 mM Tris (pH 8.0) containing 0.5% *N*-lauroylsarcosine and 0.5 mg ml^{-1} proteinase K
- *TE buffer*: 10 mM Tris·HCl (pH 8.0) and 1 mM EDTA
- *Electrophoresis loading buffer*: 10 mM EDTA, 0.25% bromophenol blue and 50% glycerol
- *TBE buffer*: 2 mM EDTA (pH 8.0), 89 mM Tris and 89 mM boric acid

PROCEDURE: MORPHOLOGICAL CHARACTERIZATION OF CELL DEATH

The two modes of cell death may be identified easily on morphological grounds. Apoptosis involves cell shrinkage, nuclear fragmentation and apoptotic bodies budding off, maintaining the ability to exclude vital dyes. Necrosis involves cell swelling, chromatin flocculation and direct cell lysis. Thus, cells undergoing necrosis rapidly lose their ability to exclude vital dyes.

Materials and equipment

- Cell fixative – a haematoxylin and eosin stain (e.g. Rapi-Diff II, Diachem Diagnostic Developments, Southport, UK)
- Trypan blue
- Cytocentrifuge

 Trypan blue is a suspected carcinogen. Gloves should be worn when using all stains.

1. Cytocentrifuge cells onto glass slides. The speed of centrifugation will depend on the cell type; a typical cytocentrifuge setting would be 300 rpm for 2 min.
2. Remove glass slides and place in fixing solution for 20 s ('A' of Rapi-Diff II stain). Remove excess fluid from slide and repeat for the cytoplasmic and nuclear stains ('B' and 'C' of Rapi-Diff II stain, respectively). Rinse with distilled water.
3. Examine under the light microscope.

The key morphological criteria for recognizing apoptotic cells are nuclear condensation and fragmentation (Figure 4.4.1). Apoptotic cells also maintain their ability to exclude vital dyes such as Trypan blue. Hence cultures with a high proportion of cells displaying nuclear condensation and fragmentation, indicative of apoptosis, should also have a relatively high number of cells with the ability to exclude vital dyes.

Necrotic cells may be recognized morphologically by cell swelling and nuclear flocculation (Figure 4.4.2). Cultures with a high proportion of necrotic cells should have a relatively low number of cells with the ability to exclude vital dyes.

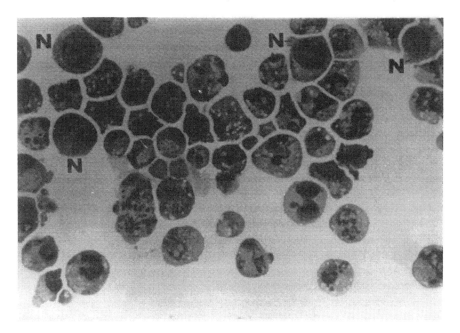

Figure 4.4.1 Morphological characteristics of apoptosis in human promyelocytic leukaemic HL-60 cells, illustrating nuclear fragmentation typical of apoptosis in HL-60 cells. N indicates normal cells. All the other cells are undergoing various phases of apoptosis. Original magnification ×500.

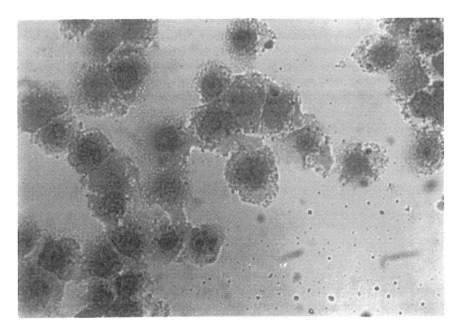

Figure 4.4.2 Morphological characteristics of necrosis in HL-60 cells. Original magnification ×500.

PROCEDURE: BIOCHEMICAL CHARACTERIZATION OF CELL DEATH

The biochemical hallmark of apoptosis is cleavage of the nuclear DNA into ~200 base pair multiples (Wyllie, 1980). This specific DNA cleavage is thought to result from the activation of an endogenous endonuclease that cleaves at the exposed linker regions between nucleosomes. Necrosis, on the other hand, is not associated with the ordered form of DNA cleavage observed during apoptosis. Hence a biochemical examination of the DNA from cells can be used to characterize the mode of cell death occurring in cell cultures.

Isolation of DNA

- Lysis buffer
- RNase A (previously heat treated to remove contaminating deoxyribonuclease activity)
- Phenol, buffered with 0.1 M Tris·HCl (pH. 7.4)
- Chloroform/isoamyl alcohol (24:1)

 Gloves should be worn throughout this procedure, both to protect the worker from phenol and chloroform and to prevent contamination of samples with deoxyribonucleases. Pipette tips and tubes should be autoclaved before use.

1. Wash cells twice in phosphate-buffered saline (PBS).
2. Resuspend cell pellets at 2×10^7 ml^{-1} in lysis buffer and incubate in a 50°C water-bath for 1 h.
3. Add RNase A to a concentration of 0.25 mg ml^{-1} and continue incubation at 50°C for 1 h.
4. Extract the crude DNA preparations twice with an equal volume of phenol (retain the aqueous layer after each extraction).
5. Extract twice with an equal volume of chloroform/isoamyl alcohol (retain the aqueous layer after each extraction).
6. Centrifuge the DNA preparations at 13 000 *g* for 15 min to separate intact from fragmented chromatin.
7. Place the supernatant (containing the fragmented chromatin) into a separate tube.
8. Precipitate the DNA in two volumes of ice-cold ethanol overnight at −70°C. The DNA may be stored in this condition for long periods.

Electrophoresis of DNA

Materials and equipment

- TE buffer
- Electrophoresis loading buffer
- Agarose
- TBE buffer

- Ethidium bromide
- UV source (Transilluminator (UV: 302 nm), UVP Inc., San Gabriel, CA, USA)

 Ethidium bromide is an irritant and known mutagen. Gloves should be worn throughout this procedure.

1. Recover the DNA precipitates by centrifugation at 13 000 g for 15 min. Allow the tube to air-dry at room temperature for 10 min and resuspend in TE buffer. Store at 4°C.
2. Add electrophoresis loading buffer in a 1:5 ratio. Place samples into a 65°C water-bath for 10 min and then maintain on ice.
3. Place samples into wells of a 1% agarose gel.
4. Electrophoresis is carried out in TBE buffer at 6 V cm^{-1} of gel.
5. After electrophoresis, soak the gel in TBE buffer containing 1 mg ml^{-1} ethidium bromide. Destain briefly in TBE buffer.

 Visualize the DNA by UV fluorescence. Care should be taken not to expose skin or eyes to UV light.

The DNA isolated from apoptotic cells separates into bands of 200 or multiples of 200 base pairs (Figure 4.4.3, lanes 3 and 4), whereas DNA isolated from control cells shows relatively little degradation (Figure 4.4.3, lanes 1 and 2). The DNA isolated from necrotic cells shows no ordered DNA fragmentation.

Note: The use of DNA molecular size markers enables an estimate of the size of the fragmented DNA.

SUPPLEMENTARY PROCEDURE: PURIFICATION OF APOPTOTIC CELLS

If apoptosis is only occurring at relatively low levels in cultures, it may be necessary to obtain a purified population of apoptotic cells prior to DNA isolation and electrophoresis. This may be achieved by exploiting the fact that apoptotic cells are more dense than normal cells. Hence it is possible to purify apoptotic cells by isopycnic centrifugation. Percoll can be used to create solutions of different densities. The precise Percoll densities used for isolation of apoptotic cells will depend on the cell type under investigation.

1. Prepare Percoll solutions of various densities. These densities should typically range from 1.05 to 1.08 in the case of most mammation cells.
2. Carefully layer the Percoll solutions sequentially into a test tube. Wash cells in PBS, resuspend at a high concentration (typically 20×10^6 ml^{-1}) in PBS and place on top of the Percoll column.
3. Centrifuge at 400 g for 30 min in a swing-out rotor. Cells distribute to their isopycnic points, creating bands representing normal, dead and apoptotic cells.
4. Elute the different bands from the Percoll gradients. Morphological examination of these bands indicates which band is enriched for apoptotic cells. This provides an estimate for the density of the apoptotic cells.

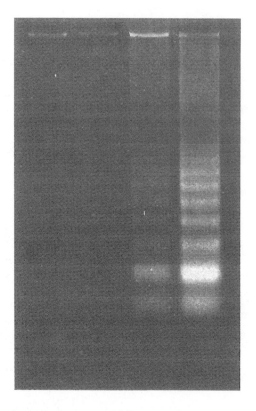

Figure 4.4.3 Cleavage of DNA into 200 or multiples of 200 base pairs during apoptosis in HL-60 cells (lanes 3 and 4). Control cells do not display any DNA degradation (lanes 1 and 2).

5. Adjust the Percoll gradients, placing the solution with the estimated density of apoptotic cells at the bottom of the tube.
6. Repeat the above procedure until apoptotic cells pellet to the bottom of the test tube after centrifugation. It is possible to obtain a purified population of apoptotic cells (>90%) by this method.

Figure 4.4.4 shows a typical Percoll gradient for isolation of apoptotic cells from a human promyelocytic leukaemic cell culture (HL-60).

DISCUSSION

Apoptosis was first described in 1972 (Kerr *et al.*, 1972). Since then it has become apparent that this mode of cell death plays a role in a number of important physiological processes. Thus it is advantageous to be able to recognize when this mode

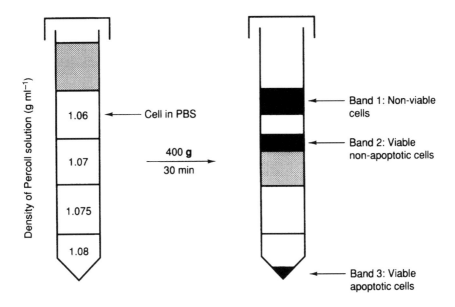

Figure 4.4.4 Percoll gradient for isolation of apoptotic cells from an LH-60 cell culture.

of death is occurring. Investigations into the mechanistic aspects of apoptosis have revealed a regulatory role of different ions in this process. For example, apoptosis occurs in certain human cell lines under conditions of zinc deficiency (Martin et al., 1991). Likewise, apoptosis may be inhibited by the addition of zinc to the culture medium (Martin & Cotter, 1991) or by removal of calcium (McConkey et al., 1989). Hence it is apparent that different culture conditions may activate or inhibit this mode of cell death. The above morphological and biochemical procedures should be used to complement each other in identifying the mode of cell death occurring in cells.

REFERENCES

Kerr JFR, Wyllie AH & Currie AR (1972) Apoptosis: a basic biological phenomenon with wide ranging implications in tissue kinetics. *British Journal of Cancer* 26: 239–257.

Lennon SV, Martin SM & Cotter TG (1991) Dose-dependent induction of apoptosis in human tumour cell lines by widely diverging stimuli. *Cell Proliferation* 24: 203–214.

Martin SJ & Cotter TG (1991) Ultraviolet B irradiation of human leukaemia HL-60 cells *in vitro* induces apoptosis. *International Journal of Radiation Biology* 59: 1001–1016.

Martin SJ, Bradley JG & Cotter TG (1990) HL-60 cells induced to differentiate towards neutrophils subsequently die via apoptosis. *Clinical Experimental Immunology* 79: 448–453.

Martin SJ, Mazdai C, Strain JJ, Cotter TG & Hannigan BM (1991) Programmed cell death (apoptosis) in lymphoid and myeloid cell lines during zinc deficiency. *Clinical and Experimental Immunology* 83: 338–343.

McConkey DJ, Hartzell P, Amador-Perez JF, Orrenius S & Jondal M (1989) Calcium-dependent killing of immature thymocytes by stimulus via the CD3/T cell receptor complex. *Journal of Immunology* 145: 1801–1806.

Nieto MA & Lopez-Rivas A (1989) IL-2 protects T lymphocytes from glucocorticoid-induced DNA fragmentation and cell death. *Journal of Immunology* 143: 4166–4170.

Savill JS, Wyllie AH, Henson JE, Walport MJ, Henson PE & Haslett C (1989) Macrophage phagocytosis of aging neutrophils in inflammation: programmed cell death in the neutrophil leads to its recognition by macrophages. *Journal of Clinical Investigation* 83: 865–871.

Wyllie AH (1980) Glucocorticoid-induced thymocyte apoptosis is associated with endogenous endonuclease activation. *Nature (London)* 284: 555–556.

4.5 DETOXIFICATION OF CELL CULTURES

When the growth of a cell culture slows down at the end of the exponential growth phase, the probable reason for this is either that the cells have consumed essential growth factors or nutrients, or that the cells are producing inhibitory or toxic metabolites. In those cases where the latter of these alternatives is predominant, it might be useful to remove the toxic metabolites in order to obtain higher cell densities or a prolonged culture lifetime. In cultures that produce a cellular product (e.g. monoclonal antibody or recombinant DNA product), the purpose of removing toxic metabolites might be to increase product concentration in the medium.

Procedures are given for two different methods: detoxification by dialysis and detoxification by gel filtration.

PROCEDURE: DETOXIFICATION BY DIALYSIS

Materials and equipment

- Growth medium with all supplements *except* serum, at least ×10 of the volume of the culture medium to be dialysed (preferably ×50)
- Dialysis tubing with molecular weight cut-off 10 000 (for preparation see below)
- Sterile flask or bottle that may contain the above medium
- Magnetic bar and stirrer

Preparation of dialysis tubing

1. Wash tubing twice for 1 h in 1 l of 50% ethanol.
2. Wash tubing twice for 1 h in 10 mM NaHCO$_3$.
3. Wash tubing twice for 1 h in 1 mM EDTA.
4. Wash tubing twice for 1 h in distilled water.
5. Store in distilled water with 10% acetic acid at 4°C.

Detoxification

1. Make two knots in one end of the dialysis tubing and boil it in distilled water for 2 h. Cool before use. All work from this stage should be sterile.
2. Remove the cell-free medium from the cell culture, leaving a minimum amount of medium for the cells to survive for approximately 2 h. If the cells are not

adherent, or immobilized by other means, they may be removed by, for example, centrifugation.
3. Take the tubing in one hand with sterile gloves, pipette the medium into the tubing and immerse it in a bottle containing fresh growth medium without serum.
4. Put a magnetic bar (sterile) into the bottle and place it on a magnetic stirrer at room temperature for 2 h.
5. With a pipette, withdraw the medium from the tubing and return it to the cell culture from which it was taken.
6. If sterility is not guaranteed, the medium must be filtered through a sterile filter (0.2 µm) before returning it to the cells.

ALTERNATIVE PROCEDURE: DETOXIFICATION BY GEL FILTRATION

Materials and equipment

- Column packed with Sephadex G-25, bed volume ×3 of the volume of the medium to be treated
- Growth medium with all supplements *except* serum, at twice the bed volume of the column
- Equipment for running the medium through the column and collecting the eluted fractions (pump, UV monitor, recorder and fraction collector)
- Sterile filter (0.2 µm)

Detoxification

1. Equilibrate the column with one bed volume growth medium without serum.
2. Remove the cell-free medium from the culture, leaving a minimum amount of medium for the cells to survive for approximately 2 h. If the cells are not adherent, or immobilized by other means, they must be removed by, for example, centrifugation.
3. Apply the cell-free medium to the column and start the recorder and fraction collector.
4. Elute growth medium without serum (one bed volume) and collect suitable fractions (e.g. 1/10 to 1/5 of the volume applied to the column).
5. Pool the fractions containing the proteins (peak adsorption at 280 nm) to give a total volume equal to the volume applied to the column.
6. Filter the medium through a sterile filter (0.2 µm) and return to the cell culture that it was taken from.
7. Wash the column with suitable buffer (e.g. PBS); if the column is kept, the buffer must contain a preservative (e.g. thimerosal or sodium azide).

DISCUSSION

Background information

Both the described methods are based on the observation that the toxic metabolites are usually low molecular weight compounds. The rationale is to remove these low molecular weight compounds and retain the macromolecular fraction containing growth factors and potential cell product. Several methods have been employed to remove toxic metabolites. In addition to the two methods described above, ion exchange chromatography has been used. Another approach is to detoxify the medium by adding complex binding substances. Serum appears to protect against toxic effects because low-serum cultures are often more sensitive to toxic metabolites than cultures with higher serum concentration. Albumin often functions as a complex-forming agent; however, at present no other detoxicants are known, and one has to use physicochemical methods in order to remove the toxic metabolites.

An alternative to batch treatment, such as the two methods described, is continuous removal by dialysis. Anchorage-independent cells may be grown inside dialysis tubing in roller bottles with cell-free medium outside the tubing, or in hollow-fibre cartridges. When larger volumes of medium are to be dialysed (e.g. in connection with a bioreactor), ultrafiltration membranes can be used. The medium (with or without cells) is circulated through the ultrafiltration unit and dialysed against fresh medium, usually lacking serum, which is circulated on the other side of the membrane.

Time considerations

Detoxification of the cell culture medium with one of the above-described methods takes 1–2 h. This should be taken into consideration when planning the experiments and care should be taken over storage of the cells whilst treating the medium.

REFERENCES

Adamson SR, Fitzpatrick SL, Behie LA, Gaucher GM & Lesser BH (1983) *In vitro* production of high titer monoclonal antibody by hybridoma cells in dialysis culture. *Biotechnology Letters* 5: 573–578.

Knazek RA, Gullino PM, Kohler PO & Dedrick RL (1972) Cell culture on artificial membranes. Approach to tissue growth *in vitro*. *Science* 178: 65–67.

Rønning ØW, Schartum M, Winsnes A & Lindberg G (1991) Growth limitation in hybridoma cell cultures: the role of inhibitory or toxic metabolites. *Cytotechnology* 7: 15–24.

Sjøgren-Jansson E & Jeansson S (1985). Large scale production of monoclonal antibodies in dialysis tubing. *Journal of Immunological Methods* 84: 359–364.

4.6 OXYGENATION

The control and measurement of oxygen concentration in culture media are essential elements in *in vitro* cell culture. Oxygen is often the first component to become limiting at high cell densities because of the low solubility of oxygen in aqueous media. In addition, the supply of oxygen may also limit the productivity in cell culture. Therefore, the ability to deliver sufficient oxygen to the culture must be incorporated in a successful experimental procedure. The dissolved oxygen concentration must also be controlled, because high oxygen concentrations may result in cellular damage through the generation of oxygen free radicals. The rate of cellular oxygen consumption can be used to provide information on the status of a culture; it correlates well with the cell density in bioreactors (Fleischaker & Sinskey, 1981; Lydersen *et al.*, 1985; Shirai *et al.*, 1988; Yamada *et al.*, 1990) as well as with protein production (Nicholson *et al.*, 1991). The oxygen consumption rate can also serve as a rapid and sensitive indicator for problems developing in the culture because a rapid decrease can occur in response to changes in pH or temperature or depletion of essential medium components.

Units

The units used for oxygen concentration in media are: percentage air saturation, percentage dissolved oxygen, oxygen tension, P_{O_2} or molar units. Percentage air saturation (identical to percentage dissolved oxygen, % DO) is represented by:

$$\% \text{ air sat.} (\% \text{ DO}) = (C_L/C_{sat}) \times 100\% \quad (4.6.1)$$

where C_L is the actual oxygen concentration in the bulk medium and C_{sat} is the oxygen concentration in equilibrium with air (21% oxygen). Oxygen tension or P_{O_2} (mmHg) is the partial pressure of oxygen in the liquid phase in equilibrium with the partial pressure of oxygen in the gas phase. At equilibrium, the partial pressure of oxygen in the liquid phase (corrected for the partial pressure of water) is equal to the partial pressure of oxygen in the gas phase. Thus, $P_{O_2} = 0.21(760 - 47 \text{ mmHg}) = 150$ mmHg at 100% air saturation, where the partial pressure of water is 47 mmHg at 37°C. The oxygen concentration in culture medium (RPMI 1640) in equilibrium with air has been reported to be 0.2 mM (Oller *et al.*, 1989; Wohlpart *et al.*, 1990) and with water-saturated air to be 0.19 mM (Oller *et al.*, 1989). Oxygen concentrations for other oxygen tensions are plotted in Figure 4.6.1. The equilibrium oxygen concentration in media can also be estimated by correcting the solubility of oxygen in water for the presence of electrolytes (Schumpe *et al.*, 1982). Chemical methods to determine the oxygen concentration are given in Schumpe *et al.* (1978), Slininger *et al.* (1989) and Wohlpart *et al.* (1990).

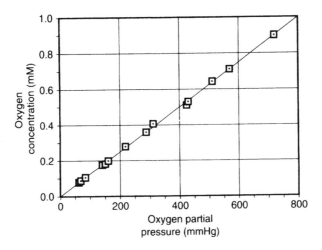

Figure 4.6.1 Equilibrium concentration (mM) of dissolved oxygen in RPMI medium at different oxygen partial pressures (mmHg). Based on data of Oller *et al.* (1989) by permission of *Journal of Cell Science*.

Effect of oxygen concentration on cells

Oxygen, while it is required for growth, can be toxic at high concentrations. Effects of varying oxygen concentrations (reported as partial pressure of oxygen, mmHg) on various properties of cultured cells are presented in Table 4.6.1. Established cell lines appear to function well over a wide range of oxygen tensions (~15–90% DO). Primary cell types, however, appear to survive best at lower oxygen tensions, which better simulate the *in vivo* environment.

Typical values of the specific oxygen consumption rate

The specific oxygen consumption rate, q, has been reported to be affected by cell type (Yamada *et al.*, 1990), cell density (Wohlpart *et al.*, 1990; Yamada *et al.*, 1990). Proliferative state of the culture (Yamada *et al.*, 1990), glucose concentration (Kilburn *et al.*, 1969; Wohlpart *et al.*, 1990) and glutamine concentration (Wohlpart *et al.*, 1990). The specific oxygen consumption rate is fairly constant over a wide range of dissolved oxygen concentrations (Fleischaker & Sinskey, 1981; Wohlpart *et al.*, 1990), but was found to decrease with P_{O2} below 5–10 mmHg (Miller *et al.*, 1987). Oxygen consumption rates for a variety of cell types are listed in Table 4.6.2.

PROCEDURE: MEASUREMENT OF OXYGEN TRANSFER COEFFICIENT AND OXYGEN UPTAKE RATE

Values for $K_L a$ (in units of inverse time), which is the product of K_L, the overall mass transfer coefficient, and a, the gas–liquid interfacial area per unit of reactor liquid volume, can be determined experimentally by the 'dynamic gassing-out

Table 4.6.1 Effects of oxygen partial pressures (mmHg) on cultured cells[a]

Cell type	Positive effects	negative effects	No effect	Reference
TK6		>510 (G)	80–425 (G)	Oller et al. (1989)
Vero		>310 (G)	80–215 (G)	Oller et al. (1989)
SP2/0 hybridoma		>425 (G)	80–310 (G)	Oller et al. (1989)
	Max. at 1.5 (V)		15–150 (V)	Miller et al. (1988)[b]
	Max. at 75 (Ab)			Miller et al. (1988)[b]
Murine hybridoma		143 (G)	7.5–135 (G)	Meilhoc et al. (1990)[b]
	Min. 30–75 (D)	<30 (G)	30–120 (G)	Ozturk & Palsson (1990)[b]
			22.5–150 (Ab)	Ozturk & Palsson (1990)[b]
Murine/human hybridoma		<1.6, >320 (G)	12–160 (G)	Kilburn et al. (1969)[b]
Porcine aorta	Max. 60–75 (G)			Friedl et al. (1989)[b]
Human umbilical vein	Max. 105–120 (G)			Friedl et al. (1989)[b]
Rat hepatocytes		90 (V)		Suleiman & Stevens (1987)[c]
Human haemopoietic	Gas phase 37.5 (V)	Gas phase 150 (V)		Koller et al. (1992a,b)
Human macrophage progenitor	Gas phase 37.5 (V)	Gas phase 150 (V)		Broxmeyer et al. (1990)
Murine haemopoietic	Gas phase 3–5.25 (V)	Gas phase 7.5 (V)		Rich (1986)
Blood monocytes	Gas phase 9 (V)	Gas phase ≥15 (V)		Lin & Hsu (1986)

[a]G = growth rate; D = death rate; V = cell viability; Ab = antibody production (per cell)
[b]Values originally reported as % air sat., conversion based upon 100% air sat. = 150 mmHg.
[c]Paper reports O_2 conc. at air sat. as 0.22 mM, conversion based upon 0.22 mM = 150 mmHg.

Table 4.6.2 Reported oxygen consumption rates of cells in culture[a]

Cell type	Specific oxygen consumption rate ($\mu M\ h^{-1}\ 10^{-6}$ cells)	Reference
CHO	0.15	Lin & Miller (1992)
BHK-21	0.2	Radlett et al. (1972)
Human foreskin (FS-4)	0.05	Fleischaker & Sinskey (1981)
Human granulocytes	0.03–0.26	Van Dissel et al. (1986)
Murine hybridoma	0.05–0.2	Miller et al. (1988)
	0.03–0.37	Ramirez & Mutharasan (1990)
	0.110	Meilhoc et al. (1990)
	0.078–0.086	Ozturk & Palsson (1990)
	0.12–0.48	Wohlpart et al. (1990)
Murine/human hybridoma	0.267–0.452	Shirai et al. (1988)
Murine myeloma	0.057	Meilhoc et al. (1990)
Murine macrophages	0.06–0.07	Van Dissel et al. (1986)
Vero	0.24	Oller et al. (1989)

[a]CHO = Chinese hamster ovary; BHK = baby hamster kidney. Additional values are given in Fleischaker & Sinskey (1981).

OXYGENATION

method'. Equation 4.6.1 describes the dynamics of the dissolved oxygen concentration in liquid medium without cells (no oxygen consumption):

$$\frac{dC_L}{dt} = K_L a(C^* - C_L) \qquad (4.6.2)$$

where C_L and C^* are the actual and dissolved oxygen concentration in equilibrium with the oxygen concentration in the bulk gas phase, respectively. To measure $K_L a$, the vessel is first equilibrated with nitrogen until the oxygen concentration in the medium is negligible. The oxygen concentration is then recorded as a function of time after starting the flow of the supply gas. The value of $K_L a$ is then determined by integration of Equation 4.6.2 (with the initial condition $C_L = 0$ at time $t = 0$), which yields:

$$-K_L at = \ln\left(1 - \frac{C_L}{C^*}\right) \qquad (4.6.3)$$

where C_L is the recorded output and C^* is the final equilibrium output. Using Equation 4.6.3, the slope of the line $\ln[1 - (C_L/C^*)]$ versus time is numerically equal to $-K_L a$. Measurements of $K_L a$ need to be performed under conditions identical to those of the culture, because $K_L a$ is a function of the method of oxygenation (i.e. surface aeration, direct sparging, etc.), liquid volume, power input, the ionic strength and viscosity of the medium (Moo-Young & Blanch, 1981; Schugerl, 1981), temperature (Matsuoka et al., 1992) as well as antifoam concentration (Lavery & Nienow, 1987; Johnson et al., 1990). The equilibrium concentration of oxygen, C^*, is also a function of the ionic strength of the medium, as well as the culture temperature and partial pressure of oxygen in the supply gas (Schumpe et al., 1982). Although this method is simple and convenient, it contains some inherent sources of error due to the oxygen probe response time (time required for the output to measure a percentage of a step change). Ideally, the oxygen probe response time, τ_p, should be much smaller than the mass transfer response time, $1/K_L a$. This condition would apply for most animal cell bioreactors where a typical Clark-type (polarographic) electrode has a 95% response time of ~15 s (Lee & Tsao, 1979), which is much smaller than most of the inverses of $K_L a$ values reported in Table 4.6.3. Other sources of error include the development of liquid films about the probe (important in viscous media or with slow agitation), and as a result of an initial variation in the concentration of oxygen in the gas phase (C^*) due to the initial concentration of nitrogen in the gas space (important in highly agitated sparged reactors where the main gas residence times are long). Where applicable, there are several papers that review corrections to the dynamic gassing-out procedure (Lee & Tsao, 1979; Van't Riet, 1979; Ruchti et al., 1981).

Once $K_L a$ is known, the cellular oxygen consumption rate, qX, can be measured under culture conditions. The mass balance for oxygen in a batch reactor with cells is given by:

$$\frac{dC_L}{dt} = K_L a(C^* - C_L) - qX \qquad (4.6.4)$$

where X is the cell concentration and q is the specific oxygen consumption rate (μmol O_2 consumed h^{-1} cell^{-1}). In a bioreactor with DO control ($dC_L/dt = 0$), Equation 4.6.4 can be rearranged to yield:

$$q = \frac{K_L a(C^* - C_L)}{X} \qquad (4.6.5)$$

In reactors without DO control, a value for dC_L/dt can be determined from a C_L versus t plot.

It is possible to construct a small vessel (respirometer, ~2–50 ml) for the specific purpose of measuring the cellular oxygen consumption rate. Equipment descriptions and/or procedures to measure q are given by Noll et al. (1986), Van Dissel et al. (1986), Shirai et al. (1988), Wohlpart et al. (1990) and Yamada et al. (1990). The vessel is typically a glass-jacketed vessel with two ports, one for a DO probe and the other for sample injection. A magnetically coupled flea (small stirring bar) can be used to suspend the cells. After injecting the cell sample, the vessel is sealed (ground-glass joints work well) and the decrease in DO is recorded over a 5–10-min period. The slope of the DO profile is the consumption rate, because Equation 4.6.4 reduces to:

$$\frac{dC_L}{dt} = -qX \qquad (4.6.6)$$

for a vessel where the oxygen transfer term can be neglected. The most critical step is the removal of all air bubbles from the vessel. The value of q measured by the respirometer is independent of $K_L a$, which was needed to determine q by Equation 4.6.5.

SUPPLEMENTARY PROCEDURE: OXYGENATION METHODS

The concentration below which oxygen becomes limiting to the culture (Table 4.6.1) is known as the critical oxygen concentration. Below this level, the cells may be limited by energy (ATP) supply and rely more on glycolysis (Miller et al., 1987). Thus the method of oxygen delivery must be able to maintain oxygen levels above the critical oxygen concentration. The oxygen balance over the bioreactor (Equation 4.6.4) implies that oxygen transfer can be increased by increasing the mass transfer coefficient K_L, increasing the area available for mass transfer, a, or by increasing the driving force for oxygen transfer, $C^* - C_L$. The methods used to oxygenate laboratory-scale reactors include surface aeration, membrane aeration and direct gas sparging. For larger and production-scale bioreactors, direct gas sparging is necessary. Overall mass transfer coefficients for these various aeration systems are listed in Table 4.6.3.

Surface aeration

Surface aeration is the simplest form and is sufficient for most laboratory applications. However, this is limited to laboratory-size reactors and becomes

Table 4.6.3 Overall transfer coefficients for oxygenation using different aeration systems

Aeration system	$K_L a (s^{-1})$	Reference
Static liquid surface	10^{-6}	Spier & Griffiths (1984)
Stirred liquid surface	10^{-6}–4×10^{-3}	Sinskey et al. (1981); Lavery & Neinow (1987); Katinger & Scheirer (1985); Aunins et al. (1989); Johnson et al. (1990)
Silastic membrane aeration	10^{-4}–10^{-2}	Aunins et al. (1986); Dorrington et al. (1985)
Sparged with agitation	7×10^{-3}–2.5	Spier & Griffiths (1984); Matsuoka et al. (1992)
Gas exchange impeller	4×10^{-3}–3×10^{-2}	Johnson et al. (1990); Shi et al. (1992)

Modified from Table 2 of Bliem & Katinger (1988) by permission of *Trends in Biotechnology*.

inadequate for large systems because the surface area for oxygen transfer per unit reactor volume decreases with increasing volume. For example, $K_L a$ was found to have an approximately inverse square relationship with culture volume (Fleischaker & Sinskey, 1981). Oxygen transfer by surface aeration is limited by the degree of mixing at the gas/liquid interface. Thus, increasing agitation intensity and the use of a larger impeller (Aunins et al., 1989) have both been shown to increase $K_L a$. Fleischaker & Sinskey (1981) found increases in $K_L a$ to be proportional to increases in agitation intensity. Unfortunately, there are upper limits to the level of agitation that can be used in bioreactors due to the fragility of mammalian cells (see section 4.7). The position of the impeller in the vessel can also influence $K_L a$: $K_L a$ was essentially constant with impeller position in a 500-ml spinner flask until the impeller was positioned within 1 cm of the liquid surface; at that point $K_L a$ increases dramatically (approximately four times the submerged value) (Aunins et al., 1989). Positioning the impeller near the liquid surface, however, would result in poor mixing even in most laboratory-size vessels. Thus, a second impeller could be attached to the impeller shaft and positioned at the liquid surface. This partially submerged second impeller, known as a surface aerator, was seen to increase the mass transfer coefficient, K_L, from 6.4 to 26.2 cm h^{-1} in a 1-l bioreactor and from 3.3 to 13.5 cm h^{-1} in an 8-l bioreactor by increasing mixing at the air/liquid interface (Hu et al., 1986). Oxygenation with air is difficult because the optimal DO levels (15–90% air sat.) are close to the equilibrium concentration of oxygen in the medium. Thus, the driving force for oxygen transfer ($C^* - C_L$) is small but this can be increased by supplementing with pure oxygen. In those cases where a lower oxygen concentration is desired (Table 4.6.1), the driving force for oxygen transfer can be decreased by mixing the supply air with nitrogen.

Sparger aeration

Oxygen can be supplied to the culture by directly sparging the supply gas into the bioreactor. The basic concepts describing mass transfer in gas-sparged bioreactors are reviewed in Moo-Young and Blanch (1981). Sparger aeration offers the advantages of very high oxygen transfer rates (Table 4.6.3) because of the large interfacial area, a, for bubbles. Disadvantages arise from the fact that animal cells are susceptible to damage from direct sparging and foaming of the culture medium (Papoutsakis, 1991a). Nevertheless, many different cell types have been successfully grown in air-sparged bioreactors. There are two main types of sparged reactors: bubble columns and sparged reactors with agitation.

Bubble column bioreactors

The attachment of cells to bubbles and the related shear forces associated with the rise of bubbles through the culture medium are not detrimental to cells (Tramper et al., 1988; Handa-Corrigan et al., 1989; Jobses et al., 1991); nor is there significant cell death at the sparger site (Jobses et al., 1991). Cell damage in sparged reactors is thought to occur primarily at the top of the reactor due to bubble break-up or shearing in thin films of collapsing foams (Papoutsakis, 1991a). Thus, parameters that affect cell death are those that affect the exposure of cells to bursting bubbles and unstable foams, as well as the frequency of bubble bursts. The height of the sparged reactor affects cell viability; increasing the column height reduces cell damage (Tramper et al., 1988; Handa-Corrigan et al., 1989; Jobses et al., 1991) due to the decrease in the surface area for bubble disengagement per unit reactor volume. The maximum viable cell density approached that achieved in surface-aerated and agitated reactors, as the liquid height/column diameter ratio exceeded 10:1 (Handa, 1986). Increased column height also increases the bubble residence time in the reactor. While high gas flow rates are desirable, because $K_L a$ is proportional to the ratio of superficial gas velocity to bubble size (Moo-Young & Blanch, 1981), there is a direct relationship between increasing the specific gas flow rate (increasing bubble frequency) and increased cell death in bubble columns at a fixed column height (Handa, 1986; Tramper et al., 1988; Jobses et al., 1991). At present, there is no definite correlation between bubble size and cell death. Tramper et al. (1988) found cell damage to be independent of bubble diameter for bubbles larger than 2 mm, while Handa-Corrigan et al. (1987) studied bubble diameters less than 1.68 mm and found cells to be more sensitive to smaller bubble sizes (0.98 and 0.2 mm).

It is suggested that bubble columns are operated with large height/diameter ratios (>5) to decrease the exposure of the cells to bursting bubbles. The gas flow rate should be kept to the minimum level allowable. However, the gas flow rate needs to be sufficiently large to keep the cells in suspension and to maintain an adequate DO level. If it is not possible to keep the gas flow rates below levels at which there is significant cell damage, the effect of smaller bubble sizes to increase oxygen transfer should be tested. The bubble diameter is affected by the pore size of the air-bubble distributor, the liquid properties of the medium and gas flow rates (only at higher gas flow rates). Perforated metal plates can be used to achieve larger bubble sizes (4–6 mm) while sintered glass filters can be used to achieve smaller diameters. Cell viability in sparged reactors is also affected by medium components. 'Shear protectants' are commonly used in both stirred-tank reactors and in gas-sparged vessels to protect cells (see section 4.8 and Papoutsakis (1991b)). For example, at high air sparger rates, the non-ionic surfactant Pluronic

F-68 (BASF Corp., Parsippany, NJ) at concentrations in the range 0.1–0.2% (w/v) had a strong protective effect on hybridoma cells (Handa-Corrigan et al., 1987; Jobses et al., 1991). To prevent foam formation in sparged bioreactors, silicon antifoams have been used at concentrations of 6–100 ppm (Handa-Corrigan et al., 1989; Croughan & Wang, 1991; Jobses et al., 1991) and in this concentration range there was little effect on cell growth.

Sparged bioreactors with agitation

Mechanical agitation is necessary for suspending the cells, circulating the medium and increasing oxygen transfer rates. Thus, lower gas velocities are more necessary in sparged bioreactors with agitation than in bubble column reactors. Bliem et al (1991) reported values for gas flow rates of 1 VVh (volume gas per volume reactor per hour) for bubble columns and 0.1–1 VVh for sparged reactors with mechanical agitation. These lower gas flow rates may reduce the level of antifoams and shear protectants required.

The agitator can be located above the sparger, but it may be desirable to place it under the sparger because there is some evidence that this results in lower shear forces. In addition to perforated metal plates or sintered glass filters, gas-exchange impellers (Johnson et al., 1990; Shi et al., 1992) and hollow-fibre membranes (Matsuoka et al., 1992) have been used as air-bubble distributors. Gas exchange impellers sparge the supply gas through a steel mesh screen attached to a hollow agitation shaft. In contrast to bubble column reactors, in this system the height/diameter ratio is not relevant in minimizing injury from shear forces. Bubbles are retained longer due to agitation-induced fluid motion. However, to minimize hydrodynamic cell damage, it is necessary to reduce or eliminate vortex formation around the impeller shaft, minimize bubble entrainment and, finally, minimize or eliminate bubble break-up at the free liquid surface. The latter can be accomplished by operating the reactor without a gas headspace and allowing bubbles to migrate through a medium-filled tube attached to a port at the reactor headplate. The outlet of the bubble disengagement tube should be connected to a sterile glass vessel to allow for the rise of a foam layer inside the disengagement tube.

Microcarrier cultures

There have been few reports of anchorage-dependent cells being cultured on microcarriers in sparged reactors. Cells grown on the surface of microcarriers are likely to be more susceptible to bubble-associated damage. In particular, foam formation is detrimental because microcarriers concentrate in the foam layer (Fleischaker & Sinskey, 1981) and deposit on the vessel wall due to bubble bursting (Croughan & Wang, 1991). Even with antifoam (Medical Emulsion AF, Dow Corning, Michigan), the concentration of CHO cells in a sparged microcarrier reactor at low superficial gas velocities (0.01 cm s^{-1}) was ~50% lower than that in surface-aerated cultures (also using antifoam) (Croughan & Wang, 1991).

The culture of cells on microcarriers in sparged reactors requires a conservative approach, including low gas flow rates, small bubble diameter for high oxygen transfer, low-intensity mechanical agitation and the use of antifoams. In addition, it may prove beneficial to use 'shear protectants' (such as Pluronic F-68 or polyethylene glycols (Papoutsakis, 1991b) to minimize microcarrier flotation (i.e. aggregation of microcarriers at the free liquid surface due to attachment to bubbles). It has been found recently that these additives reduce the attachment of freely suspended cells to bubbles, and this may also be the case for microcarriers.

Membrane aeration

Bubble-free oxygenation by membrane aeration provides efficient oxygen transfer with minimal shear damage and minimal foaming. Design equations for membrane aeration are presented in Aunins *et al.* (1986) and Su *et al.* (1992). In stirred-tank reactors, the oxygen supply gas is pumped through gas-permeable, hydrophobic, autoclavable tubing (silicon, polypropylene, polytetrafluoroethylene) wound about a fixed wire cage or a modified agitator (Lehmann *et al.*, 1988; Wagner & Lehmann, 1988). The main factors that affect oxygen transfer are the oxygen concentration in the supply gas, the membrane porosity and the surface area of the tubing, as well as the agitation rate because oxygen transfer occurs through convection. The reactor oxygen concentration can be controlled by adjusting the mixtures of O_2, N_2 and CO_2 (for pH control) in the supply gas. The tubing should be connected to the reactor through a sterile filter prior to the inlet and through a sterile trap at the outlet. The system should be tested for gas leaks both before and after it is autoclaved. Membrane aeration has limitations in large reactor systems, primarily due to the design complexity and the large membrane area required. Vorlop and Lehmann (1989) used 2.5–3.0 m of tubing per litre of medium for a 100-l reactor, thus using 250–300 m of tubing. Su *et al.* (1992) found that increased tubing length increases the oxygen transfer rate only up to a certain length, and above this oxygen transfer was not significantly affected. The 'critical length' of tubing needed decreases with smaller tubing diameter. It was also found that dividing the tubing into parallel segments supplied by a gas distributor improved oxygen transfer. The difficulty involved in cleaning and sterilizing membrane reactors between runs in an additional disadvantage.

REFERENCES

Aunins JG, Croughan MS, Wang DIC & Goldstein JM (1986) Engineering developments in homogeneous culture of animal cells: oxygenation of reactors and scale-up. *Biotechnology and Bioengineering Symposium* 17: 699–723.

Aunins JG, Woodson BA, Hale TK & Wang DIC (1989) Effects of paddle impeller geometry and mass transfer in small-scale animal cell culture vessels. *Biotechnology and Bioengineering* 34: 1127–1132.

Bliem R & Katinger H (1988) Scale-up engineering in animal cell technology: Part II. *Trends in Biotechnology* 6: 224–230.

Bliem R, Konopitzky K & Katinger H (1991) Industrial animal cell reactor systems: aspects of selection and evaluation. *Advances in Biochemical Engineering* 44: 2–26.

Broxmeyer HE, Cooper S, Lu L, Miller ME, Langefeld CD & Ralph P (1990) Enhanced stimulation of human bone marrow macrophage colony formulation *in vitro* by recombinant human macrophage colony-stimulating factor in agarose medium at low oxygen tension. *Blood* 76: 323–329.

Croughan MS & Wang DIC (1991) Hydrodynamic effects on animal cells in microcarrier bioreactors. In: Ho CS & Wang DIC (eds) *Animal Cell Bioreactors*, pp. 213–249. Butterworth–Heinemann, Stoneham, MA.

Dorrington KL, Ralph ME, Bellhouse BJ, Gardaz JP & Sykes MK (1985) Oxygen and CO_2 transfer of a polypropylene dimpled membrane lung with variable secondary flow. *Journal of Biomedical Engineering* 7: 88–99.

Fleischaker RJ & Sinskey AJ (1981) Oxygen demand and supply in cell culture. *European Journal of Applied Microbiology and Biotechnology* 12: 193–197.

Friedl P, Chang JJ & Tatje D (1989) Different oxygen sensitivities of vascular endothelial cells from porcine aorta and from human veins during fermentation in a stirred bioreactor. In: Spier RE, Griffiths JB, Stephenne J & Crooy PJ (eds) *Advances in Animal Cell Biology and Technology for Bioprocesses*, pp. 233–237. Butterworth, London.

Handa A (1986) Gas–liquid interfacial effects on the growth of hybridoma and other suspended mammalian cells. *PhD Thesis*. University of Birmingham, Birmingham, UK.

Handa-Corrigan A, Emery EN & Spier RE (1987) On the evaluation of gas–liquid interfacial effects on the hybridoma viability in bubble column reactors. *Developments in Biological Standardization* 66: 241–253.

Handa-Corrigan A, Emergy EN & Spier RE (1989) Effect of gas–liquid interfacial effects on the growth of suspended mammalian cells: mechanisms of cell damage by bubbles. *Enzyme and Microbial Technology* 11: 230–235.

Hu WS, Meier J & Wang DIC (1986) Use of surface aerator to improve oxygen transfer in cell culture. *Biotechnology and Bioengineering* 28: 122–125.

Jobses I, Martens D & Tramper J (1991) Lethal events during gas sparging in animal cell culture. *Biotechnology and Bioengineering* 37: 484–490.

Johnson M, Andre G, Chavarie C & Archambault J (1990) Oxygen transfer rates in a mammalian cell culture bioreactor equipped with a cell-lift impeller. *Biotechnology and Bioengineering* 35: 43–49.

Katinger H & Scheirer W (1985) Mass cultivation and production of animal cells. In: Spier RE & Griffiths JB (eds) *Animal Cell Biotechnology*, vol. 1, pp. 167–193. Academic Press, London.

Kilburn DG, Lilly MD, Self DA & Webb FC (1969) The effect of dissolved oxygen partial pressure on the growth and metabolism of mouse LS cells. *Journal of Cell Science* 4: 25–37.

Koller MR, Bender JG, Papoutsakis ET & Miller WM (1992a) Beneficial effects of reduced oxygen tension and perfusion in long-term hematopoietic cultures. *Annals of the New York Academy of Sciences* 665: 105–116.

Koller MR, Bender JG, Miller WM & Papoutsakis ET (1992b) Reduced oxygen tension increases hematopoiesis in long-term culture of human stem and progenitor cells from cord blood and bone marrow. *Experimental Hematology* 20: 264–270.

Lavery M & Nienow AW (1987) Oxygen transfer in animal cell culture medium. *Biotechnology and Bioengineering* 30: 368–373.

Lee YH & Tsao GT (1979) Dissolved oxygen electrodes. *Advances in Biochemical Engineering* 13: 35–86.

Lehmann J, Vorlop J & Buntemeyer H (1988) Bubble-free reactors and their development for continuous culture with cell recycle. In: Spier RE & Griffiths JB (eds) *Animal Cell Biotechnology*, vol. 3, pp. 222–237. Academic Press, Orlando, FL.

Lin AA & Miller WM (1992) CHO cell responses to low oxygen: regulation of oxygen consumption and sensitization to oxidative stress. *Biotechnology and Bioengineering* 40: 505–516.

Lin H & Hsu S (1986) Modulation of tissue mononuclear phagocyte clonal growth by oxygen and antioxidant enzymes. *Experimental Hematology* 14: 840–844.

Lydersen BK, Pugh GG, Paris MS, Sharma BP & Noll LA (1985) Ceramic matrix for large scale animal cell culture. *Bio/Technology* 3: 63–67.

Matsuoka H, Fukada S & Toda K (1992) High oxygen transfer rate in bubble aeration using hollow fiber membrane: a proposal of a new aeration system. *Biotechnology and Bioengineering* 40: 346–352.

Meilhoc E, Wittrup KD & Bailey JE (1990) Influence of dissolved oxygen concentration on growth, mitochondrial function and antibody production of hybridoma cells in batch culture. *Bioprocess Engineering* 5: 263–274.

Miller WM, Blanch HW & Wilke CR (1987) The effects of dissolved oxygen concentrations on hybridoma growth and metabolism in continuous culture. *Journal of Cellular Physiology* 132: 524–530.

Miller WM, Wilke CR & Blanch HW (1988) Transient responses of hybridoma metabolism to changes in the oxygen supply rate in continuous culture. *Bioprocess Engineering* 3: 103–111.

Moo-Young M & Blanch HW (1981) Design of biochemical reactors: mass transfer criteria for simple and complex systems. *Advances in Biochemical Engineering* 19: 1–69.

Nicholson ML, Hampson BS, Pugh GG & Ho CS (1991) Continuous cell culture. In: Ho CS & Wang DIC (eds) *Animal Cell Bioreactors*, pp. 269–303. Butterworth-Heinemann, Stoneham, MA.

Noll T, de Groot H & Wissemann P (1986) A computer-supported oxystat system maintaining steady-state O_2 partial pressures and simultaneously monitoring O_2 uptake in biological systems. *Biochemical Journal* 236: 765–769.

Oller AR, Buser CW, Tyo MA & Thilly WG (1989) Growth of mammalian cells at high oxygen concentrations. *Journal of Cell Science* 94: 43–49.

Ozturk SS & Palsson BO (1990) Growth, metabolic, and antibody production kinetics of hybridoma cell cultures: III. Effect of dissolved oxygen concentration. *Biotechnology Progress* 6: 437–446.

Papoutsakis ET (1991a) Fluid-mechanical damage of animal cells in bioreactors. *Trends in Biotechnology* 9: 427–437.

Papoutsakis ET (1991b) Media additives for protecting freely suspended animal cells against agitation and aeration damage. *Trends in Biotechnology* 9: 316–324.

Radlett PJ, Telling RC, Whitside JP & Maskell MA (1972) The supply of oxygen to submerged cultures of BHK 21 cells. *Biotechnology and Bioengineering* 14: 437–445.

Ramirez OT & Mutharasan R (1990) Cell cycle- and growth phase-dependent variations in size distribution, antibody productivity, and oxygen demanding hybridoma cultures. *Biotechnology and Bioengineering* 36: 839–848.

Rich I (1986) A role for the macrophage in normal hemopoiesis. II. Effect of varying physiological oxygen tensions on the release of hemopoietic growth factors from bone-marrow-derived macrophages in vitro. *Experimental Hematology* 14: 746–751.

Ruchti G, Dunn IJ & Bourne JR (1981) Comparison of dynamic oxygen electrode methods for the measurement of K_La. *Biotechnology and Bioengineering* 23: 277–290.

Schügerl K (1981) Oxygen transfer into highly viscous media. *Advances in Biochemical Engineering* 19: 71–174.

Schumpe E, Adler I & Deckwer W (1978) Solubility of oxygen in electrolyte solutions. *Biotechnology and Bioengineering* 20: 145–150.

Schumpe E, Quicker G & Deckwer WD (1982) Gas solubilities in microbial culture media. *Advances in Biochemical Engineering* 24: 1–38.

Shi Y, Ryu DDY & Park SH (1992) Performance of mammalian cell culture bioreactor with a new impeller. *Biotechnology and Bioengineering* 40: 260–270.

Shirai Y, Hashimoto K, Yamaji H & Kawahara H (1988) Oxygen uptake rate of immobilized growing hybridoma cells. *Applied Microbiology and Biotechnology* 29: 113–118.

Sinskey AJ, Fleischaker M, Tyo MA, Giard DJ & Wang DIC (1981) Production of cell-derived products: virus and interferon. *Annals of the New York Academy of Sciences* 369: 47–59.

Slininger PJ, Petroski RJ, Bothast RJ, Ladisch MR & Okos MR (1989) Measurement of oxygen solubility in fermentation media: a colorimetric method. *Biotechnology and Bioengineering* 33: 578–583.

Spier RE & Griffiths B (1984) An examination of the data and concepts germane to the oxygenation of cultured animal cells. *Developments in Biological Standardization* 55: 81–92.

Su WW, Caram HS & Humphrey AE (1992) Optimal design of the tubular microporous membrane aerator for shear-sensitive cell cultures. *Biotechnology Progress* 8: 19–24.

Suleiman SA & Stevens JB (1987) The effect of oxygen tension on rat hepatocytes in short-term culture. *In Vitro Cellular and Developmental Biology* 23: 332–338.

Tramper J, Williams JB & Joustra D (1986) Shear sensitivity of insect cells in suspension. *Enzyme and Microbial Technology* 8: 33–36.

Tramper, J, Smit D, Straatman J & Vlak JM (1988) Bioreactor bubble-column design for growth of fragile insect cells. *Bioprocess Engineering* 3: 37–41.

Van Dissel JT, Olievier KN, Leijh PCJ & Van Furth R (1986) A reaction vessel for the measurement of oxygen consumption by small numbers of cells in suspension. *Journal of Immunological Methods* 92: 271–280.

Van't Riet K (1979) Review of measuring methods and results in nonviscous gas–liquid mass transfer in stirred vessels. *Industrial Engineering Chemistry, Process Design, and Development* 18: 357–363.

Vorlop J & Lehmann J (1989) Oxygen transfer and carrier mixing in large-scale membrane stirrer cell culture reactors. In: Spier RE, Griffiths JB, Stephenne J & Crooy PJ (eds) *Advances in Animal Cell Biology and Technology for Bioprocesses*, pp. 366–369. Butterworth, London.

Wagner R & Lehmann J (1988) The growth and productivity of recombinant animal cells in a bubble-free aeration system. *Trends in Biotechnology* 6: 101–104.

Wohlpart D, Kirwan D & Gainer J (1990) Effects of cell density and glucose and glutamine levels on the respiration rates of hybridoma cells. *Biotechnology and Bioengineering* 36: 630–635.

Yamada K, Furushou S, Sugahara T, Shirahata S & Murakami H (1990) Relationship between oxygen consumption rate and cellular activity of mammalian cells cultured in serum-free medium. *Biotechnology and Bioengineering* 36: 759–762.

4.7 MIXING

Mixing is required in animal cell bioreactors to provide a homogeneous environment throughout the bioreactor and to increase the mass transfer rates of oxygen and other nutrients to the cells. It is well known that intense agitation can be detrimental to mammalian cells in culture. The current understanding of the damage mechanisms for both attached and suspended cells in bioreactors has been reviewed recently (Croughan & Wang, 1991; Papoutsakis, 1991a).

ASSESSING CELL DAMAGE

There are several assays available to quantitate cell death in bioreactors. The simplest assay uses Trypan blue dye exclusion from intact cells to determine live and dead cell concentrations. Cells that have died and lysed, however, cannot be accounted for using this assay (see Chapter 2, section 2.2).

Cell lysis, as determined by the concentration of intracellular proteins present in the culture supernatant, is a valuable indicator of cell damage due to excessive agitation. The cytoplasmic enzyme lactate dehydrogenase (LDH) is frequently used as a measure of cell damage for both suspension and attached cells in a variety of culture systems (see Chapter 2, section 2.5). Lactate dehydrogenase activity as a measure of cell injury should be used with care, however, because several problems may complicate interpretation of the results. The concentration of LDH in the culture supernatant can very likely be correlated with cell damage because intracellular LDH levels are constant during the exponential growth phase. However, intracellular LDH has been seen to decline towards the end of batch culture. Culture conditions may also affect the level of intracellular LDH. For example, the activity of LDH has been shown to increase in oxygen-limited cultures (Geaugey *et al.*, 1990). In one culture system, LDH was found to be relatively stable in the culture supernatant, with the loss in activity not exceeding 5% per day (Wagner *et al.*, 1992). The generality of this finding has not yet been established.

Cell damage under different agitation conditions can be assessed by monitoring the increase in LDH release. To equate the LDH content in culture supernatant to the number of cells that have lysed, a standard curve must be established for each cell type and (where applicable) for different environmental conditions to which the cells were exposed. The procedures for the Trypan blue dye exclusion assay and the LDH assay are described in Chapter 2, sections 2.2 and 2.5.

PARAMETERS USED TO CORRELATE CELL DAMAGE DUE TO AGITATION AND/OR AIR SPARGING

Literature reports have used the following reactor parameters to correlate the effects of 'agitation intensity' with cell injury in bioreactors: agitator rpm, impeller tip speed, integrated shear factor and Kolmogorov eddy size. Additional parameters have been used for microcarrier bioreactors (discussed below). All correlations of cell injury with a bioreactor parameter should be used only qualitatively. These correlations are, at present, indicative of various trends or mechanistic hypotheses and should not be used for quantitative bioreactor scale-up. In addition, such correlations are applicable to the specific cell type, because different cell types are likely to exhibit different responses to fluid forces.

Correlations of cell damage with 'agitator rpm' are specific to a single reactor configuration because different bioreactor designs will result in different mechanical stresses being experienced by the cells. The mechanical stresses likely to be experienced by cells will depend on the impeller design and diameter, the vessel design and diameter and the fraction of the reactor occupied by the liquid phase, in addition to the impeller rpm. Thus, results in terms of impeller rpm are relevant only for reporting trends that can be generalized to different reactor types (i.e. the protective effect of medium additives, damaging effects of certain impellers, etc.).

The 'integrated shear factor' (ISF), which is assumed to be (incorrectly, strictly speaking) a measure of the strength of the shear field between the impeller and the vessel wall, and may be somewhat more useful for scale-up purposes, is defined as:

$$\text{ISF} = \frac{2\pi N D_i}{(D - D_i)} \qquad (4.7.1)$$

where N is the impeller rpm, D_i is the impeller diameter and D is the vessel diameter. Cell growth was found not to correlate well with impeller tip speed ($2\pi N D_i$) or the ISF, and thus the correlation of cell growth with ISF is not suited for incorporation into mechanistic models of hydrodynamic cell damage in agitated bioreactors (Tramper & Vlak, 1988). Information concerning damage in terms of impeller tip speed or ISF would not be transferable to different culture systems.

Correlation of cell damage with 'Kolmogorov eddy size' represents attempts to correlate cell damage with more fundamental fluid dynamic parameters (Papoutsakis, 1991a). The large eddies formed from the movement of the impeller through the medium eventually break down into eddies of progressively smaller sizes, forming a range of eddy sizes that decrease in size down to the smallest turbulent eddies in the dissipation range of the turbulent spectrum. The smallest eddies transfer their kinetic energy as thermal energy to the surrounding fluid through viscous dissipation. The characteristic size of the eddies in the viscous dissipation range of the spectrum of isotropic turbulence, the Komogorov eddies (η, in cm), can be estimated using Kolmogorov's isotropic turbulence theory:

$$\eta = (\nu^3/\varepsilon)^{1/4} \qquad (4.7.2)$$

where ν (in cm^2 s^{-1}) is the kinematic viscosity of the medium and ε (in cm^2 s^{-3}) is the viscous dissipation rate per unit fluid mass, given by:

$$\varepsilon = P/\rho_f V \tag{4.7.3}$$

and:

$$P = N_p \rho_f n^3 D_i^5 \tag{4.7.4}$$

where P is the mixing-power input (in g cm^2 s^{-3}), N_p is the dimensionless power number, ρ_f is the fluid density (in g cm^{-3}), n is the agitation rate (rps), D_i is the impeller diameter (in cm) and V is the liquid volume (in cm^3) in which the energy dissipation takes place. A good estimate for the dissipation volume, V, is D_i^3, i.e. the volume of highest turbulence around the impeller. Thus, Equation 4.7.3 becomes:

$$\varepsilon = N_p n^3 D_i^2 \tag{4.7.5}$$

where N_p is a function of impeller geometry, impeller Reynolds number ($Re_i = nD_i^2/\nu$, a dimensionless number) and vessel characteristics, and will be either provided by the manufacturer of the bioreactor vessel or measured directly, and can be estimated from correlations. Correlations for N_p versus Re_i for various vessels and impeller designs are given by Nagata (1975) or specifically for spinner flasks by Aunins et al. (1989).

CULTURES OF FREELY SUSPENDED CELLS

In most cases, cell damage to suspension cells in agitated bioreactors is primarily the result of air entrainment and bubble break-up (Papoutsakis, 1991a). For example, cell damage to hybridoma cells in a 2-l surface-aerated bioreactor was found to begin at agitation rates that promote air-bubble entrainment and bubble break-up near the bottom of the vortex (Kunas & Papoutsakis, 1990). Cell damage became significant at 180–200 rpm in the 2-l surface-aerated bioreactor when vortex formation and bubble entrainment were not controlled. However, in the absence of vortex formation and gas entrainment the 2-l bioreactors were agitated at rates up to 700 rpm without significant cell damage. Without vortex formation and bubble entrainment, Kunas and Papoutsakis (1990) found cell damage likely to occur at agitation rates where the predicted Kolmogorov eddy size (Equation 4.7.2) approached the size of a single hybridoma cell (10–15 μm). More recent data (Michaels et al., 1992) suggest that the main source of cell injury in agitated and sparged bioreactors is bubble break-up at the free gas/liquid interface in the bioreactor. Other interactions (such as bubble coalescence and break-up in the bulk liquid or turbulent stresses in the bulk liquid) probably do not contribute significantly to cell injury (except for some very fragile cells), and can be ignored for all practical purposes until one reaches agitation intensities where the Kolmogorov eddy size (Equation 4.7.2) becomes approximately equal to the cell size.

Consequently, for the case of surface-aerated bioreactors, the most important condition to avoid is the formation of a vortex that can lead to the entrainment of large, unstable bubbles that will break up at the free gas/liquid interface. When high agitation rates are required, baffles can be installed to decrease vortex formation and increase surface aeration and mixing. If feasible, gas entrainment and bubble break-up at the free gas/liquid interface can be eliminated by completely filling the reactor and operating it without a gas headspace. This will require that either direct sparging of air or oxygen or gas-permeable membranes (such as thin-walled silicon tubing (see section 4.6) be used for oxygenating the reactors. For direct sparging, gas has to be removed from the bioreactor, and this can be accomplished by an overflow vessel (above the bioreactor headplate) connected through one or multiple tubes to the headplate. The tubes and overflow vessel allow for controlled bubble break-up and foam dissipation without damage to the cells (Michaels *et al.*, 1992). Vessels with large height/diameter ratios (of at least 2–3) are desirable for culturing freely suspended cells if the gas headspace in the bioreactor cannot be eliminated completely. This is because these reactors reduce the ratio of the free liquid interface to the reactor volume, thus reducing cell injury due to bubble break-up at the free gas/liquid interface. Vessel geometry is important for effective mixing. The reactor vessel should have a hemispherical or a round bottom to eliminate corners where cells could settle out. Vessels can be modified with a hump located on the bottom ('profiled' bottom), directly underneath the impeller, to eliminate a dead spot beneath the impeller and improve mixing. Typically, pitched-blade, marine-type or hydrofoil impellers with two to four blades are used for agitation in mammalian cell bioreactors, in contrast to bacterial or yeast fermenters, which use Rushton impellers (vertical blades) for vigorous mixing. Rushton impellers may be perfectly suitable for agitation if completely filled bioreactors are used, as discussed above. For vessels with large height/diameter ratios, even high agitation rates may not provide sufficient mixing. In such cases, additional impellers can be attached to the agitation shaft to increase the number of mixing zones inside the vessel, resulting in increased mixing at lower agitation rates. Helical ribbon impellers have been used successfully for agitation of highly shear-sensitive cultures, such as insect cells (Kamen *et al.*, 1991).

Small-volume cultures (or starter cultures) are commonly established in spinner flasks (see Chapter 5, section 5.3). Spinner flasks (e.g. Bellco, Vineland, NJ, USA) are small glass vessels (20–5000 ml) with two ports for aeration and sampling. Agitation is provided by a magnetically driven stir bar suspended by a shaft. The stir bar must be suspended in the culture medium to avoid grinding the cells against the vessel bottom. The magnetic stir plates must be able to agitate the cultures at a constant speed (i.e. the agitation should not increase as the stir plate warms up and agitation should be free of pulsing) in the range of 30–350 rpm.

Shear protectants have been successfully used for the protection of cells at high agitation or mixing intensities in both agitated and bubble column or airlift bioreactors (see section 4.8 and Papoutsakis (1991b)). For reviews and illustrations of impellers and other mixing devices, see Prokop & Rosenburg (1988), Patel (1988) and Griffiths (1988).

ANCHORAGE-DEPENDENT CELLS (MICROCARRIER CULTURES)

Anchorage-dependent cells are grown in suspension culture by growing the cells attached to small microcarrier beads (100–300 μm) suspended in the culture medium by agitation. There are two types of microcarriers available (Chapter 5, sections 5.8 and 5.9; Griffiths, 1990): the non-porous microcarriers whereby cells are growing only on the outer surface of the beads, and porous microcarriers whereby cells are growing predominantly in the internal porous structure of the microcarrier beads.

For porous microcarriers (Chapter 5, section 5.9), the essential mixing issue pertains to what is needed for good suspension of the beads and how much agitation/mixing the beads will endure without being mechanically damaged. The cells in the porous beads are well protected from mechanical forces, and obviously no cells will be able to grow on the external surface of the beads if the agitation intensity is high. For porous microcarriers, it is desirable to use the highest possible agitation or mixing in order to increase the mass transfer of nutrient and metabolites in and out of the porous structure of the beads. The best strategy is to test the beads without cells at various agitation/mixing intensities before using them, and examine the beads microscopically for mechanical damage.

For non-porous microcarriers, in contrast to suspension cultures of anchorage-independent cells, damage to cells attached to microcarriers occurs at agitation intensities below which vortex formation and bubble entrainment and break-up can take place. Thus, the maximum agitation intensities should be much lower for microcarrier cultures than for suspension cultures. As a starting point, the agitation rate that will just keep the microcarriers in suspension should be chosen. When excessive shear damages cells, it will eventually lead to detachment of cells from the microcarrier surface. This is an especially critical problem when cells on microcarriers are used for prolonged recombinant protein expression in a serum-free medium that does not allow good cell growth. In such a case, it can be seen (e.g. Figure 4.7.1) that the rate of cell detachment from confluent microcarriers in 100-ml spinner flasks increases as the agitation intensity increases. These agitation intensities are far below levels where damage was seen for a comparable cell line in suspension culture and below which bubble entrainment begins. Earlier data on cellular injury of cells in microcarrier cultures have been reviewed by Papoutsakis (1991a).

There is no substantial literature on direct sparging of non-porous microcarrier cultures. As is discussed in section 4.6, the difficulty is that the presence of bubbles induces bead flotation, i.e. attachment of beads to bubbles, and the formation of large bead–bubble aggregates that tend to rise and accumulate at the surface of the culture vessel, which is a highly undesirable characteristic. Nevertheless, it is possible slowly to sparge microcarrier cultures without undue cellular injury if suitable surfactants/antifoams (e.g. Pluronic F-68 or Medical Emulsion AF; see section 4.6) are used.

For non-sparged cultures of non-porous microcarriers, cellular injury is likely to be the result of three distinct mechanisms (Papoutsakis, 1991a): interactions

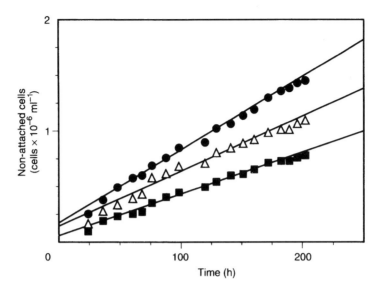

Figure 4.7.1 Cumulative concentrations of non-attached CHO cells in a serum-free medium from confluent microcarriers in spinner flasks at (■) 80, (△) 115 and (●) 150 rpm. Cumulative values are the sum of suspended cells by nuclei counting in the supernatant. Microcarrier concentration was 5 g l^{-1}. Based on data of Borys and Papoutsakis (1992) by permission of *Cytotechnology*.

(collisions) between microcarrier beads, interactions of beads with the turbulent liquid (turbulent eddies) and interactions of beads with bioreactor internals, i.e. the impeller and probes. The first two are most likely to contribute to cellular injury, and the most significant source of cellular injury is likely to be the bead–bead interactions. Although models have been proposed and used to relate these mechanistic events to cellular injury, predictions remain uncertain and somewhat cumbersome for routine calculations (Papoutsakis, 1991a). A rough estimate of conditions that may lead to agitation-induced cell injury is based on the Kolmogorov eddy length concept (Equation 4.7.2). Specifically, because both bead–bead collisions and bead–eddy interactions become significant when the Kolmogorov eddy length (which decreases with increasing mixing intensity) becomes similar to the bead size, the detrimental effects of agitation become quite severe as the Kolmogorov eddy lengths approach the size of the microcarriers (150–200 μm). However, it is not possible to specify precisely the smallest ratio of Kolmogorov eddy length to bead diameter (η/d) below which cellular injury becomes significant. This is because: cellular injury commences at a higher value of this ratio when higher bead concentrations are used; there is substantial error in estimating η (Equation 4.7.2) in various reactors; and there is substantial variability in the amount of shear that various cells can tolerate. An η/d ratio of 2–2.5 (calculated according to Equations 4.7.2–4.7.5) is a good conservative estimate. Dextran is presently the only medium additive that has been shown to protect cells from mechanical damage in microcarrier cultures by increasing the medium viscosity (see 4.8). When the medium viscosity is increased, the eddy length (Equation 4.7.2) increases, and thus a culture can be agitated at higher rates without cellular injury (Papoutsakis, 1991a). The beneficial effect of increased medium viscosity is more pronounced at higher agitation rates (Lakhotia & Papoutsakis, 1992).

Beyond the aforementioned simple criterion based on the Kolmogorov eddy length, several reactor (vessel) design characteristics should be considered with respect to mixing. Selection of the proper impeller is important both for proper mixing and to minimize cell damage. The discharge and trailing vortex velocities are considerably higher from radial flow impellers (vertical blade) than from axial flow impellers (pitched blade) for identical impeller tip speeds (Chang et al., 1981). The higher velocities of the trailing vortexes mean higher maximum local (i.e. in the area around the impeller) shear rates, and also smaller local Kolmogorov eddy lengths. Thus, it is recommended that pitched-blade impellers be used for agitation. Spinner flasks used for microcarrier culture are typically modified with a profiled bottom (raised hump on the vessel bottom) to eliminate a dead spot where microcarriers could accumulate. The use of a profiled bottom is more important in microcarrier culture than in suspension culture because microcarriers settle out much faster than individual cells. It is very important for the stir bar to be positioned at least 1 cm above the hump in spinner flasks to avoid grinding of the microcarriers between the stir bar and the hump. For microcarrier culture the stir drive must be able to deliver constant agitation rates, especially for the low range of agitation rates. When cells are first being attached to the microcarriers, it is critical to have slow, pulse-free agitation.

A common occurrence in microcarrier culture is the formation of large microcarrier aggregates in which the microcarriers are joined by cellular bridges. Microcarrier aggregates made up of as many as 10 or more microcarriers are not uncommon. Microcarrier bridging occurs mainly during the growth phase of the culture, with little additional bridging occurring after cell growth has ceased (Borys & Papoutsakis, 1992). This study also showed that there is an inverse relationship between the rate of microcarrier bridging and agitation intensity. Thus, it may be of interest to operate at higher agitation intensities during the growth phase of the culture to minimize microcarrier aggregation, and to slow down the agitation as cell growth slows in order to minimize cell detachment during the later stages of the culture. In certain cases, such as to promote bead-to-bead transfer of cells to bare microcarriers, low agitation rates would be desirable during the culture growth phase.

REFERENCES

Aunins JG, Woodson BA, Hale TK & Wang DIC (1989) Effects of paddle impeller geometry and mass transfer in small-scale animal cell culture vessels. *Biotechnology and Bioengineering* 34: 1127–1132.

Borys MC & Papoutsakis ET (1992) Formation of bridges and large cellular clumps in CHO-cell microcarrier cultures: effects of agitation, dimethyl sulfoxide and calf serum. *Cytotechnology* 8: 237–248.

Chang TPK, Tatterson GB & Dickey DS (1981) Liquid dispersion mechanisms in agitated tanks: Part II. Straight blade and disc style turbines. *Chemical Engineering Communications* 10: 215–222.

Croughan MS & Wang DIC (1991) Hydrodynamic effects on animal cells in microcarrier bioreactors. In: Ho CS & Wang DIC (eds) *Animal Cell Bioreactors*, pp. 213–249. Butterworth–Heinemann, Stoneham, MA.

Geaugey, V, Pascal F, Engasser JM & Marc A (1990) Influence of the culture oxygenation on the release of LDH by hybridoma cells. *Biotechnology Techniques* 4: 257–262.

Griffiths JB (1990) Advances in animal cell immobilization technology. In: Spier RE & Griffiths JB (eds) *Animal Cell Biotechnology*, vol. 4, pp. 72–121. Academic Press, London.

Kamen AA, Tom RL, Caron AW, Chavarie C, Massie B & Archambault J (1991) Culture of insect cells in a helical ribbon bioreactor, *Biotechnology and Bioengineering* 38: 619–628.

Kunas KT & Papoutsakis ET (1990) Damage mechanisms of suspended animal cells in agitated bioreactors with and without bubble entrainment. *Biotechnology and Bioengineering* 36: 476–483.

Lakhotia S & Papoutsakis ET (1992) Agitation induced cell injury in microcarrier cultures. The protective effect of viscosity is agitation-intensity dependent: experiments and modeling. *Biotechnology and Bioengineering* 39: 95–107.

Michaels JD, Kunas KT & Papoutsakis ET (1992) Fluid-mechanical damage of freely-suspended cells in agitated bioreactors: effects of dextran, derivatized celluloses and polyvinyl alcohol. *Chemical Engineering Communications* 118: 341–360.

Mossman T (1983) Rapid colorimetric assay for cellular growth and survival: application to proliferation and cytotoxicity assays. *Journal of Immunological Methods* 65: 55–63.

Nagata S (1975) *Mixing – Principles and Applications*. Halsted Press, New York.

Papoutsakis ET (1991a) Fluid-mechanical damage of animal cells in bioreactors. *Trends in Biotechnology* 9: 427–437.

Papoutsakis ET (1991b) Media additives for protecting freely suspended animal cells against agitation and aeration damage. *Trends in Biotechnology* 9: 316–324.

Tramper J & Vlak JM (1988) Bioreactor design for growth of shear-sensitive mammalian and insect cells. In: Mizrahi A (ed.) *Advances in Biotechnological Processes*, vol. 7, pp. 199–228. Alan R. Liss, New York.

Wagner A, Marc A, Engasser JM & Einsele A (1992) The use of lactate dehydrogenase (LDH) release kinetics for the evaluation of death and growth of mammalian cells in perfusion reactors. *Biotechnology and Bioengineering* 39: 320–326.

4.8 MECHANICAL PROTECTION

Mixing is required in animal cell bioreactors to provide a homogeneous environment throughout the bioreactor and to increase the mass transfer rates of oxygen and other nutrients to the cells. Intense agitation and/or aeration (air or oxygen sparging) can be detrimental to animal cells in culture. The current understanding of the damage mechanisms in bioreactors, for both freely suspended cells and cells on microcarriers, has been reviewed recently (Croughan & Wang, 1991; Papoutsakis, 1991a). In most studies, cell damage has been assessed on the basis of cell viability and cell death, or reduced cell growth (see section 4.7). Although such assessments may be adequate for cell culture applications (such as in vaccine production or the production of cells for cellular immunotherapies or gene therapies) are likely to require more sophisticated assessments of the effects of fluid-mechanical forces on cultured cells. Several such effects have been investigated in the biomedical engineering literature (Papoutsakis, 1991a) and in the biotechnology literature (Lakhotia *et al.*, 1992). Such effects include alterations in cell metabolism, cell cycle kinetics and DNA synthesis, alterations in protein expression and effects on the concentration of surface proteins.

Cell damage and mechanisms of protection by additives

Freely suspended cells

In most practical situations, damage of freely suspended cells in agitated and/or aerated bioreactors is due to the interactions of cells with bubbles and rearranging gas/liquid interfaces (Papoutsakis, 1991a). Specifically, cellular injury results from the forces released on the cells that have been collected near the bubble surface during the break-up of bubbles at the free liquid surface, or alternatively from forces released on cells entrapped in thin liquid films generated during the collapse of foams at the liquid surface. Several additives have been used since the early 1950s to protect cells from fluid-mechanical damage (Papoutsakis, 1991b). Serum and the pluronic family of non-ionic surfactants are the best documented and most widely studied. Other additives include several derivatized celluloses and starches, cell-derived fractions and proteins. Additives that protect freely suspended cells from fluid-mechanical injury either decrease the fragility of the cells by nutritional or other biological mechanisms, or alter the magnitude and/or frequency of forces released on cells due to interactions with gas/liquid interfaces by physicochemical mechanisms. Michaels *et al.* (1991) have proposed the following terminology regarding the nature of the protection effect or mechanism of an additive:

- A 'biological' protection mechanism or effect implies that the additive changes the cell itself to make it more shear resistant
- A 'physical' protection, mechanism means that the cell resistance to shear remains unchanged, but that the factors that affect the level or frequency of transmitted shear forces to the cell in a given culturing system have changed so that less cell damage is observed
- The 'biological' effect may be the result of alterations that take place either very fast and without requiring metabolic events, or requiring changes that demand metabolic events brought about by the additive. For the latter case, a longer exposure (at least a substantial fraction of the cell cycle time) to the additive would be required. The former will be referred to as a 'fast-acting biological' mechanism, and the latter as a 'metabolic biological' mechanism

Microcarrier cultures

In microcarrier cultures, cell injury is due to the interactions (collisions) between microcarrier beads on which cells are attached, interactions between microcarriers and small turbulent eddies and interactions between microcarriers and bioreactor internals, such as the impeller and various probes (Papoutsakis, 1991a). To protect cells against injury in microcarrier bioreactors, an increased medium viscosity has been the only documented and studied medium alteration (Croughan *et al.*, 1989; Papoutsakis, 1991a; Lakhotia & Papoutsakis, 1992). Dextran (Sigma) has been used to increase the medium viscosity as discussed below. This is a purely physical mechanism of protection.

PRELIMINARY PROCEDURE: ADDITIVE PREPARATION

Prepare a solution of each additive as specified by the manufacturer and/or as discussed below, filter-sterilize the solution and add to the sterile culture medium.

PROCEDURE: TESTING BEFORE USING AN ADDITIVE

Before any additive is to be used with a given cell type, it should first be tested to ensure that it has no detrimental effects on cell growth, metabolism, differentiation, protein expression, etc., and that it does indeed offer mechanical protection in agitated and/or aerated systems (bioreactors).

Parallel static cultures

1. Run cultures in well plates or T-flasks (in duplicate or triplicate) with and without the additive.
2. Use an average, and the highest reported concentration (see below), of the additive for the initial screening. If a new additive is being tested, use a range of concentrations.

3. If a small inhibitory effect is found (which is quite usual), proceed to assess whether or not the additive offers the desirable mechanical protection.
4. If the additive offers mechanical protection, check if the protection effect is present at lower non-inhibitory concentrations. Most cells can overcome a small inhibitory effect by being acclimatized to the additive.
5. To this effect, culture the cells over several (e.g. 10–20) generations in the presence of the additive in static cultures, and check if they can overcome the inhibitory effect. If they do, then propagate the cells (when preparing the culture inoculum) in the presence of the additive.

Check for mechanical protection effect

1. Decide what aeration and/or agitation conditions are desirable for the application, and run experiments to compare cultures with and without the additive under such conditions.
2. Run the experiments in duplicate or triplicate, and repeat the experiments two or three times to ascertain the protective effect.
3. If no protective effect is found, check if protection is provided under less severe agitation and/or aeration conditions.
4. If protection is found, check how good the protective effect is by comparing with parallel static or low-agitation cultures.
5. If time and resources permit, determine the highest agitation and/or aeration conditions under which the additive will provide protection.

ADDITIVES FOR FREELY-SUSPENDED CELLS

Pluronic F-68, other pluronic polyols, polyethylene glycol (PEG), polyvinyl alcohol (PVA) and polyvinylpyrrolidone (PVP)

Pluronic polyols (such as F-68, BASF, Parsippany, NJ) and PEG (Sigma) are the best and most widely studied and used additives, and should be the additives of choice for most, if not all, applications. Polyvinyl alcohol (Sigma) should be included in this list, although it has not been widely studied (Michaels *et al.*, 1991). The non-ionic surfactants Pluronic F-68 and F-88, which are block copolymer glycols of poly(oxyethylene)–poly(oxypropylene)–poly(oxyethylene), have been known to protect cells against fluid-mechanical damage in agitated and aerated bioreactors and have been used for over 30 years. These, and related, surface-active agents are used as medium additives in static, agitated and/or aerated cell cultures (Papoutsakis, 1991b). More detailed studies on the effects of various other pluronics and reverse pluronics (the order of the copolymers is reversed with poly(oxyethylene) in the middle) on cell growth and/or shear protection have been published by Mizrahi (1975) and Murhammer and Goochee (1990). Studies have also shown that pluronics have a concentration-dependent positive or negative effect on cell growth in 'static cultures'. Although it is possible that the effects observed could be due to small unknown impurities, these results suggest that these polyols may affect the growth of some cells independently of agitation or aeration (Papoutsakis, 1991b).

Two types of polyvinylpyrrolidones (PVP–Bayer, Leverkusen, Germany) have been found beneficial to cell growth in spinner cultures, and a mixed molecular weight PVP has been found to protect hybridoma cells against shear injury in a bubble column reactor (Handa, 1986).

Polyethylene glycol (which is related in chemical structure to pluronic polyols) and PVA have been tested more recently as protectants against aeration and agitation cell damage (Michaels *et al.*, 1991). Both appear to offer mechanical protection as good as F-68, without any detrimental effects on the cells they have been tested on (Chinese hamster ovary (CHO) and hybridoma cells CRL 8018).

The mechanisms through which pluronic polyols, PEG and PVA protect cells from agitation and/or aeration damage have not been firmly established and have been the subject of considerable debate and speculation (Papoutsakis, 1991b). Because F-68 and PEG (and also PVA) can protect cells from fluid-mechanical damage in bioreactors even after a short exposure, it is likely that their effect is either physical or fast-acting biological in nature. Some viscometric studies have shown that, unlike serum, PEG and F-68 do not protect cells against shear damage in viscometric flows (well-defined, laminar Couette flows) after prolonged (cells grown in the presence of the additive), short (30 min) or intermediate (1–4 h) exposure.

Viscometric studies are used to assess the shear robustness or fragility of cells by exposing them (usually for a short, 10–20 min, time period) to the reproducible and well-defined laminar flow of a viscometric device. In contrast, the (usually turbulent) flow in a bioreactor is not well defined and not easily reproducible, unless the same reactor is used.

These findings suggest that the shear-protective effects of these additives in the bioreactor are physical in nature, and specifically purely fluid-mechanical, i.e. due to changes in the interactions between bubbles, draining films and the cells. If their effect was biological, cells would have been protected in both shear environments (viscometer and bioreactor). Other experiments suggest that the protection mechanism may vary for different cell types.

For most applications, 0.1% of F-68, F-88 and various PEGs and PVAs should be sufficient for mechanical protection. Higher concentrations of pluronics (up to 0.2%) may be necessary in some cases, but this has not been well documented. Among the large variety of PEGs available, those of molecular weights between 1400 and 15 000 have been found successful for mechanical protection. Polyvinyl alcohol of only one molecular weight (10 000) has been examined, and has been found to have excellent protective properties.

Serum

Serum is a good but undefined shear protectant. It promotes better cell growth in agitated and/or aerated cultures in a dose-dependent fashion up to 10% (v/v) (Papoutsakis, 1991b). Several studies employing bioreactors and well-defined Couette flows in a viscometer have shown that the protective effect of serum is due to a physical and/or a biological mechanism (Papoutsakis, 1991b). It seems that the most significant contribution comes from the physical mechanism, so any type of serum (i.e. independently of its nutritional and hormonal value) that does not inhibit cell growth can be used. For most applications, 2–5% serum will be sufficient to provide protection. Because serum complicates protein purification, may elicit several biological responses or may become a source of undesirable variability or viral contamination, its use as a protective agent against mechanical injury should be avoided. Experimental evidence suggests that the synthetic, well-defined additives (e.g. pluronics, PEG, PVA) provide as good (or almost as good) mechanical protection as serum does.

Derivatized celluloses

Derivatized celluloses are promising but not yet reliable shear protectants. They have been used since the very early days of the suspension animal cell culture technology in the 1950s (Papoutsakis, 1991b). Bryant (1969) has reviewed the early uses of methylcelluloses (MCs), presumably as protectants against shear damage for the cultivation of a large variety of suspension cells, including human, mouse and monkey cells. Also, the most detailed study on the effects of various MC grades on various cells of variable shear fragility in both shake-flask and static cultures has been carried out (Bryant, 1969). Despite this work and the later work of several other investigators, MCs and other derivatized celluloses (such as carboxymethyl celluloses, CMCs) have not been established unequivocally as reliable shear protectants (Papoutsakis, 1991b).

The effect of MCs and CMCs as additives to protect animal cells from shear damage in agitated, bubble column or otherwise mixed cultures may depend on cell type, MC grade and MC make. In addition, MCs and CMCs may elicit biological responses from cells, either positive or negative; these responses appear also to be dependent on cell type and additive grade and make. It is not known whether the stimulatory or inhibitory effects are due to MCs and CMCs themselves or to small impurities. The effect of some MCs in causing cell aggregation for some cells has also been reported. There is little question, however, that even after almost 40 years of research very little is understood about the mechanism and generality of the protective effect of MCs.

Bryant (1969) found that low concentrations of low-viscosity MC were as effective as a protectant as the higher viscosity grades, thus concluding that the MC on insect cells derives from the higher medium viscosity. Higher concentrations or grades of MCs produce very viscous solutions, and some of the early work used MCs that produced media of very high viscosities.

Goldblum et al. (1990) have reported the effect of various MCs and (hydroxypropyl)methylcelluloses (methocels, Dow Chemical Company) and dextran in protecting insect cells (Sf9 and TN-368) in viscometric flows. All media contained 10% foetal bovine serum (FBS) in addition to the additive. They found that the best protection (58–76-fold less cell lysis compared with the unsupplemented medium) was offered by higher concentrations and higher molecular weight methocels. The best protection was offered by the additives resulting in the highest medium viscosity (4–25 times the unsupplemented medium viscosity). Similarly, dextran (MW = 476 000) offered substantial shear protection only at high concentrations (4.5%), where the medium viscosity is also high (6.6-fold higher than the unsupplemented medium). The significance of these findings in terms of protection in bioreactors is not known.

In the last 25 years, the use of MCs (in combination with other shear-protecting additives, such as serum, yeastolate or protein hydrolysates) appears to have been restricted to the cultivation of insect cells, probably because of the more widespread use of Pluronic F-68, which does not cause any of the cell-aggregation problems often associated with the use of MCs and CMCs.

Michaels et al. (1992) have recently reported the beneficial effect of methocel (MW 15 000, 50 000 and 100 000) on CHO cells in agitated and aerated bioreactors. This is the only recent and complete (with control experiments) reported study of derivatized celluloses as shear protectants in bioreactors. These findings appear very promising, but additional work will be needed before the widespread use of MCs as shear protectants.

Dextrans and modified starches

Dextrans and modified starches are not suitable for shear protection. Mizrahi (1984) has tested a large number of synthetic polymers – including hydroxyethyl starch (HES), CMC (Edifas B50), modified gelatin (Haemaccel) and several dextrans – for improved growth of human lymphocyte and lymphoblastoid cell lines in agitated cultures with low serum concentrations. Experiments were carried out in 100- or 500-ml spinner flasks at 100 rpm. It was found that the dextrans had a positive effect on the growth of two cell lines but no effect on a third line. Both HES and CMC had a positive effect on the growth of all cells tested, but the modified gelatin had a negative effect on growth. There were no static control cultures presented to test that their agitation intensity is indeed damaging to cells.

Radiolabelled polymers were used to show that CMC and HES are not metabolized by the cells. Handa (1986) found that dextran of MW 488 000 offered no protection against sparging damage in a bubble column reactor. Extensive laboratory studies have established that 1–3% (w/v) dextrans (MW = 229 000) do not protect hybridoma cells from damage due to bubble entrainment and break-up in an agitated bioreactor, but instead increase cell death under intense agitation (Michaels et al., 1992). Control static or low-agitation cultures showed that dextran does not affect the cells, at least in terms of growth rates, so its detrimental effect is due to the viscosity increase. These findings are in contrast to the viscometric studies by Goldblum et al. (1990).

In summary, the effect of dextrans can be positive or negative depending on the cell type, bioreactor or even dextran grade and make used, but all experimental evidence so far suggests that dextran offers no mechanical protection. Dextrans and modified starches *should not* be used as protectants against shear damage.

Proteins and cell extracts

For proteins and cell extracts there have been mixed results and insufficient testing. Several protein mixtures (in addition to serum) and a protein have been used as shear protectants for the cultivation of various cells. Again, their protective effect is not established unambiguously due to lack of proper control experiments. They have been reviewed in detail (Papoutsakis, 1991b) and include: peptone, lactalbumin hydrolysate, tryptose phosphate, yeast extract, yeastolate and Primatone RL (a peptic digest of animal tissue). Bovine serum albumin (BSA), a major component of bovine serum, is a widely used additive in serum-free media and has been used frequently as a shear protectant by several investigators; it was recently shown to be an effective additive against fluid-mechanical damage of hybridoma cells in an airlift bioreactor, although it has no effect on the cells in static or spinner cultures. Among all these additives, BSA is the only one recommended for use as a shear protectant (if it is also included as a medium additive for other reasons). Otherwise, even BSA should be avoided in favour of the well-defined synthetic additives, such as F-68, PEG, PVA and PVP.

ADDITIVES FOR MICROCARRIER CULTURES

As already discussed, theory and experimental evidence suggest that an increased medium viscosity will reduce damage of cells on microcarriers because of the reduced intensity and frequency of interactions that cause cell damage in microcarrier cultures.

Among the possible additives that can increase medium viscosity with no apparent detrimental effects on the cells, dextran (up to 0.3%, w/v) has been found beneficial to these cultures. A limited number of experiments have shown that MCs and CMCs are not suitable protective agents because they cause microcarrier aggregation and flotation (i.e. collection on the liquid surface in the form of aggregates).

Use dextran of high molecular weight (between 200 000 and 300 000) at 0.15% or 0.2% (w/v) or, if necessary, lower molecular weight dextrans at 0.3% (w/v).

REFERENCES

Bryant JC (1969) Methylcellulose effect on cell proliferation and glucose utilization in chemically defined medium in large stationary cultures. *Biotechnology and Bioengineering* 11: 155–179.

Croughan M, Sayre ES & Wang DIC (1989) Viscous reduction of turbulent damage in animal cell culture. *Biotechnology and Bioengineering* 33: 862–872.

Croughan MS & Wang DIC (1991) Hydrodynamic effects on animal cells in microcarrier bioreactors. In: Ho CS & Wang DEC (eds) *Animal Cell Bioreactors*, pp. 213–249. Butterworth–Heinemann, Stoneham, MA, USA.

Goldblum S, Bae YK, Hink WF & Chalmers J (1990) Protective effect of methylcellulose and other polymers on insect cells subjected to laminar shear stress. *Biotechnology Progress* 6: 383–390.

Handa A (1986) Gas/liquid interfacial effects on the growth of hybridomas and other suspended mammalian cells. *PhD Thesis*, University of Birmingham, UK.

Lakhotia S & Papoutsakis ET (1992) Agitation induced cell injury in microcarrier cultures. Protective effect of viscosity is agitation intensity dependent: experiments and modeling. *Biotechnology and Bioengineering* 39: 95–107.

Lakhotia S, Bauer KD & Papoutsakis ET (1992) Damaging agitation intensities increase DNA synthesis rate and alter cell-cycle phase distributions of CHO cells. *Biotechnology and Bioengineering* 40: 978–990.

Michaels JD, Petersen JF, McIntire LV & Papoutsakis ET (1991) Protection mechanisms of freely suspended cells (CRL 8018) from fluid-mechanical injury. Viscometric and bioreactor studies using serum. Pluronic F68 and polyethylene glycol. *Biotechnology and Bioengineering* 38: 169–180.

Michaels JD, Kunas KT & Papoutsakis ET (1992) Fluid-mechanical damage of freely-suspended animal cells in agitated bioreactors: effects of dextran, derivatized celluloses and polyvinyl alcohol. *Chemical Engineering Communications* 118: 341–360.

Mizrahi A (1975) Pluronic polyols in human lymphocyte cell line cultures. *Journal of Clinical Microbiology* 2: 11–13.

Mizrahi A (1984) Oxygen in human lymphoblastoid cell line cultures and effect of polymers in agitated and aerated cultures. *Developments in Biological Standardization* 55: 93–102.

Murhammer DW & Goochee CF (1990) Structural features of nonionic polyglycol polymer molecules responsible for the protective effect in sparged animal cell bioreactors. *Biotechnology Progress* 6: 142–148.

Papoutsakis ET (1991a) Fluid-mechanical damage of animal cells in bioreactors. *Trends in Biotechnology* 9: 427–437.

Papoutsakis ET (1991b) Media additives for protecting freely suspended animal cells against agitation and aeration damage. *Trends in Biotechnology* 9: 316–324.

CHAPTER 5

CULTURE PROCESSES AND SCALE-UP

5.1 OVERVIEW

In this chapter the subject of scale-up is reviewed, which is taking small laboratory cultures (e.g. 10 ml) to industrial-scale processes (e.g. 10 000 litre), i.e. a 1 000 000-fold scale-up! The aim of such scale-up is to provide more cells, and more cell product, in as efficient and cost-effective a manner as possible. Cell cultures have been used since 1954 for the production of human (e.g. polio, measles, mumps, rabies, rubella) and then veterinary (e.g. FMDV) vaccines (Griffiths, 1990a). Interferon was the next most important product to be developed, followed by monoclonal antibodies and a range of recombinant proteins.

The importance of cell culture as a manufacturing process is exemplified by the range of products either licensed or in trial (in addition to the classical range of vaccines and veterinary products), which are listed in Table 5.1.1.

This production activity has been made possible by the ability to grow cells in unit processes of 10^{11}–10^{13} cells. The scale-up process involves many factors, which include engineering (especially to maintain sterility in large-scale plants), physiological (supply of oxygen and nutrients and maintenance of correct environmental conditions), monitoring and control (biosensors, performance measurements, computer control) and logistic and regulatory requirements (compliance with pharmaceutical licensing and health and safety aspects when growing pathogens).

Table 5.1.1 Mammalian recombinant cell products

Monoclonal antibodies (MAbs)	OKT3/Orthoclone (1987), Centoxin (1990), Reopro (1994), Myoscint (1989), Oncoscint (1990)
tPA	Activase/Actilyse (1987)
EPO	Epogen/Procrit/Eprex (1989), Epogin/Recormon (1990)
hGH	Saizen (1989)
HBsAg	GenHevac B Pasteur (1989), HB Gamma (1990)
Interferon	Roferon (1991)
G-CSF	Granocyte (1991), Neupogen (1991)
Blood factor VIII	Recombinate (1992), Kogenate (1993)
DNase I	Pulmozyme (1993)
Glucocerebrosidase	Cerezyme (1994)
FSH	Gonal-F (1995)

rDNA Products in Development/Clinical Trial
HIV vaccines (gp120, gp160, CD4)
Herpes simplex vaccines (gB, gD)
Chimeric MAbs (her2, CD4, TNFα, CD20, CD18, TAC, leukointegrin, CF54, RSV)
Others (TSH, TNF, M-CSF, IL-6, IL-1)
In vitro diagnostic MAbs (1–200)
Tissue engineering and replacement (e.g. skin, artificial kidneys)

Cell and Tissue Culture: Laboratory Procedures in Biotechnology, edited by A. Doyle and J.B. Griffiths.
© 1998 John Wiley & Sons Ltd.

SCALE-UP FACTORS

The primary aim has been to move from multiple processes (many replicate bottles, roller cultures, spinner flasks, etc.) to a high-volume unit process. This has economical advantages in terms of manpower needed (reduction in repetitive steps for each culture) and space (particularly expensive hot-room facilities), and efficiency advantages in that it is feasible to control, reproducibly, environmental factors such as pH and O_2 in a unit process. One cannot help being aware of the huge range of alternative culture systems that are available (see Chapter 3, Figure 3.6.1; Glacken *et al.*, 1983; Griffiths, 1988; Prokop & Rosenburg, 1989). These are a result of solutions to overcoming limitations to scale-up and the need to find solutions for both suspension and anchorage-dependent cells.

The first problem on moving from a small multiple-process unit to a large unit process is usually oxygen limitation. To overcome this and supply oxygen without the damaging effects of air bubbles from sparging and the high stirring speeds needed to effect efficient mass transfer, bioreactors have been developed that oxygenate through a membrane inside the bioreactor or by external medium loops through hollow-fibre oxygenators (see Chapter 4, section 4.6). A related factor is mixing, and low-shear modifications to impellers or the airlift reactor concept have been introduced (see Chapter 4, section 4.7). A problem for anchorage-dependent cells is the limitation on surface area during scale-up. A huge range of reactors have been developed (multiple plates, spirals, ceramic matrices, glass beads; see Chapter 3, Figure 3.6.1) but the most successful has been the microcarrier (Griffiths, 1991a). By growing cells on a small bead, which can be cultured analogously to a suspension cell, the large fermenter systems used for suspension cells have been made available for anchorage-dependent cells. The next problem to be overcome is nutrient limitation and toxic metabolite build-up. Although medium changes can be used (especially for attached cells), or detoxification procedures (see Chapter 4, section 4.5), the most efficient means of overcoming these problems is continuous medium perfusion. To perfuse suspension cells, physical separation of cells and waste medium is necessary: this has been done by means of spin-filters; external loops to filters or centrifuges; or gravitational settling devices (Griffiths, 1990a). The use of perfusion resulted in the development of many immobilization procedures to prevent cell wash-out (e.g. encapsulation, hollow-fibre devices, membrane reactors, gels, fibres and porous microcarriers). Additional benefits to these developments were the provision of huge unit surface areas for cell attachment and the ability to grow cells at 50–100-fold higher densities than previously achieved in free suspension. All these developments mean that manufacturers have a choice in scale-up routes (Griffiths, 1988), as summarized in Figure 5.1.1.

SCALE-UP STRATEGIES (Griffiths, 1991b)

1. High-volume (to 10 000 l) tank fermenter systems with conventional cell densities of 1–2×10^6 ml^{-1}.

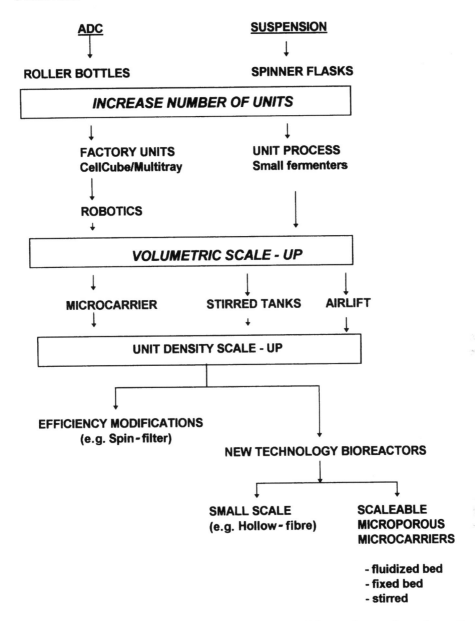

Figure 5.1.1 Scale-up options – the decision process. ADC = anchorage-dependent cells.

2. Low-volume (under 1 l) but very high density ($1-2 \times 10^8$ ml^{-1}) heterogeneous systems (e.g. hollow-fibre bioreactors).
3. Intermediate systems using a spin-filter to increase cell density to $1-2 \times 10^7$ ml^{-1} but to a volumetric scale of 500 l only.
4. High-density (1×10^8 ml^{-1}) scaleable systems (to 200 l) based on porous microcarriers.

The high-density immobilized systems have many advantages in terms of smaller process equipment, higher product concentration and reduced nutritional requirements, and perfusion allows long culture runs (thus reducing the down-time of the plant) and shorter residence times for products. They also have the advantage that they are equally suitable for suspension and anchorage-dependent cells (the Universal System?), they provide protection from shear and they have a three-dimensional structure that enhances productivity (Griffiths, 1990b, 1992a; Looby & Griffiths, 1990). Basically, the low-volume, high-density systems fill an important niche in manufacturing capability (e.g. gramme amounts of monoclonal antibodies), whilst the high-volume systems are the most widely used scale-up route for pharmaceutical products. However, there is increasing acceptance of, and need for, the scaleable high-density systems and these will be the first-choice system in future years.

GENERAL PRINCIPLES

The use of small-scale suspension cultures is described in section 5.4. At some stage a decision has to be made, if scale-up is required, of when to move from laboratory units to small-scale industrial systems (Griffiths, 1992b). This is a decisive moment because of the investment needed to set up *in situ* fermentation systems. This step should be taken at 10 l and involves:

1. Change from glass to stainless-steel vessels.
2. Change from a mobile to a static (plumbed-in) system with connection to steam for *in situ* sterilization, manipulation being carried out in the open laboratory (not a laminar flow cabinet), more sophisticated temperature control and additional vessels for medium holding and culture harvesting.
3. More sophisticated and sensitive environmental control systems.

A typical cell culture bioreactor available from all fermenter suppliers is conceptually similar to the familiar bacterial fermenter apart from modifications such as a marine (not turbine) impeller, curved or convex base for better mixing at low speeds and water-jacket – not immersion heater – temperature control (to avoid localized heating at low stirrer speeds).

MONOLAYER AND SUSPENSION CULTURE

Suspension culture is the preferred method for scale-up because it is easier and cheaper, it requires less space, cell growth can be monitored and environmental parameters can be controlled more easily. However, many cell lines have higher specific productivity when attached to the substrate. In the urge to move to suspension systems, the following advantages of anchorage-dependent systems should not be overlooked:

- They facilitate complete medium changes, which are often needed, for example, to wash out serum before adding serum-free production medium

OVERVIEW

- They are easier to perfuse
- Medium/cell ratios are changed more easily during an experiment

Microcarrier culture has most of the advantages of both suspension and anchorage-dependent systems and this is a particularly valuable asset for porous microcarrier culture (Looby & Griffiths, 1990).

CULTURE MODES (see also section 5.4)

- *Batch culture:* cells grow from seed to a final density over 4–7 days and are then harvested (typical monoclonal antibody yield is 6 mg l^{-1} per batch run) (see section 5.5)
- *Fed-batch culture:* additional media, or medium components, are added to increase the culture volume and density of cells
- *Semi-continuous batch culture:* this is in effect a batch culture that is partially harvested (e.g. 70%), topped up with fresh medium, allowed to grow up and then harvested again. Typically, three or four harvests can be made, although usually with diminishing returns at each harvest
- *Continuous-flow (chemostat) culture:* the culture is completely homogeneous for long periods of time whilst cells are in a steady state (see section 5.6). An extremely useful tool for physiological studies but not very economical for production (by definition, cells are never at maximum density)
- *Continuous perfusion culture:* differs from chemostat in that all nutrients are kept in excess and cells are retained within the bioreactor. The most productive culture system gives typical yields of 10 mg l^{-1} day^{-1} for 50–100 days (see section 5.9)

There is no definitive answer as to the best system to use: it depends upon the nature of the cell and the product, the quantity of product, downstream processing capability, licensing regulations, etc. However, a rough guide to relative costs for producing monoclonal antibodies by perfusion, continuous-flow and batch culture is the ratio 1:2:3:5. A summary of the capacity of the various systems is given in Table 5.1.2.

BIOLOGICAL FACTORS

Key factors for the successful scale-up of cell culture include using the best available cell line, inoculating cells that are in prime condition and ensuring that nutritional, physiological and physicochemical conditions are optimized.

Cells vary in their cultural characteristics. This is very noticeable when comparing clones from the same hybridization or transfection experiment. Differences in expression of the desired product are well recognized, hence the standard screening produced for high producers. However, there is also clonal variability in serum requirements, plating efficiency, growth rate, metabolism (e.g. glycolytic rate) and ability to grow in (or adapt to) suspension culture. It is thus

Table 5.1.2 Comparison of culture systems indicating volumetric size, unit and system cell yields

Scale (l)	System	Area (cm^2) (batch size)	Cells per batch (cells l^{-1})	
	Roller	1750 (× 100)	2 × 10^{10}	(2 × 10^8)
	Multitray	24 000 (× 10)	3 × 10^{10}	(3 × 10^9)
	CellCube	85 000 (× 4)	4 × 10^{10}	(9 × 10^9)
1	Hollow fibre	20 000	10^{10}	(10^{11})
20	Microporous microcarrier	2.5 × 10^6	10^{12}	(5 × 10^{10})
100	Glass sphere	10^6	10^{11}	(10^9)
200	Stacked plate	2 × 10^5	2 × 10^{10}	(1 × 10^8)
500	Microcarrier (with spin-filter)	2 × 10^{10}	10^{13}	(5 × 10^{10})
2000	Airlift		4 × 10^{12}	(2 × 10^9)
4000	Microcarrier	5 × 10^6	8 × 10^{12}	(2 × 10^9)
10 000	Stirred tank		2 × 10^{13}	(2 × 10^9)

beneficial to select a clone that has the desired characteristics for the scaled-up process that is to be used.

The quicker that cells can be harvested and re-inoculated into a new culture, the healthier they tend to be and the higher the chance of establishing a successful culture. Prolonged exposure to proteolytic enzymes (e.g. trypsin), standing in concentrated suspensions for long periods before inoculation and over-robust mixing (or pouring) whilst in the fragile post-trypsinized state must be avoided. Selecting cells in the late logarithmic phase and rapid processing into the new culture at an adequate inoculum level should allow an optimum initiation of the new culture with a short lag phase and maximum cell density. Always inoculate cells into stabilized culture conditions, because shifts in temperature, pH and oxygen levels during initiation can be damaging.

As one scales up, culture conditions become more demanding on the cells and the logistics of preparing large-cell inocula and maintaining sterility become more difficult. Therefore, do not be over-ambitious in the scale-up steps.

SUMMARY

In this overview an explanation has been given for the wide and complex range of cell culture bioreactor systems that are available. This is to put some perspective on the ones that are described in more detail in the following chapters. The selection of a suitable system follows the adage 'horses for courses', i.e. a process is selected that fulfils the particular criteria for the cell, product, scale and quantity required, plus the resources in facilities, manpower and experience that are available.

There are some additional general factors to be taken into consideration. First, upstream processing is only part of the total process, which starts at the initiation of a cell line and finishes with a purified and packaged product, and therefore should not be treated in isolation. Secondly, increasing unit productivity does not only depend upon designing a bigger, better and more efficient bioreactor. Significant increases in productivity have been, and will be, achieved by designing better media, genetically constructing more efficient and robust cell lines and more critical monitoring (by biosensors) and control (by computer) of the process.

REFERENCES

Glacken MW, Fleischaker RJ & Sinsky AJ (1983) Mammalian cell culture: engineering principles and scale-up. *Trends in Biotechnology* 1: 102–108.

Griffiths JB (1988) Overview of cell culture systems and their scale-up. In: Spier RE & Griffiths JB (eds) *Animal Cell Biotechnology*, vol. 3, pp. 179–220. Academic Press, London.

Griffiths JB (1990a) Animal cells – the breakthrough to a dominant technology. *Cytotechnology* 3: 109–116.

Griffiths JB (1990b) Advances in animal cell immobilization technology. In: Spier RE & Griffiths JB (eds) *Animal Cell Biotechnology*, vol. 4, pp. 149–166. Academic Press, London.

Griffiths JB (1991a) Cultural revolutions. *Chemistry and Industry* 18: 682–684.

Griffiths JB (1991b) Closing the culture gap. *Bio/Technology* 10: 30–32.

Griffiths JB (1992a) Animal cell culture processes – batch or continuous? *Journal of Biotechnology* 22: 21–30.

Griffiths JB (1992b) Scaling-up of animal cell cultures. In: Freshney RI (ed.) *Animal Cell Culture: a Practical Approach*, pp. 47–93, Oxford University Press, Oxford.

Looby D & Griffiths JB (1990) Immobilization of animal cells in porous carrier culture. *Trends in Biotechnology* 8: 204–209.

Prokop A & Rosenburg MZ (1989) Bioreactor for mammalian cell culture. *Advances in Biochemical Engineering/Biotechnology* 39: 29–71.

5.2 ROLLER BOTTLE CULTURE

Roller bottle culture is considered the first scale-up step for anchorage-dependent cells from stationary flasks or bottles. This is achieved by using all the internal surface for cell growth, rather than just the bottom of a bottle. The added advantages are that: a smaller volume of medium and thus a higher product titre can be achieved; the cells are more efficiently oxygenated due to alternative exposure to medium and the gas phase; and dynamic systems usually generate higher unit cell densities than stationary systems.

The basis of the roller culture system is to place multiple cylindrical bottles into an apparatus that will rotate the bottles evenly at set rotational speeds (between 5 and 60 rph). Apparatus is available that will accommodate four to hundreds of bottles, and in fact this method is used for the large-scale production of vaccines using 29 000 bottles (1-l volume) per batch. Any cylindrical bottle of the correct tissue culture grade surface (borosilicate glass or polystyrene) can be used and there are many available commercially between 250 and 2000 cm^2 in capacity. It is an advantage when using large numbers to have as small a diameter as possible in order to reduce the total volume of the culture vessels. However, sufficient headspace must be maintained to prevent oxygen limitation, and thus the medium/air volume is usually between 1:5 and 1:10.

PROCEDURE: ROLLER BOTTLE CULTURE OF ANIMAL CELLS

Materials and equipment

- Standard tissue culture media and reagents
- Roller bottles (e.g. 23 × 12-cm plastic bottle with surface area of 1400 cm^2 from, for example, Becton Dickinson, Sterilin)
- Roller bottle apparatus with speed control between 5 and 60 rph (e.g. Wheaton, Bellco, New Brunswick)

1. Add approximately 1 ml of complete culture medium at 37°C per 5-cm^2 culture area.
2. Add 5% CO_2 in air to headspace.
3. Add $1-2 \times 10^5$ cells cm^{-2}.
4. Seal bottle and place in roller culture apparatus.
5. Rotate bottle at 12–24 rph for the initial attachment phase (2–8 h, depending upon cell type). This faster speed is to get an even distribution of cells but should be reduced for cells with low attachment efficiency.
6. Reduce revolution rate to 5–10 per hour when culture is growing.

7. Cell growth can be monitored initially under an inverted microscope and later in the culture period by visual inspection.
8. A medium change can be carried out after 4–5 days if the pH becomes acid and/or maximum cell densities are required. This can also be carried out to change to a production medium, or when infecting cells, and a lower medium volume can be added to get a higher product concentration.
9. Harvest cells when confluent (5–6 days) by removing the medium and trypsinizing in the conventional way. After adding trypsin, roll the bottles at speeds of 20–60 rph until cells detach. Cell yields will be similar to or up to two-fold higher than in stationary cultures, and multilayering (non-diploid cell lines) will occur.

COMMENT

Successful roller bottle culture depends a great deal on solving various logistic problems. Larger roller bottles (1500 cm^2) are unwieldy to use and cannot be manipulated in many tissue culture cabinets. Special Class II cabinets are suitable; otherwise use the smaller size bottles. Also, do not underestimate the time involved in harvesting cells from roller bottles. If a large number of bottles are being used, trypsinize four at a time and get the cells in fresh medium with serum before beginning the next batch. Cells deteriorate very rapidly in trypsin, and also when being held at high concentrations in medium, so do not store whilst harvesting multiple batches of roller bottles, but put them in a new culture as soon as possible. Unless your laboratory is geared up for this type of work (multiple tissue culture cabinets and staff), limit the number of roller bottles to 8–16 at a time. It is very difficult to remove completely all the cells from a roller bottle without repeated washing (increases contamination risk, standing time of cells and centrifugation volume) and it is wise to assume that a 10–20% loss may occur. This is particularly important when judging how many bottles to set up as an inoculum for a larger culture.

SUPPLEMENTARY PROCEDURES

The roller bottle system is a multiple process requiring considerable staff time for repeated manipulations. Thus many modifications have been introduced that increase the efficiency or surface area capacity of roller bottles. Some examples are as follows:

1. Perfusion. Specially modified swivel caps have been developed that allow continuous perfusion of roller bottles (e.g. Bellco Autoharvester, New Brunswick apparatus).
2. Spira-Cell Multi-Surface Roller Bottle (Bibby Sterilin Ltd). This bottle contains a spiral polystyrene cartridge on which cells grow on both sides of the film; 3000-cm^2, 4500-cm^2 and 6000-cm^2 sizes are available.

3. Extended Surface Area Roller Bottle (ESRB) (Bibby Sterilin Ltd). This polystyrene bottle has a ribbed or corrugated surface, doubling the surface area of the 850-cm^2 flask to 1700 cm^2.
4. Bellco-Corbeil Culture System (Bellco). A roller bottle is packed with a cluster of small glass tubes (arranged in parallel and separated by silicone spacer rings). Models are available in 1000-cm^2, 10 000-cm^2 and 15 000-cm^2 sizes. Medium is perfused through the vessel and the bottles rotate 360° in alternate directions to avoid tube twisting and the need for special caps.
5. If hundreds of roller bottles need to be handled, then a robotic culture system (Cellmate, The Automation Partnership, Royston, UK) could be considered, which reduces manning by 10-fold as well as giving improved consistency and reproducibility.

DISCUSSION

The roller bottle technique is a well-established and successful culture method widely used for the production of cells and products. One reason for this is that a single contamination event does not mean that the whole batch is lost, as with a single unit process. However, roller culture needs a considerable financial investment both in the apparatus itself and in incubator/hot-room facilities. Also, some cell lines (particularly epithelial) may not be as successfully grown in roller bottles as in stationary bottles. Common problems are streaking, clumping or inadequate spreading over the total surface (e.g. non-locomotory cell lines). An alternative scale-up route is to use multisurface stationary systems such as the Nunclon Cell Factory (Nunc) (see section 5.7) or the CellCube (Costar).

BACKGROUND READING

Griffiths JB (1992) Scaling-up of animal cell cultures. In: Freshney RI (ed.) *Animal Cell Culture: a Practical Approach*, pp. 47–93. IRL Press, Oxford.

Kruse PF, Keen LN & Whittle WL (1970). Some distinctive characteristics of high density perfusion cultures of diverse cell types. *In Vitro* 6: 75–81.

Panina GF (1985) Monolayer growth systems: multiple processes. In: Spier RE & Griffiths JB (eds) *Animal Cell Biotechnology*, vol. 1, pp. 211–242. Academic Press, Orlando.

Polatnick J & Bacharach HL (1964) Production and purification of milligram amounts of foot-and-mouth disease virus from baby hamster kidney cell culture. *Applied Microbiology* 12: 368–373.

5.3 SPINNER FLASK CULTURE

The first scale-up step for cells growing in suspension either naturally or after adaptation, or anchorage-dependent cells on microcarriers, is the spinner flask. This technique originated by placing a silicone or Teflon-coated magnet with a central ring into a glass vessel that is placed on a magnetic stirrer. Although this simplistic approach is still used, it is preferable to use a specially constructed flask where the magnet is suspended (essential for microcarrier culture, because otherwise the cells and microcarriers will be destroyed by the grinding action of the magnet) just above the bottom of the flask. A typical spinner flask is shown in Figure 5.3.1, and has a stirrer shaft containing the magnet fitted to the bottle top. In addition, side arms are fitted in order to add cells, change medium or gas with oxygen or CO_2-enriched air (in which cases filters should be fitted). It is preferable to have a slightly convex surface under the stirrer bar to aid mixing. There are many variations on this theme from different suppliers, mostly with regard to size, number and positioning of the stirrer bar.

The spinner flask is a glass vessel, usually intended to be used for replicate cultures, on a multibased magnetic stirrer. Sizes range from a few millilitres to 20 l, but for ease of handling, safety and physicochemical reasons it is advisable to consider 10 l as a maximum practical size. The units are sterilized by autoclaving. A very important factor is the quality of the magnetic stirrer. This should be able to give an extremely stable speed of rotation between 10 and 300 rpm, have an accurate tachometer, restart after power failure at the set speed, not overheat on the top surface and operate reliably for long periods of time in a 37°C environment. Thus high-quality stirrers are essential with a strong magnetic field. For unit scales over 10 l, and definitely for those over 20 l, *in situ* fermentation systems should be used (see sections 5.1 and 5.5) rather than spinner flasks.

PROCEDURE: CULTURE OF SUSPENSION CELLS IN A SPINNER FLASK

Reagents and solutions

Growth medium

Standard tissue culture media can be used with various supplements. Serum at 10% can cause foaming at fast stirring speeds, so the serum level is usually reduced to as low a level as possible (0.5–5%). The use of HEPES buffer to stabilize the pH during the setting-up procedure is beneficial.

Cell and Tissue Culture: Laboratory Procedures in Biotechnology, edited by A. Doyle and J.B. Griffiths.
© 1998 John Wiley & Sons Ltd.

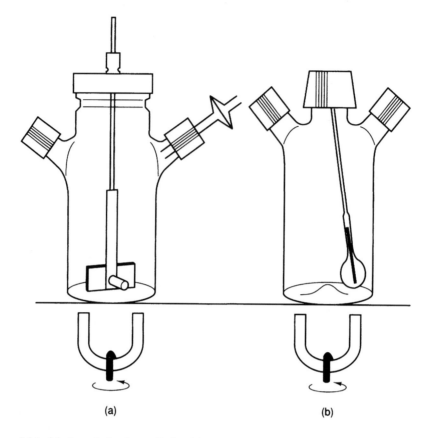

Figure 5.3.1 (a) A typical spinner flask with magnetic spinner bar. (b) A Techne flask with asymmetric stirring rod.

Medium supplements

Pluronic F-68 (polyglycol) (BASF, Wyandot) can be used at 0.1% to protect cells against mechanical damage, especially at reduced serum concentrations. Carboxymethyl cellulose (CMC) (15–20 cP) can also be used to protect cells from mechanical damage at 0.1% concentration. Antifoam (6 ppm) (Dow Chemical Co.) is recommended at serum concentrations above 2–3%. For more information, see Chapter 4, section 4.8.

Materials and equipment

- Spinner flask

1. Add 200 ml of growth medium to a 1-l spinner flask, gas with 5% CO_2 and warm to 37°C.
2. Add cells from a logarithmically growing seed culture at $1-2 \times 10^5$ cells ml^{-1}.
3. Place spinner flask on stirrer and stir at 100–250 rpm (this is variable depending

upon the cell type and geometry of the stirrer bar and vessel – a guideline is to use the minimum speed, which gives, by visual examination, complete homogeneity).

4. Monitor cell growth at least daily by taking a small sample from the side arm (remove flask to a tissue culture cabinet) and carrying out a cell count (Trypan blue stain and a haemocytometer).
5. Monitor the pH. In closed systems (i.e. all ports closed with no filters) the pH will become acidic and need adjusting by day 2–3. Remove the vessel to a tissue culture cabinet and: regas with 5% CO_2 in air; add sodium bicarbonate (5.5%); or allow cells to settle, remove around 50% of the medium and replace with fresh prewarmed medium.
6. After 3–5 days the culture density will typically reach $1–2 \times 10^6$ ml^{-1} (for many hybridomas, $0.8–1.5 \times 10^6$ ml^{-1} only) and can be harvested, or the culture prolonged for extra cell growth with twice-daily 50% medium changes. Suspension cells tend to have only a limited stationary phase and it is thus important to monitor growth more closely than in monolayer culture to ensure that healthy, rather than dying/dead, cells are harvested.

If cells are prone to clumping, or adhering to surfaces, then a medium with reduced Ca^{2+} and Mg^{2+} concentration should be used (e.g. suspension MEM). Additionally it is good practice to siliconize the glass vessels with Dow Corning 1107 (which has to be baked on, in an oven) or Repelcote (dimethyldichlorosilane) (Hopkins and Williams).

There is a tendency to subculture suspension cells by dilution into fresh medium. However, some cells do not grow optimally under this regime and should be centrifuged (800 g for 5 min) and resuspended in new medium.

There is a wide range of spinner vessels available:

- Conventional vessels (Figure 5.3.1a) from Bellco or Wheaton
- Conventional vessels but with large-bladed paddles attached to the magnet or base of the stirrer shaft (Bellco and Cellon μ spinner flasks). These vessels give more efficient stirring with better mixing at the lower speeds and are thus primarily aimed at microcarrier culture
- The Techne system uses a radial stirring action, which is designed for good mixing at low stirring speeds and to minimize mechanical stress to cells (Figure 5.3.1b)
- The Bellco dual overhead drive culture system uses a similar stirring system to the Techne and additionally permits continuous medium perfusion
- The Techne BR-06 Bioreactor uses a floating impeller, which again gives a gentle stirring action. An advantage of this system is that the vessel can be used equally effectively at various working volumes from 500 ml to 3 l. This allows an *in situ* build-up of cell seed by just adding extra medium as cell growth occurs.

If both side arms are fitted with filters, then continuous aeration through the headspace can be carried out, giving better control of pH and higher cell densities.

DISCUSSION

Suspension cells, e.g. hybridomas and BHK, have been the main consideration in this module because microcarrier culture is described in section 5.8. However, most of the principles for suspension cells are applicable to microcarrier culture. The main difference is stirring speed. Microcarrier culture should be initiated at a very low speed (15–30 rpm) whilst cell attachment occurs (stir just fast enough to keep cells and carriers in suspension). Do not aim for homogeneity at this stage. In fact, it aids attachment to have all cells and carriers in the lower 30–50% portion of the medium. Once cell attachment is complete (3–8 h), increase stirring speed to achieve complete homogeneous mixing (40–80 rpm). As the culture develops, and the carriers get increasingly loaded with cells, the stirring speed can be increased further (70–120 rpm) to prevent excessive clumping of carriers. Also, the vessels with large paddles or the Techne stirrer must be used – a magnetic bar alone is insufficient.

BACKGROUND READING

De Bruyne NA (1988) The design of bench-scale reactors. In: Spier RE & Griffiths JB (eds) *Animal Cell Biotechnology*, vol. 3, pp. 142–178. Academic Press, London.

Griffiths JB (1988) Overview of cell culture systems and their scale-up. In: Spier RE & Griffiths JB (eds) *Animal Cell Biotechnology*, vol. 3, pp. 179–220. Academic Press, London.

Griffiths JB (1990) Scale-up of suspension cells and anchorage-dependent cells. In: Pollard JW & Walker JM (eds) *Methods in Molecular Biology*, vol. 5, pp. 49–63. Humana Press, Clifton, NJ.

Griffiths JB (1992) Scaling-up of animal cell cultures. In: Freshney RI (ed.) *Animal Cell Culture: a Practical Approach*, pp. 47–93. IRL Press, Oxford.

5.4 PILOT-SCALE SUSPENSION CULTURE OF HYBRIDOMAS – AN OVERVIEW

The most common cell culture systems developed for pilot- and commercial-scale production of monoclonal antibodies (MAbs) are hollow-fibre and ceramic matrix modules, stirred bioreactors and airlift fermenters. These systems allow cultivation of cells in batch, fed-batch, continuous or perfusion mode. The selection of a culture system and culture mode for the large-scale production of a particular MAb should take into account the growth and antibody-production characteristics of the particular hybridoma line. This module therefore presents an overview of the important characteristics of these systems. Detailed descriptions with accompanying results and a large collection of cited literature are given elsewhere (Seaver, 1987; Mizrahi, 1989; sections 5.1 and 5.9).

PILOT- AND LARGE-SCALE *IN VITRO* SYSTEMS FOR HYBRIDOMAS

Hybridomas can be cultured in either non-homogeneous or homogeneous cultivation systems (Reuveny & Lazar, 1989). In non-homogeneous systems the cells are immobilized in a growth chamber. Most systems have growth chambers with hollow fibres (Von Wedel, 1987) or a porous ceramic matrix (Putnam, 1987). In homogeneous systems cells are grown in bioreactors with either mechanical stirring or air sparging to keep cells in suspension (Reuveny & Lazar, 1989). Short descriptions with lists of characteristics of these cultivation systems are outlined below.

Hollow-fibre and ceramic matrix modules

In hollow-fibre modules cells grow in the extracapillary space. The medium is pumped through capillaries and the pH and oxygen saturation are adjusted after each passage. The fibres are made of semi-permeable membranes, allowing diffusion of oxygen, low molecular weight nutrients and waste products but not the high molecular weight MAbs, which are harvested in a concentrated form from the extracapillary space.

The principle of ceramic matrix modules is the same as for hollow-fibre systems, with the exception that cells are not separated from the medium circulation by a membrane. Cells are held in the highly porous matrix and are supplied with nutrients by circulating medium through small squared channels in the matrix. Some characteristics of these two systems are:

- High cell density of >10^8 ml^{-1} (Andersen & Gruenberg, 1987; Handa-Corrigan, 1988)
- High product concentration – in hollow fibres: 400 mg l^{-1} HIgM, 1100 mg l^{-1} HIgG (Andersen & Gruenberg, 1987); in ceramic matrix modules: 300 mg l^{-1} (Putram, 1987)
- Nutrient, oxygen and waste product gradients in the growth chamber
- Long diffusion distance for nutrients and oxygen in hollow-fibre modules
- Product harvesting with no, or few, cells
- Estimation of cell density in the growth chamber only via oxygen consumption
- No access to representative cell samples for viability and morphology check during cultivation
- No hydrodynamic shear forces in hollow-fibre modules
- Hollow-fibre membrane may clog
- Scale-up limited because the tube length cannot exceed a certain length and only the diameter of the modules or number of installed modules can be increased
- Transfer of cells, grown in small systems, into large systems is difficult

Stirred bioreactor

Different agitation systems, e.g. turbine-type or marine-type impeller, vibromixer and cell lift, were developed in order to prevent exposure of animal cells, cultured in stirred bioreactors, to high shear forces (Reuveny & Lazar, 1989). In large-scale or high-cell-density perfusion cultures, air sparging or oxygen-permeable silicone/polypropylene tubing will provide adequate aeration (Shevitz et al., 1989).

- Homogeneous cell suspension; even supply of substrate in bioreactor
- Controlled cell environment (pH, oxygen, nutrients)
- Easy access to representative cell samples for viability/morphology check
- Easy transfer of cells to the next scale-up step
- Oxygen supply is the most critical parameter in large bioreactors or high-cell-density cultures. Long oxygen-permeable silicone/polypropylene tubing will provide bubble-free aeration. The tubing may need to be anything from 0.5 m to 2 m per unit of liquid volume, depending upon the oxygen requirements (Shevitz et al., 1989). Oxygen transfer by air or oxygen sparging can be used (Backer et al., 1988) but may be detrimental for shear-sensitive hybridoma clones
- Cells are exposed to hydrodynamic shear forces

Airlift bioreactor

Air or oxygen sparged into the bioreactor at the bottom drives circulation to keep cells in suspension, instead of a mechanical stirring device. To improve circulation and oxygen input and minimize cell damage, the height to diameter ratio is normally high in airlift fermenters (Handa et al., 1987).

- No need for long silicone tubing for aeration
- Cell damage due to shear forces produced by disintegration of air bubbles
- Addition of non-ionic detergent may be necessary to prevent foaming, especially in media with a high protein content
- Cell circulation may be enhanced with a mechanical stirring device to minimize the necessary sparging rate at low cell densities to a level sufficient to maintain dissolved oxygen tension at the desired level

CULTIVATION MODES

The following cultivation modes have been described for production of MAbs in *in vitro* large-scale systems: batch, fed-batch, chemostat and perfusion (Reuveny & Lazar, 1989). Batch and fed-batch processes are the most common methods. More demanding are chemostat and perfusion modes, which allow a continuous production of MAbs. All methods are applicable in homogeneous and, with the exception of chemostat, non-homogeneous systems.

In non-homogeneous systems and homogeneous perfusion systems, cells remain in the growth chamber and culture supernatant is collected intermittently or continuously. The cell concentration is not affected by harvesting, in contrast to homogeneous systems where cells are harvested together with the antibody. In the following, characteristics of homogeneous systems are described.

Batch and fed-batch mode

Cells are inoculated, grown up to a desired cell or product concentration and harvested. The duration of batch culture depends on inoculation density, cell line and characteristics such as growth rate and antibody-production kinetics of the cell line. After every run the bioreactor must be cleaned and autoclaved for the next run.

In fed-batch mode, cells are fed either continuously or intermittently and parts of the cell suspension are withdrawn periodically.

- Low cell density compared with perfusion or non-homogeneous systems (at the end, less than 5×10^6 cells ml^{-1})
- Product concentration normally less than 100 mg l^{-1}, but up to 200 mg l^{-1} in batch and 500 mg l^{-1} in fed-batch is possible (Reuveny & Lazar, 1989)
- Low demand for process control
- Variable substrate and (waste) product concentration
- Bad ratio of production time versus regeneration time in batch mode

Chemostat

Cells are inoculated and grown to a certain density, as in the batch mode. Subsequently the medium is continually pumped into the bioreactor at a constant rate and the cell suspension containing antibodies is withdrawn at the same rate.

The dilution rate can be varied within the range of zero (batch) and maximal growth rate; higher dilution rates lead to wash-out of the cells. Cell density increases until one or several substrates (e.g. glucose, nitrogen) or waste products (e.g. ammonia) limit cell growth and steady-state conditions are reached. Antibody production may be continued for weeks or even months.

- Cell density approaches the final concentration reached in batch cultures ($1-5 \times 10^6$ cells ml^{-1})
- Product concentration up to 300 mg ml^{-1} (Reuveny et al., 1986)
- Constant cultivation conditions
- Optimization for antibody production but not for biomass production is possible
- Optimization of parameters under steady-state conditions with high accuracy
- Higher product yields per reactor volume and unit time (Reuveny et al., 1986)
- Flow rate of medium and liquid level in bioreactor must be controlled

Chemostat culture is described more fully in section 5.6.

Perfusion

Perfusion is initiated in the same way as chemostat culture but, instead of the cell suspension being withdrawn, the cells are continuously separated by filtration or sedimentation and recycled into the bioreactor. The increasing cell population is fed by increasing the flow rate of medium pumped into the bioreactor. Continuous harvesting of antibody containing supernatant via a filtration device increases proportionally. The two most widespread filtration systems are rotating filter cages, located within the bioreactor or in an external loop, and tangential flow filtration units (Reuveny & Lazar, 1989).

- High cell density ($>10^7$ ml^{-1})
- Product concentration up to 1000 mg l^{-1} (Velez et al., 1987)
- High yield per unit time and reactor volume
- Better utilization of medium
- Smaller reactor volumes for the same amount of monoclonal antibody
- Sophisticated instrumentation is necessary to control the process
- Clogging of cell separation filters may limit culture times

REFERENCES

Andersen BG & Gruenberg ML (1987) Optimization techniques for the production of monoclonal antibodies utilizing hollow-fiber technology. In: Seaver SS (ed.) *Commercial Production of Monoclonal Antibodies. A Guide for Scale Up*, pp. 175–195. Marcel Dekker, New York.

Backer MP, Metzger LS, Slaber PL, Nevitt KL & Boder GL (1988) Large scale production of monoclonal antibodies in suspension culture. *Biotechnology and Bioengineering* 32: 993–1000.

Handa A, Emery AN & Spier RE (1987) Detrimental effects of sparger aeration on suspended mammalian cell cultures and their prevention. *Proceedings of the 4th European Congress on Biotechnology* 3: 601–604.

Handa-Corrigan A (1988) Large scale *in vitro* hybridoma culture: current status. *Biotechnology* 6: 784–786.
Mizrahi A (ed.) (1989) *Monoclonal antibodies: production and application*, Advances in Biotechnological Processes, vol. 11. Alan R. Liss, New York.
Putnam JE (1987) Monoclonal antibody production in a ceramic matrix. In: Seaver SS (ed.) *Commercial Production of Monoclonal Antibodies. A Guide for Scale Up*, pp. 119–138. Marcel Dekker, New York.
Reuveny S & Lazar A (1989) Equipment and procedures for production of monoclonal antibodies in culture. In: Mizrahi A (ed.) *Monoclonal Antibodies: Production and Application*, Advances in Biotechnological Processes, vol. 11, pp. 45–80. Alan R. Liss, New York.
Reuveny S, Velez D, Miller L & Macmillian JD (1986) Comparison of cell propagation methods for their effect on monoclonal antibody yield in fermentors. *Journal of Immunological Methods* 86: 61–69.
Seaver SS (ed.) (1987) *Commercial Production of Monoclonal Antibodies. A Guide for Scale Up*. Marcel Dekker, New York.
Shevitz J, Reuveny S, LaPorte TL & Cho GH (1989) Stirred tank perfusion reactors for cell propagation and monoclonal antibody production. In: Mizrahi A (ed.) *Monoclonal Antibodies: Production and Application*, Advances in Biotechnological Processes, vol. 11, pp. 81–106. Allan R. Liss, New York.
Velez D, Reuveny S, Miller L & Macmillan JD (1987) Effect of feeding rate on monoclonal antibody production in a modified perfusion-fed fermentor. *Journal of Immunological Methods* 102: 275–278.
Von Wedel RJ (1987) Mass culture of mouse and human hybridoma cells in hollow-fibre culture. In: Seaver SS (ed.) *Commercial Production of Monoclonal Antibodies. A Guide for Scale Up*, pp. 159–173. Marcel Dekker, New York.

5.5 PILOT-SCALE SUSPENSION CULTURE OF HUMAN HYBRIDOMAS

The optimization of culture parameters and the scale-up of a human heterohybridoma in a stirred bioreactor are described in this section. From the viewpoint of scale-up and handling, a stirred bioreactor is chosen as the most practical approach for industrial-scale production of monoclonal antibodies (MAbs). Furthermore, stirred bioreactors are very flexible with regard to the optimization of culture parameters, i.e. oxygen supply (bubble free, air sparging) and culture mode (e.g. batch, fed-batch, chemostat and perfusion). They also give easy access to cell samples at any time of culture, and keep cells homogeneously supplied with nutrients and oxygen.

PROCEDURE: OPTIMIZATION OF CULTURE PARAMETERS AND SCALE-UP

The optimization of culture parameters is an important step in scale-up. The growth rate, the viability and the antibody production of hybridoma cells largely depend on culture conditions. Optimization studies were carried out in a 1.5-l bioreactor, equipped with pH and pO_2 regulators, that provides reproducible culture conditions and minimizes medium requirements. The results were then applied to a 12-l bioreactor.

Media

- **Low serum-containing medium:** Iscove's modified Dulbecco's medium (IMDM) (Sigma Chemical Company, St Louis, MO, USA) supplemented with 2% foetal bovine serum and 5×10^{-5} M β-mercaptoethanol
- **Serum-free media:** evaluation of serum-free media for human heterohybridomas has been described (Lang *et al.*, 1990b).

Cell line

The heterohybridoma (human × human × mouse) 4–8KH15, used for scale-up, secretes human monoclonal IgM specific for *Pseudomonas aeruginosa* lipopolysaccharide of Habs serotype 4. The generation and characterization of human hybridomas is described by Lang *et al.* (1989, 1990a).

Cell and Tissue Culture: Laboratory Procedures in Biotechnology, edited by A. Doyle and J.B. Griffiths.
© 1998 John Wiley & Sons Ltd.

Spinner flasks

Spinner flasks (Bellco) of 500 ml volume are placed in an incubator aerated with air enriched with 5% CO_2. Agitation speed was set at 100 or 150 rpm.

KLF 2000 bioreactor

The bioreactor KLF 2000 (Bioengineering AG, Wald, Switzerland) with a capacity of 1.5 l was used to optimize growth and antibody production conditions of the hybridoma 4–8KH15. This flat-bottomed bioreactor is equipped with a marine impeller and a draft tube with diameters of 4.7 cm and 4.9 cm, respectively. The height to diameter ratio is 2:4. Temperature is regulated to 37°C with a 800-W heat finger. Dissolved oxygen tension and pH are measured on-line with heat-sterilizable electrodes (Ingold Messtechnik AG, Urdorf, Switzerland). Cultures are aerated bubble free with air and O_2 to maintain the dissolved oxygen concentration at 40%, and the pH is adjusted with CO_2 and NaOH. The bioreactor is sterilizable *in situ* by heating water, filled into the bioreactor, to 120°C.

The flat-bottomed design of the KLF 2000 bioreactor affords a higher stirrer speed to prevent sedimentation of cells compared with the round-bottomed NLF 22 bioreactor (see below). For cultivation of shear-sensitive cells, round-bottomed bioreactors may be necessary. To avoid high temperatures near the heat finger, a power control unit is used to restrict the power to the range 800–200 W but as low as possible to maintain a temperature of 37°C. In addition to this, heating is restricted to intervals of a few seconds only and the glass part of the bioreactor is insulated.

NLF 22 bioreactor

The 12-l NLF 22 bioreactor (Bioengineering AG, Wald, Switzerland) is equipped in the same way as the KLF 2000. The impeller and draft tube diameters are 7.9 cm and 10.9 cm, respectively. The bottom of the NLF 22 is spherically shaped. The height to diameter ratio is 1:95. The temperature is regulated through an integrated water-jacket. The NLF 22 bioreactor is steam sterilizable *in situ* by heating water, filled into the bioreactor, to 120°C. Cultures are aerated bubble free at low cell density, and at high cell density are intermittently sparged with air/O_2 to maintain the dissolved oxygen saturation at 40%. Bubble size and sparging rate were not measured.

INOCULUM PREPARATION AND OPTIMIZATION OF PARAMETERS

1. Grow hybridoma cells in stationary cultures in T-flasks and passage every 2–3 days in a ratio of 1:3 to 1:10.
2. From T-flasks, transfer 100 ml of cell suspension to a 500-ml spinner flask and add 200 ml of fresh medium. Initial cell density is about 10^5 cells ml^{-1}. Two days later, add an additional 200 ml of fresh medium.

3. From the spinner flask, use 300–500 ml of cell suspension at a concentration of $3\text{–}6 \times 10^5$ cells ml^{-1} as inoculum for the 1.5-l KLF 2000 bioreactor for optimization experiments (Table 5.5.1).
4. For cultures in the 12-l NLF fermenter, apply the same parameters as optimized for the 1.5-l KLF 2000 bioreactor (Table 5.5.1). Only the stirrer speed, which is strongly dependent on the vessel geometry, has to be re-evaluated in the 12-l bioreactor.
5. Transfer 1500 ml of cell suspension containing $7\text{–}8 \times 10^5$ cells ml^{-1} from the KLF 2000 to the NLF 22 bioreactor. The dilution rate of spent to fresh medium should be 1:8 and the cell density after inoculation should be about 1×10^5 cells ml^{-1}.
6. The scale-up protocol is summarized in Table 5.5.2.

Table 5.5.1 Parameters applicable to optimization in stirred bioreactors

Batch	Chemostat	Perfusion
Inoculation density	pH	Perfusion rate
Stirrer speed	pO_2	Circulation rate
	Growth rate μ	Cell recycle rate
	Oxygen supply	Oxygen supply

Table 5.5.2 Scale-up of human hybridoma 4-8KH15 to the 12-l stirred bioreactor[a]

Culture system	Volume (ml)	Cultivation time (days)	Parameters measured off-line	Parameters controlled on-line
T-flask	100	8–11	Cell density Viability MAb conc.	–
Spinner flask	500	11–15	Cell density Viability MAb conc. Glucose pH	–
KLF	1500	14–18	Cell density Viability MAb conc. Glucose Lactate	pH pO_2 rpm Temperature
NLF	12 000	18–22	Cell density Viability MAb conc. Glucose Lactate	pH pO_2 rpm Temperature

[a]Cells are thawed into culture medium and continuously processed from T-flasks to the NLF 22 bioreactor. At each level the parameters measured and the cumulative cultivation time required are shown. The volume of cell suspension required to proceed to the next step is given.

DISCUSSION

Background information

Perfusion culture results in the highest yield of MAb per reactor volume and time and the highest product concentration achievable in stirred bioreactors (Reuveny et al., 1986). Despite the need for sophisticated instrumentation to control the process, this mode seems to be the most favourable for large-scale antibody production.

Independent of the chosen culture method, cells are inoculated at low density into stirred bioreactors and initially grown in batch mode. Inoculum density, the ratio of fresh to conditioned medium and stirrer speed are the first parameters to be optimized in batch cultures. Normally, confluent cell cultures are passaged by transferring a portion of the cell suspension into new medium at a certain ratio. Alternatively, cells can be centrifuged and resuspended in completely fresh medium. Most cells grow better in partially conditioned medium as compared with totally fresh medium and at a cell density below $1-2 \times 10^5$ cells ml^{-1}. Therefore, low dilution ratios and consecutively a high ratio of conditioned medium may be advantageous, e.g. when the culture system is changed from stationary T-flask to stirred bioreactor. On the other hand, high dilution ratios are necessary to minimize scale-up steps. The stirrer speed has to be optimized in order to prevent cell sedimentation and cell damage due to hydrodynamic shear forces.

Parameters such as pH, pO_2 and growth rate can be optimized in chemostat cultures. First, cells are grown to a steady state (normally 3–4 liquid volume changes) with a chosen set of parameters. Second, the parameter of interest is changed to a new value and cell density, viability and antibody secretion are monitored. With this strategy different sets of culture parameters can be defined. One set maintains a fast-growing cell population and a high viability during scale-up. Alternatively, culture parameters can be adjusted for a high cell density and a high specific antibody secretion in a production-scale bioreactor.

In perfusion cultures, the perfusion rate for a given cell density, the circulation rate in the cell-retaining filter and the cell recycling rate should be evaluated. A complete cell recycling rate may lead to steadily increasing cell concentration and accumulation of dead cells and cell debris in the bioreactor. To operate the perfusion culture in a steady state, the recycling rate can be set to a value below 100% (Seamans & Hu, 1990) or filters with a large pore size should be used to prevent accumulation of cell debris.

Critical parameters

The oxygen supply of mammalian cell cultures is the most critical parameter (Shevitz et al., 1989; Chapter 4, section 4.6). In high-density cultures or in large bioreactors, oxygen supply may be insufficient. Factors influencing oxygen transfer into the medium are the oxygen concentration in the inlet gas, the stirrer speed and the area of the gas/liquid interface. The area of the gas/liquid interface can be increased by air or oxygen sparging directly into the culture. Alternatively, bubble-free aeration by oxygen-permeable silicon or polypropylene tubing submerged in the medium may be used. The sparging method can affect cell viability and can cause foaming in serum-supplemented media with a high protein content. Oxygen-permeable tubing in the range of 0.5 m l^{-1} to more than 2 m l^{-1} culture volume seems to be impractical. Another critical parameter in stirred bioreactor cultures comprises shear forces produced by agitation or disintegration of air bubbles (Reuveny & Lazar, 1989; Shevitz et al., 1989; Chapter 4, section 4.7). The sensitivity to shear forces is clone specific and varies within a wide range. Different low-shear mixing systems have been developed and may avoid cell destruction (Reuveny & Lazar, 1989). The detrimental effect on cells produced by air sparging can be reduced by the addition of serum, non-ionic detergent (Pluronic F-68) or increasing height to diameter ratio of the bioreactor (Handa-Corrigan et al., 1988).

Results

In stirred batch cultures after 3–4 days a maximum cell density of $7–10 \times 10^5$ cells ml^{-1} with a viability above 85% was achieved. Optimal stirrer speeds were 100–150, 300 and 100–150 rpm in the spinner flasks, the KLF 2000 bioreactor and the NLF 22 bioreactor, respectively. The high stirrer speed of 300 rpm in the flat-bottomed KLF 2000 is necessary to prevent cell sedimentation. In the NLF 22, with a spherically shaped bottom, a reduced stirrer speed is sufficient. Accordingly, the impeller tip speed was 0.8 m s^{-1} in the KLF 200 bioreactor and 0.8–1.2 m s^{-1} in the NLF 22, values that were not detrimental to cells (Backer et al., 1988). Up to 40 µg ml^{-1} IgM was produced in batch cultures in 3–5 days. The cell-doubling time of about 25 h and the antibody production in stirred cultures were similar to those achieved in stationary cultures. The 4–8KH15 cells grow in serum-free medium under the same conditions as in low serum (2%)-containing media. Maximum cell density achieved in continuous culture was 1.5×10^6 cells ml^{-1} and viability was 80–90%. Without any change in medium composition, optimization of growth rate and pH in chemostat cultures resulted in a threefold increase of the specific IgM production rate, whereas IgM yield per reactor volume and time increased fourfold.

At cell densities exceeding 5×10^5 ml^{-1} in the NLF 22 bioreactor, the dissolved oxygen tension had to be maintained by intermittently sparging O_2 directly into the culture and raising the stirrer speed from 100 rpm (impeller tip speed 0.8 m s^{-1}) to 150 rpm (tip speed 1.2 m s^{-1}). No negative effects on cell growth and viability were observed.

Time considerations

A batch experiment takes 3–5 days to grow cells to confluency, up to 5 more days to investigate antibody production during decline phase and 1–2 days for cleaning and sterilizing. Optimization of parameters in chemostat experiments requires 5–12 days per selected value. The time required depends essentially on the dilution rate. Provided that the 12-l NLF 22 bioreactor is the final scale for production,

18–22 days are required to grow cells from the working cell bank to production scale. To reduce the time to start a new culture in the production bioreactor, it is important to maintain a culture at the preliminary stages.

REFERENCES

Backer MP, Metzger LS, Slaber PL, Nevitt KL & Boder GL (1988) Large scale production of monoclonal antibodies in suspension culture. *Biotechnology and Bioengineering* 32: 993–1000.

Handa-Corrigan A, Emery AN & Spier RE (1988) Effect of gas–liquid interfaces on the growth of suspended mammalian cells. Mechanisms of cell damage by bubbles. *Enzyme and Microbial Technology* 11: 230–235.

Lang AB, Fürer E, Larrick JW & Cryz SJ Jr (1989) Isolation and characterization of a human monoclonal antibody that recognizes epitopes shared by *Pseudomonas aeruginosa* immunotype 1, 3, 4 and 6 lipopolysaccharides. *Infection and Immunity* 57: 3851–3855.

Lang AB, Fürer E, Senyk G, Larrick JW & Cryz SJ Jr (1990a) Systematic generation of antigen specific human monoclonal antibodies with therapeutical activities using active immunization. *Human Antibodies and Hybridomas* 1: 96–103.

Lang AB, Schürch U & Cryz SJ Jr (1990b) Optimization of growth and secretion of human monoclonal antibodies by hybridomas cultured in serum free media. *Hybridoma* 10: 401–409.

Reuveny S & Lazar A (1989) Equipment and procedure for production of monoclonal antibodies in culture. In: Mizrahi A (ed.) *Monoclonal Antibodies: Production and Application*, Advances in Biotechnological Processes, vol. 11, pp. 45–80. Alan R. Liss, New York.

Reuveny S, Velez D, Miller L & Macmillian JD (1986) Comparison of cell propagation methods for their effect on monoclonal antibody yield in fermentors. *Journal of Immunological Methods* 86: 61–69.

Seamans TC & Hu WS (1990) Kinetics of growth and antibody production by a hybridoma cell line in a perfusion culture. *Journal of Fermentation Bioengineering* 70: 241–245.

Shevitz J, Reuveny S, LaPorte TL & Cho GH (1989) Stirred tank perfusion reactors for cell propagation and monoclonal antibody production. In: Mizrahi A (ed.) *Monoclonal Antibodies: Production and Application*, Advances in Biotechnological Processes, vol. 11, pp. 81–106. Alan R. Liss, New York.

5.6 CHEMOSTAT CULTURE

Determining the effect of individual parameters on cell physiology in batch cultures is complicated by the exposure of cells to a continuously changing environment. This is because a simple batch culture is a closed system in which cell growth is accompanied by the depletion of nutrients and the accumulation of end products. However, in an open system where nutrient consumption is balanced by input of medium, and cell and metabolite accumulation are balanced by removal of spent medium and cells, it is possible to achieve a steady state where cell and metabolite concentrations remain constant. This can be achieved by the use of a chemostat, as demonstrated independently by Monod (1950) and Novick and Szilard (1950). The theory of the chemostat is based upon the principle of Monod, which states that, at growth rates less than the maximum, the specific growth rate of a culture (μ) is determined by the concentration of a single growth-limiting substrate (s). Monod described the relationship between the specific growth rate and the limiting substrate concentration using an equation that takes a similar form to the Michaelis–Menten model of enzyme kinetics:

$$\mu = \mu_{max} [s/(s + K_s)]$$

The term μ_{max} represents the maximum specific growth rate that occurs when the growth-limiting substrate is in excess, i.e. when s is very large and $\mu = \mu_{max}$. The saturation constant (K_s) is equal to the substrate concentration at $\mu_{max}/2$. Both K_s and μ_{max} may be determined experimentally by Lineweaver–Burk-type analysis, plotting $1/\mu$ versus $1/s$ and extrapolating to obtain $-1/K_x$ at the x-axis intercept and $1/\mu_{max}$ at the y-axis intercept. Alternatively, a more accurate estimation of these parameters may be made by fitting the Monod model to the experimental data using suitable curve-fitting software.

The specific growth rate (μ) has units of reciprocal time (h^{-1}) and can be related to the cell-doubling time (t_d) by the following:

$$t_d = \ln 2/\mu$$

In the chemostat the culture volume is kept constant by removing spent medium and cells at the same rate as the addition of fresh medium. The rate of dilution of the culture (D) is a function of the culture volume (V) and the flow rate (F), where $D = F/V$. The rate of change of cell concentration (dx/dt) is therefore a function of the specific growth rate of the cell (μx) and the dilution rate (Dx):

$$dx/dt = \mu x - Dx$$

Provided that the dilution rate is less than the maximum specific growth rate, the cells will continue to grow until the substrate concentration in the fermenter

---

Cell and Tissue Culture: Laboratory Procedures in Biotechnology, edited by A. Doyle and J.B. Griffiths.
© 1998 John Wiley & Sons Ltd.

becomes limiting. At this point, further cell growth is determined by the rate at which substrate is added to the fermenter, and at a constant dilution rate a steady state is achieved. Because at steady state the cell concentration is constant (i.e. $dx/dt = 0$), it follows that $\mu = D$. Thus, at steady state the specific growth rate can be controlled by varying the dilution rate.

However, it is important to note that the specific growth rate is only equal to the dilution rate when all cells in the population are viable. Below 100% viability, the precise relationship between μ and D is more complex than the simple model described above, and a number of studies have shown that the actual specific growth rate deviates from D, particularly at low dilution rates. This results from an increase in the proportion of dead cells in the population, which in turn can be related to an increase in the specific rate of cell death (μ_d) at low D. The specific death rate at a particular dilution rate is related to the proportion of viable (x_v) and non-viable cells (x_d), where:

$$\mu_d = D\,(x_d/x_v)$$

Taking into account the contribution from the specific death rate, the actual relationship between μ and D becomes:

$$\mu = D + \mu_d$$

Whereas the specific growth rate is determined by the rate of addition of the growth-limiting substrate, the cell concentration in the fermenter is determined by the concentration of the limiting substrate in the feed medium. At steady state the cell yield (Y) on the growth-limiting substrate can be represented by:

$$Y = x/(s_R - s)$$

where s_R is the substrate concentration in the feed medium, s is the residual substrate concentration in the fermenter and x is the total cell concentration ($x_v + x_d$).

At steady state, the concentrations of cells and metabolites are constant and thus calculation of specific utilization or production rates for each metabolite is relatively simple. For example, the specific rate of utilization of a nutrient q_n at steady state is given by:

$$q_n = D(n_R - n)/x_v$$

where n_R is the nutrient concentration in the feed medium and n is the nutrient concentration in the chemostat.

Similarly, the specific rate of production for a cell product (q_p) is given by:

$$q_p = D(p - p_R)/x_v$$

In summary, it can be seen that the chemostat enables the study of steady-state cell physiology at predetermined growth rates or biomass concentrations.

EQUIPMENT

Fermenter vessel

Most fermenter vessels suitable for suspension cell culture can be readily adapted for chemostat culture. A device for maintaining a constant culture volume will be required, the simplest and most reliable of which is a weir that allows the culture to overflow to a collection vessel. Alternatively, a dip tube that draws medium from the surface of the culture can be mounted in the headplate of the vessel and the medium pumped to the collection vessel. Most manufacturers supply modified tubes for chemostat operation of their fermenters.

Pumps

Peristaltic pumps are most convenient because they can be used in conjunction with sterile pump tubing (e.g. silicone or Viton). Constant wear of the tubing can lead to changes in the flow rate or tubing failure, so provision should be made to change the tubing regularly using sterile connectors. Alternatively, by using a long section of pump tubing that can be advanced through the pump head when necessary, excessive wear of the tubing can be avoided.

The flow rate should be checked regularly and this can be achieved by fitting a graduated pipette on the reservoir side of the pump. Medium can be drawn into the pipette using a syringe, and the flow rate calculated from the time taken for a set volume of medium to be withdrawn.

Media and collection vessels

The feed reservoir and collection vessel are connected to the inlet and outlet lines using sterile connectors so that they may be replaced as necessary. A sampling port should be fitted to the feed reservoir or at some point in the inlet line to enable sampling of the feed medium. The feed reservoir should be stored at 4°C to reduce degradation of labile medium components such as glutamine. If product is to be harvested, the collection vessel should also be refrigerated. Figure 5.6.1 shows a typical set-up for chemostat culture.

METHOD

Initiating a chemostat culture

Cells can be inoculated at a normal inoculum level, e.g. $1-2 \times 10^5$ cells ml^{-1}, and grown as a batch culture until the mid-exponential phase of growth. The supply of medium from the feed reservoir can then be started at a flow rate that gives the required dilution rate. In order that the cells are not washed out of the fermenter, the initial dilution rate should be below the maximum specific growth rate of the cells, although too low a dilution rate may result in low cell viability

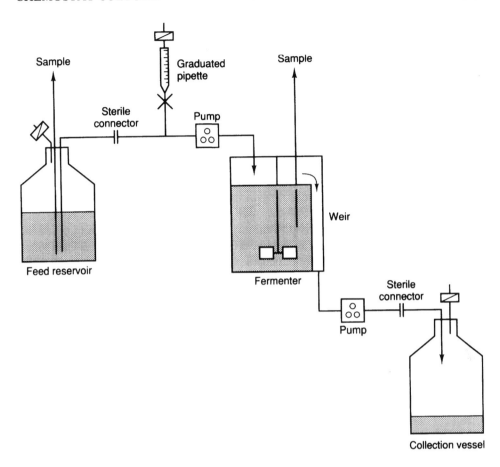

Figure 5.6.1 Chemostat culture apparatus.

and oscillations in cell numbers. A dilution rate of between 50% and 75% of μ_{max} is usually a suitable starting point.

An indication of the value of μ_{max} for a particular cell line may be derived from batch culture experiments, and for most mammalian cell lines it is likely to fall in the range 0.02–0.06 h^{-1}. If the dilution rate exceeds μ_{max}, then cells will be washed out of the fermenter at a rate that is a function of the dilution rate and maximum specific growth rate. Indeed, the maximum specific growth rate can be determined experimentally by increasing D until wash-out occurs and then measuring the rate of cell wash-out. The maximum specific growth rate can then be determined using the following equation:

$$\mu_{max} = \frac{(\ln x_t - \ln x_0)}{t} + D$$

where x_0 and x_t are the respective cell concentrations at the start and end of the time interval and t is the time interval.

Growth-limiting substrate

The chemostat is a particularly powerful tool for examining the effect of different nutrient limitations on the physiology of mammalian cells. Selection of the growth-limiting substrate will depend on a number of factors, but most importantly it must be a nutrient essential for cell growth, e.g. essential amino acids, glucose, oxygen. The choice of substrate will also depend on cell type, because some nutrients will be essential for some cells but not for others. For example, glutamine can be regarded as an essential nutrient for most hybridomas because cells of lymphoblastoid origin lack glutamine synthetase and are unable to synthesize glutamine from glutamate. Conversely, Chinese hamster ovary cells express glutamine synthetase and will switch to glutamine synthesis in response to glutamine limitation.

The concentration of the growth-limiting substrate in the feed medium should be set to that it becomes limiting while other nutrients are still present in relative excess. Small-scale batch culture experiments in which the cell yield is determined at different initial substrate concentrations will give some indication of a suitable substrate feed concentration (s_R). Once at steady state, it can be confirmed that the cells are limited by the chosen substrate because a change in the substrate feed concentration should give a proportionate change in the cell concentration.

Characterization of the steady state

Steady state is achieved when cell (as measured by cell number, biomass, DNA or protein content) and metabolite (e.g. glucose, lactate, ammonia, amino acids, product) concentrations remain stable over a period of time. The attainment of steady state should be confirmed by measuring several parameters, because the stability of a single parameter (e.g. cell number) is not necessarily indicative of a steady state.

The time taken for a culture to adjust to a new steady state will depend on the dilution rate, because it is advisable to allow five culture volume changes following a change. Thus it takes 100 h to reach steady state when $D = 0.05$ h^{-1}, but 500 h when $D = 0.01$ h^{-1}.

Sampling

Once at steady state, several samples should be collected over a period covering at least three culture volume changes, to ensure that the data generated are representative of the steady state. The volume of sample removed for analysis will depend on the number and type of analyses to be performed and the frequency of sampling. Removing large samples from the fermenter should be avoided, because this changes the volume of the culture and thus affects the dilution rate. Sample volumes of no more than 1–2% of the culture volume are recommended and the frequency of sampling should be reduced at low dilution rates. Samples should also be removed from the feed reservoir to check the composition of the feed medium.

DISCUSSION

More extensive discussions of the theoretical aspects of chemostat culture have been published elsewhere (e.g. Herbert *et al.*, 1956; Pirt, 1975; Bailey & Ollis, 1977). Many of the theoretical principles applied to microbial chemostat cultures can be applied successfully to animal cell cultures, although there are a number of instances where the behaviour of animal cells has been reported to deviate from the models used to describe microbial growth kinetics (Tovey, 1980; Boraston *et al.*, 1984; Miller *et al.*, 1988).

Some of these differences may be ascribed to the complex nutritional requirements of animal cells, which precludes the use of single carbon or nitrogen sources as limiting substrates. For this reason, the growth-limiting substrate is often not defined in animal cell chemostat studies, although a number of studies using defined growth-limiting substrates have been published. Glucose has been used as a limiting substrate in many chemostat studies on animal cells (e.g. Pirt & Callow, 1964; Moser & Vecchio, 1967; Tovey & Brouty-Boyé, 1976; Frame & Hu, 1991; Hayter *et al.*, 1992a). Phosphate-, oxygen-, and glutamine-limited chemostats have also been described in the literature (Kilburn & Van Wezel, 1970; Boraston *et al.*, 1983; Birch *et al.*, 1985; Hayter *et al.*, 1992b).

Chemostat cultures have been employed in studies on the effect of the specific growth rate on mammalian cell physiology, gene expression and culture productivity (e.g. Griffiths & Pirt, 1967; Birch *et al.*, 1985; Miller *et al.*, 1988; Shaughnessy & Kargi, 1990a; Frame & Hu, 1991; Linardos *et al.*, 1991; Hayter *et al.*, 1993; Banik *et al.*, 1996). Chemostats have also been used to study the effect of specific nutrients and metabolites on cell physiology (Miller *et al.*, 1989a,b; Shaughnessy & Kargi, 1990b; Vriezen *et al.*, 1997) and protein glycosylation (Hayter *et al.*, 1992a). The effects on cell growth of other parameters, such as pH (Miller *et al.*, 1988; Goergen *et al.*, 1993), interferon (Tovey, 1981), growth factors (Martial *et al.*, 1991), Pluronic F-68 (Al-Rubeai *et al.*, 1992) and gas sparging (Jobses *et al.*, 1991), have been investigated at steady state using chemostat cultures. Others have used chemostats with partial cell retention, which enables the study of cell behaviour at steady states that are difficult to achieve in a completely open chemostat system, e.g. high cell densities and low specific growth rates (see, for example, Hiller *et al.*, 1993; Banik & Heath, 1995).

REFERENCES

Al-Rubeai M, Emery AN & Chalder S (1992) The effect of Pluronic F68 on hybridoma cells in continuous culture. *Applied Microbiology and Biotechnology* 37: 44–45.

Bailey JE & Ollis DF (1977) *Biochemical Engineering Fundamentals*. McGraw-Hill, New York.

Banik GG & Heath CA (1995) Partial and total cell retention in a filtration-based homogeneous perfusion reactor. *Biotechnology Progress* 11: 584–588.

Banik GG, Todd PW & Kompala DS (1996) Foreign protein expression from S phase specific promoters in continuous cultures of recombinant CHO cells. *Cytotechnology* 22: 179–184.

Birch JR, Thompson PW, Lambert K & Boraston R (1985) The large scale cultivation of hybridoma cells producing monoclonal antibodies. In: Feder J & Tolbert WR (eds) *Large-scale Mammalian Cell Culture*, pp. 1–18. Academic Press, New York.

Boraston R, Garland S & Birch JR (1983) Growth and antibody production by mouse hybridoma cells growing in oxygen-limited chemostat culture. *Journal of Chemical Technology and Biotechnology* 33B: 200.

Boraston R, Thompson PW, Garland S & Birch JR (1984) Growth and oxygen requirements of antibody producing mouse hybridoma cells in suspension culture. *Developments in Biological Standardisation* 55: 103–111.

Frame KK & Hu W-S (1991) Kinetic study of hybridoma growth in continuous culture. 1. A model for non-producing cells. *Biotechnology and Bioengineering* 37: 55–64.

Goergen JL, Marc A & Engasser JM (1993) Determination of cell lysis and death kinetics in continuous cultures from the measurement of lactate dehydrogenase release. *Cytotechnology* 11: 189–195.

Griffiths JB & Pirt SJ (1967) The uptake of amino acids by mouse cells (strain LS) during growth in batch culture and chemostat culture: the influence of growth rate. *Proceedings of the Royal Society of London, Series B* 168: 421–438.

Hayter PM, Curling EMA, Baines AJ, Jenkins N, Salmon I, Strange PG, Tong JM & Bull AT (1992a) Glucose-limited chemostat culture of Chinese hamster ovary cells producing recombinant interferon-γ. *Biotechnology and Bioengineering* 39: 327–335.

Hayter PM, Kirkby NF & Spier RE (1992b) Relationship between hybridoma growth and monoclonal antibody production. *Enzyme and Microbial Technology* 14: 454–461.

Hayter PM, Curling EMA, Gould ML, Baines AJ, Jenkins N, Salmon I, Strange PG & Bull AT (1993) The effect of the dilution rate on CHO cell physiology and recombinant interferon-γ production in glucose-limited chemostat culture. *Biotechnology and Bioengineering* 42: 1077–1085.

Herbert D, Elsworth RE & Telling RC (1956) The continuous culture of bacteria: a theoretical and experimental study. *Journal of General Microbiology* 14: 601–622.

Hiller GW, Clark DS & Blanch HW (1993) Cell retention–chemostat studies of hybridoma cells – analysis of hybridoma growth and metabolism in continuous suspension culture in serum-free medium. *Biotechnology and Bioengineering* 42: 185–195.

Jobses I, Martens D & Tramper J (1991) Lethal events during gas sparging in animal cell culture. *Biotechnology and Bioengineering* 37: 484–490.

Kilburn DG & Van Wezel AL (1970) The effect of growth rate in continuous flow cultures on the replication of rubella virus in BHK cells. *Journal of General Virology* 9: 1–7.

Linardos TI, Kalogerakis N, Behie LA & Lamontagne LR (1991) The effect of specific growth rate and death rate on monoclonal antibody production in hybridoma chemostat cultures. *Canadian Journal of Chemical Engineering* 69: 429–438.

Martial A, Nabet P, Engasser JM & Marc A (1991) Kinetic effect of growth factors on batch and continuous cultures. In: Spier RE, Griffiths JB & Meignier B (eds) *Production of Biologicals from Animal Cells in Culture*. pp. 606–608. Butterworth-Heinemann, London.

Miller WM, Blanch HW & Wilke CR (1988) A kinetic analysis of hybridoma growth and metabolism in batch and continuous suspension culture: effect of nutrient concentration, dilution rate and pH. *Biotechnology and Bioengineering* 32: 947–965.

Miller WM, Wilke CR & Blanch HW (1989a) Transient responses of hybridoma cells to nutrient additions in continuous culture: 1. Glucose pulse and step changes. *Biotechnology and Bioengineering* 33: 477–486.

Miller WM, Wilke CR and Blanch HW (1989b) Transient responses of hybridoma cells to nutrient additions in continuous culture: 1. Glutamine pulse and step changes. *Biotechnology and Bioengineering* 33: 487–499.

Monod J (1950) La technique de culture continué: théorie et applications. *Annals de l'Institut Pasteur* 79: 390–410.

Moser H & Vecchio G (1967) The production of stable steady states in mouse ascites mast cell cultures maintained in the chemostat. *Experientia* 23: 1–10.

Novick A & Szilard L (1950) Description of the chemostat. *Science* 112: 715–716.

Pirt SJ (1975) *Principles of Cell and Microbe Cultivation*. Blackwell, Oxford.

Pirt SJ & Callow DS (1964) Continuous flow culture of the ERK and L types of mammalian cells. *Experimental Cell Research* 33: 413–421.

Shaughnessy TS & Kargi F (1990a) Growth and product inhibition kinetics of T-cell hybridomas producing lymphokines in batch and continuous culture. *Enzyme and Microbial Technology* 12: 669–675.

Shaughnessy TS & Kargi F (1990b) Transient behaviour of T-cell hybridomas in response to changes in metabolite concentrations in continuous culture. *Enzyme and Microbial Technology* 12: 676–684.

Tovey MG (1980) The cultivation of animal cells in the chemostat: applications to the study of tumour cell multiplication. *Advances in Cancer Research* 33: 1–37.

Tovey MG (1981) Use of the chemostat culture for study of the effect of interferon on tumour cell multiplication. In: Pestka S (ed.) *Methods in Enzymology*, vol. 79, pp. 391–404. Academic Press, New York.

Tovey MG & Brouty-Boyé D (1976) Characteristics of the chemostat culture of murine leukemia L1210 cells. *Experimental Cell Research* 101: 346–354.

Vriezen N, Romein B, Luyben KCAM & van Dijken JP (1997) Effects of glutamine supply on growth and metabolism of mammalian cells in chemostat culture. *Biotechnology and Bioengineering* 54: 272–286.

5.7 GROWTH OF HUMAN DIPLOID FIBROBLASTS FOR VACCINE PRODUCTION MULTIPLATE CULTURE

For many years human diploid cells have been utilized for the large-scale manufacture of viral vaccines. With the most recent developments in biotechnology, these cells are also capable of producing a variety of protein products. The history of establishment, growth and storage of human diploid cell lines has been extensively investigated and well documented. The absence of spontaneous transformation and adventitious viruses, a stable diploid karyotype and support of growth of a wide range of viruses are some of the reasons why these cells have been the substrate of choice for the production of biologicals. The growth of these cell lines has been greatly facilitated by the development of microcarriers. In microcarrier cultures, cells grow as monolayers on the surface of small spherical beads that are suspended in a suitable medium and in a vessel with constant stirring. The advantage of using this type of cell culture methodology is a homogeneous, well-controlled cellular environment (see section 5.8).

In this section growth in plastic tissue culture vessels with scale-up to multiplate units is described.

PROCEDURE: PROPAGATION AND SUBCULTIVATION OF HUMAN DIPLOID CELLS IN 150-cm² PLASTIC CULTURE VESSELS

This method describes the establishment and subcultivation of 150-cm² cell cultures originating from a human diploid working cell bank. The harvests of these cultures will be used to seed further similar vessels or microcarrier culture systems.

Materials and equipment

- Ampoule/vial of MRC-5/WI-38
- Trypsin in phosphate-buffered saline (PBS)
- Dulbecco's modified Eagle's medium (DMEM) with 10% foetal bovine serum (FBS)
- 150-cm² plastic culture vessels

- Harvest vessel

1. Place frozen ampoule/vial of MRC-5/WI-38 from a working cell bank (3.0×10^6 cells at a predetermined population doubling level (PDL)) in warm water (37–40°C).
2. Shake gently until the contents are entirely thawed.
3. Remove thawed ampule/vial and swab with 70% ethanol.
4. Transfer contents aseptically, with a pipette and pro-pipetter, into a 150-cm^2 flask containing 100 ml of DMEM with 10% FBS.
5. Incubate at 37°C in a 5% CO_2/95% air incubator.
6. Fluid change the medium the following day and observe cultures macroscopically and microscopically during the incubation period.
7. Once the desired level of confluency is achieved (4–6 days, $25–30 \times 10^6$ cells), transfer the culture(s) from the incubator to a tissue culture cabinet.
8. Pour off spent medium into a discard bottle, then wash cell sheet with 20 ml of PBS and discard immediately.
9. Add 10 ml of warm trypsin solution using a 10-ml plastic pipette and pipetter.
10. Gently roll the solution over the cell sheet and pour off the trypsin into the discard bottle.
11. Incubate culture(s) at 37°C until cells have loosened and separated.
12. Transfer the culture(s) from the incubator to the cabinet and break cells up with 10 ml of DMEM with 10% serum by pipetting up and down. Transfer the suspension to a cell harvest vessel.
13. Rinse culture flask(s) with an additional 10 ml of the cell-suspending medium and transfer the suspension to the harvest vessel. Pipette up and down to ensure homogeneity.
14. Remove 1 ml of cell suspension into a disposable culture tube for a cell count.
15. Determine the cell concentration of the suspension by direct counting and monitor the viability using the Trypan blue exclusion test.
16. If the following transfer is into 150-cm^2 flasks, seed a minimum of 3.0×10^6 viable cells per flask. Incubate at 37°C in a CO_2 environment (3–5 days).

PROCEDURE: SEEDING, CULTIVATION, TRYPSINIZATION AND INFECTION OF A NUNC 6000-cm^2 MULTIPLATE UNIT

The following method describes the seeding, cultivation, trypsinization and inoculation of human diploid cells in Nunc 6000-cm^2 multiplate (MP) units. These MP units may be used to provide human diploid cells for vaccine production in essentially two formats: cells may be provided as monolayer cultures in the MP units for subsequent viral infection; or cells may be trypsinized out of the MP units and used to seed microcarrier culture systems or other MP vessels.

Materials and equipment

- MRC-5/WI-38 cells in exponential growth phase
- DMEM with 10% FBS
- Viral inoculum
- Viral growth medium
- Trypsin in PBS
- Nunc 6000-cm^2 multiplate unit with a 0.2-μm disk filter unit
- Cell harvest vessel with connection/air filter and magnetic bar
- Seeding vessel with connections, air filter and magnetic bar
- Trypsin vessel (500 ml) with connections and air filter
- Glass bell-end attachment
- Multiple-end connection
- Laboratory stand
- Cautery pump
- Spring or screw clamps

1. Remove MP unit from its plastic packing and inspect visually for cracks, faulty seals, etc.
2. Under sterile conditions in a tissue culture cabinet, remove the seal from one of the adapter caps and insert a sterile disk filter (0.2 μm) in the cap. Remove the seal from the second adapter cap and insert a 5/16-inch stainless-steel connector into the second cap. The stainless-steel connector is part of a multiplate-end connection (Figure 5.7.1) designed to facilitate transfer of ingredients and cells, as well as final harvest. Each and every end of this connection contains sterile wrapping and should be clamped individually using spring or screw clamps.
3. For seeding, each unit requires 2 l of diluted cell suspension, prepared in a sterile 2-l Erlenmeyer flask equipped with a No. 10 rubber stopper, air filter and silicone tubing (Figure 5.7.2).
4. Secure the rubber stopper on the Erlenmeyer flask with a piece of tape. Then, aseptically connect the open end of the silicone tubing (Figure 5.7.2) to one of the 5/16-inch stainless-steel connectors of the multiplate-end connection (Figure 5.7.1), and elevate the vessel on an overhead platform.
5. In order to facilitate the flow of all solutions by means of gravity, place the multiplate cell factory unit in a position with the supply tube to the bottom. To start the flow, a minimal amount of air pressure is applied on the unit using a cautery pump (connected to filter in Figure 5.7.2). Excessive air pressure may result in damage to the plastic seals.
6. Allow the cell suspension to flow in and ensure that an even level is achieved in each and every chamber. Clamp off the supply.
7. Rotate unit through 90° in the plane of the monolayer onto the short side away from the inlet.
8. Rotate unit through 90° perpendicular to the plane of the monolayer so that it now lies flat on its base with the culture surfaces horizontal.
9. Incubate MP unit culture for 3–5 days at 37°C in a 5% CO_2/95% air incubator.
10. When manipulating a multiplate cell factory unit, always lift from the sides or the bottom, never by the top plate.

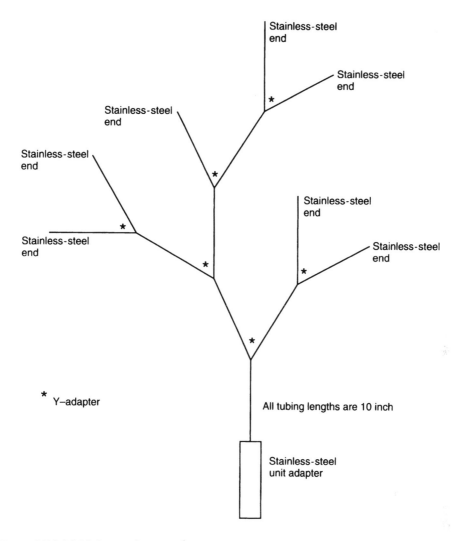

Figure 5.7.1 Multiplate-end connection.

11. Prior to harvest or viral infection, visually inspect the multiplate cell factory for any signs of contamination and/or leaks.
12. Connect four containers to the multiplate-end connection aseptically. The first vessel contains the trypsin solution and the second contains the growth medium required to dilute the cells and inactivate the trypsin. The third and fourth vessels are used to capture the spent medium (supernatant) and the cell suspension, respectively. The recommended procedure for connecting the vessels to the multiplate-end connection is described in detail in Steps 2–4.
13. Open the clamp of the supply tube and drain the spent growth medium by gravity. When the entire unit is empty, clamp off the portion of the tube leading into the discard vessel.

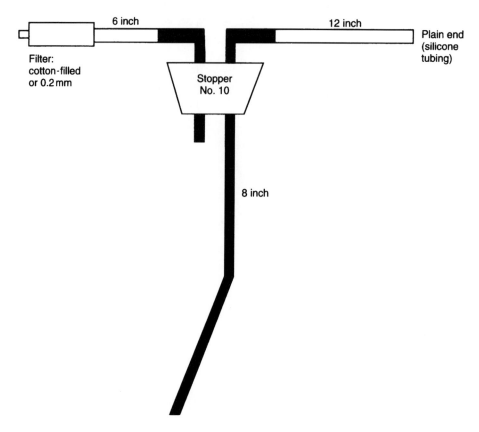

Figure 5.7.2 Stopper connection plus one plain end.

14. Fill the multiplate cell factory unit with 170–200 ml of trypsin and rinse the cell growth surface of each and every chamber. Discard the excess trypsin, leaving a small volume that is just enough to provide superficial moisture. Clamp off the portion of the tube leading into the trypsin vessel. *Note*: Some operators prefer to wash the monolayers with PBS (Ca^{2+}-free and Mg^{2+}-free) prior to the addition of trypsin. When working with a large number of MP units, this additional step becomes impractical.
15. Incubate the multiplate cell factory unit at 37°C until the cells are observed to slide off the growth surface.
16. Once cell detachment is obtained, open the clamp of the tube leading from the vessel containing the growth medium and pump 300–500 ml of cell growth medium inside the cell factory unit. Rotate the unit so that the medium will dislodge as many cells as possible. Close the clamp of the cell growth medium vessel and open the clamp of the portion of the tubing leading into the cell harvest vessel. The cell harvest vessel is equipped with a magnetic bar. For better cell dispersion and homogeneous cell suspension, the magnetic bar must be activated as soon as the first volume of cell suspension reaches the harvest vessel.

17. Repeat Step 16 with an equal amount of medium, making sure that each and every chamber is properly rinsed.
18. Close off the tubing leading into the cell harvest vessel and let the cell suspension mix thoroughly for a few more minutes.
19. Connect a glass bell-end attachment to the multiplate-end connection. Pressurize the harvest vessel and, via the glass bell-end, remove a small volume of cell suspension for a viable cell count. An MP unit seeded with a minimum of 70×10^6 cells will yield approximately $0.8-1.6 \times 10^6$ human diploid cells. For operations involving viral infection instead of trypsinization, the following steps should be considered.
20. Proceed with Step 13.
21. Aseptically, connect the viral seeding vessel to the multiplate-end connection. The viral seeding vessel will contain the desired viral inoculum in viral growth medium.
22. Open the clamp of the tube leading from the viral seeding vessel and pump the contents into the MP unit. Clamp off and incubate the infected MP unit.
23. Following viral growth, proceed with downstream processing of supernatant.

DISCUSSION

Background information

Human diploid cell lines have been utilized classically as *in vitro* hosts for the propagation of polio, mumps, rubella, cytomegalovirus, varicella-zoster, rabies, hepatitis A, respiratory syncytial virus, parainfluenza and many other viruses. Other uses in the biotechnology industry include large-scale cultivation for the production of various cellular products, such as human interferon beta.

For the production of viral vaccines, the cultivation of cells and virus is primarily carried out in a batch mode. Cells are expanded on a compatible growth surface to a desired confluency and then subcultivated repeatedly until a sufficient inoculum is achieved for large-scale cultivation. In order for the microcarrier cultivation to be an efficient process, critical inoculation and growth parameters need to be identified (Lindner *et al.*, 1985; section 5.8). Due to the limited *in vitro* lifespan of human diploid cells and for practical reasons of experimental reproducibility, a cell banking system is utilized. Each experimental run is initiated from cells frozen in liquid nitrogen, for a fixed number of subcultivations. In order to establish a small working cell bank, 3.0×10^6 cells per cryogenic vial (1×10^6 cells ml^{-1}) are frozen in DMEM containing 10% FBS and 7.5% dimethylsulphoxide, at a predetermined population doubling level (PDL) of 20–30. It has been established that microcarrier culture performance is largely influenced by the seeding inoculum density (Hu *et al.*, 1985). For human diploid cells the maximum growth rate is achieved with lower cell concentrations. Conversely, high cell concentrations result in an extended lag phase and poor microcarrier culture growth. This phenomenon appears to be related to the synthesis of intrinsic growth inhibitors and is cell type dependent.

The use of Nunc cell factory technology for large-scale cell culture and viral vaccine production offers a variety of advantages. When utilized properly, the MP units require less than half the incubating and storage space compared to the T-flask or roller bottle system. There is also a reduced possibility for contamination. Once connected correctly, the units are a closed system and do not require multiple closures or openings. If a unit is found to be contaminated, it can be segregated easily from the remainder of the batch. An MP unit is processed in a fraction of the time required to process the $40 \times 150 \text{ cm}^2$ T-flasks that it replaces For certain applications the units can be re-used, making the activity extremely cost-effective. However, the fact that the cell factory cannot be observed microscopically may be the only drawback. This technical difficulty can be overcome by preparing a single plate unit in conjunction with the multiplate cell factory. By carefully examining the single plate, the operator can determine the growth pattern and cell morphology.

Troubleshooting

Suboptimal quality of the raw materials may be detrimental to the growth of human diploid cells. The most obvious component is serum. Sera other than FBS (newborn, calf and adult) appear to support the growth of MRC-5/W1–38, at least

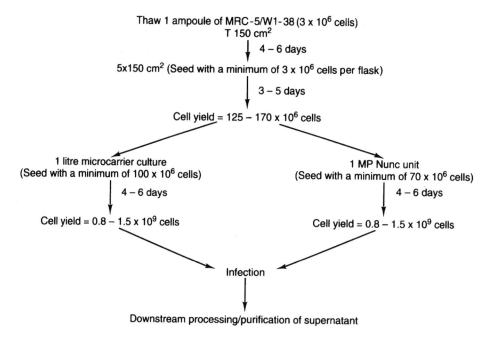

Figure 5.7.3 The expected results and time frames when following a microcarrier or Nunc multiplate (MP) culture path. The microcarrier culture system requires one more extra day of equipment preparation compared with the MP pathway. One should also note that WI-38 will require an additional 24-h growth period in order to achieve comparable yields to that of MRC-5.

for the short term (two or three passages). Longer term passaging requires FBS supplementation. It is important to prescreen your serum, because not all FBS lots will support the growth of human diploid cells. Attention should also be given to the water purity and washing practices of the cultureware and to the substratum (Varani et al., 1988).

REFERENCES

Hu WS, Meier J & Wang DJC (1985) A mechanistic analysis of the inoculum requirement for the cultivation of mammalian cells on microcarriers. *Biotechnology and Bioengineering* 27: 585–595.

Lindner E, Arvidsson AC, Wergeland I & Billig D (1985) Subpassaging cells on microcarriers: the importance for scaling up to production. *Developments in Biological Standardization* 66: 299–305.

Varani J, Bendelow MJ & Hillegas WJ (1988) The effect of substrate on the production of infectious virus by cells in culture. *Journal of Biological Standardization* 16: 333–338.

BACKGROUND READING

Hayflick L (1963) Human diploid cell strains as hosts for viruses. *Perspectives in Virology* 3: 213–237.

Hayflick L (1963) An analysis of the potential oncogenicity of human virus vaccine cell substrates. In: *Proceedings and Symposium on Oncogenicity of Virus Vaccines*, pp. 39–46. Permanent Section of Microbiological Standardization, International Association of Microbiological Societies, Opatija, Yugoslavia.

Hayflick L (1968) Cell substrates for human virus vaccine preparation general comments. *National Cancer Institute Monograph* 29: 83–91.

Hayflick L (1969) *A Consideration of the Cell Substrates used in Human Virus Vaccine Preparation*, Progress in Immunobiological Standardization 3. Karger, Basel.

Hayflick L & Jacobs JP (1968) Suggested methods for the management and testing of a diploid cell culture used for virus vaccine production. In: *Proceedings and Symposium on Oncogenicity of Virus Vaccines*, pp. 181–185. Permanent Section of Microbiological Standardization, International Association of Microbiological Societies, Opatija, Yugoslavia.

Hayflick L, Plotkin SA, Norton TW & Koprowski H (1962) Preparation of poliovirus vaccines in a human fetal diploid cell strain. *American Journal of Hygiene* 75: 240–258.

Pharmacia Fine Chemicals AB (1981) *Microcarrier Cell Culture: Principles and Methods*, Technical Booklet Series. Pharmacia, Uppsala, Sweden.

von Seefried A, Chun JM, Grant JA, Letvenuk L & Pearson EW (1984) Inactivated poliovirus vaccine and test development at Connaught Laboratories Ltd. *Reviews of Infectious Diseases* 6: 345–349.

5.8 MICROCARRIERS – BASIC TECHNIQUES

Microcarrier culture, i.e. the growth of anchorage-dependent cells on small particles (usually spheres) 100–300 μM in size suspended in stirred culture medium, has made a tremendous impact on upstream processes (Griffiths, 1991). During the 1960s, the availability of human diploid cell (HDC) lines allowed a rapid expansion in the manufacture of human vaccines (section 5.7). However, large-scale production was restricted to using many replicate small cultures (flask and roller bottles) because HDC lines were anchorage-dependent, a restriction that did not apply to veterinary vaccines, which were being produced in large scalable fermenter processes (a suspension cell line, BHK, was licensed for the production of veterinary vaccines). A similar unit scale-up process was a goal being sought for surface-growing cells; hence the large range of process bioreactors described in section 5.1 and Chapter 3, Figure 3.6.1. The opportunity came when van Wezel (1967) showed that cells would grow on dextran beads in stirred bioreactors. However, the chromatography-grade dextran (Sephadex A-50, Pharmacia) being used was unsuitable for consistent and reliable growth of cell lines, particularly HDC. After considerable developmental work by van Wezel and Pharmacia, a range of suitable microcarriers became available (the Cytodex range, Pharmacia). The first industrial process using microcarriers was described by Meignier *et al.* (1980) for foot and mouth disease virus (FMDV) vaccine production. Subsequently a whole range of microcarriers based on gelatin, collagen, polystyrene, glass, cellulose, polyacrylamide and silica have been manufactured to meet all situations (Table 5.8.1). The key criteria were to get the surface chemically and electrostatically correct for cell attachment, spreading and growth. To put this development in full perspective the following facts illustrate the sheer scale of opportunity that this method gives the cell culturist:

1. 1 g of Cytodex microcarrier has a surface area of 6000 cm^2 and, used at a modest concentration of 2 g l^{-1} gives 12 000 cm^2 l^{-1}. This is equivalent to 8 large or 15 small roller bottles.
2. Suspension culture systems, unlike bottles, etc., can be environmentally controlled and were optimized in the late 1960s to a minimum 100-l capacity, i.e. a 1 × 100-l fermenter was equivalent to 800–1500 roller bottles. The scale-up potential is 10 000 l (for suspension cells), although currently only 4000 l has been used for microcarrier culture.

The availability of microcarriers has not only opened up industrial opportunities but also allowed the laboratory worker to produce substantial quantities of cells and cell products for research and development purposes.

Cell and Tissue Culture: Laboratory Procedures in Biotechnology, edited by A. Doyle and J.B. Griffiths.
© 1998 John Wiley & Sons Ltd.

Table 5.8.1 Examples of microcarriers

Product name	Manufacturer	Type	Surface area ($cm^2\ g^{-1}$)	Specific gravity
Biosilon	Nunc	Polystyrene	255	1.05
Bioglas	Solihull Engineering	Glass/latex	350	1.03
Bioplas	Solihull Engineering	Polystyrene	350	1.04
Cytodex 1, 2	Pharmacia Biotech	Dextran	6000	1.03
Cytodex 3	Pharmacia Biotech	Collagen	4600	1.04
Dormacell	Pfeifer & Langen	Dextran	7000	1.05
Gelibead	Hazelton Laboratories	Gelatin	3800	1.03
Micarcel G	Reactifs IBF	Polyacrylamide	5000	1.03
Microdex	Dextran Products	Dextran	250	1.03
Ventregel	Ventrex Laboratories	Gelatin	4000	1.03

General principles

1. Cells differ in their attachment requirements, and thus a range of microcarriers should be assessed for suitability (see Table 5.8.1).
2. It is critical to have suitable culture equipment. Spinner flasks should have a special impeller (e.g. Bellco, Wheaton and Techne flasks are specifically designed for microcarriers) because a magnetic bar is unsuitable. The shape of the vessel is important for homogeneous mixing, as is the stirring unit, which must operate smoothly without vibration at low revolutions (15–100 rpm). Culture vessels must be siliconized to prevent microcarriers sticking to the glass, as well as the bottles used for preparing and storing microcarriers (see Preliminary Procedure below).
3. Microcarriers can be purchased as dry powders or sterilized solutions. The dry powders are swelled in Ca^{2+}/Mg^{2+}-free phosphate-buffered saline (PBS) (3 h), the supernatant is decanted and discarded and then the residual powder is washed and sterilized in the same buffer (50 ml g^{-1}) by autoclaving (15 lb in^{-2}, 15 min, 115°C). *Warning*: do not overheat. The powder microcarriers are 'softer' than glass and plastic and can be used in higher concentrations, thus giving vastly greater unit surface areas (e.g. Cytodex, 6000 $cm^2\ g^{-1}$; glass/polystyrene, 300 $cm^2\ g^{-1}$). Some microcarriers are available at different specific gravities (usually between 1.01 and 1.05), offering a choice depending upon the stirring (or mixing) system – roller bottles, shake flasks or airlift bioreactors can be used – and process requirements (e.g. if multiple medium changes are to be made, the heavier beads settle out more quickly and efficiently).
4. Consideration may have to be given to using a supplemented medium during the initial stages of culture to aid cell attachment and to offset the effects of low cell density, which will be more critical in microcarrier than stationary culture. The supplementation may be simple, e.g. non-essential amino acids, pyruvate (0.1 mg ml^{-1}), adenine (10 μg ml^{-1}), hypoxanthine (3 μg ml^{-1}) and thymidine (10 μg ml^{-1}). Other supplements include tryptose phosphate broth (1 mg ml^{-1}), HEPES (5 mM), transferrin (10 mg l^{-1}) and fibronectin (2 μg ml^{-1}). Serum, unless serum factors are added, may have to be used at 5–10% initially before being reduced after 1–2 days of culture.

5. The conventional cell-counting procedure of trypsinizing cells and then counting in a haemocytometer with Trypan blue can be tedious or even inaccurate (due to problems in removing cells from the microcarrier sludge). A better method is to use the nuclei-counting method developed by Sanford *et al.* (1951). The microcarriers are suspended in 0.1 M citric acid containing 0.1% crystal violet. The mixture can be vortex mixed or left at 37°C for at least 1 h. An advantage is that samples can be stored for long periods (at least 1 week) at 4°C before being counted. However, it is a total, not viable, cell count. Cells can also be fixed and stained on microcarriers for microscopic examination using standard procedures. Haematoxylin is the most widely used stain.
6. It is possible to re-use some microcarriers using good washing procedures and re-equilibration in PBS. Except for glass microcarriers this is not recommended, especially for more than one re-cycle, because performance drops significantly. There are reports of *in situ* re-colonization of microcarriers (Crespi & Thilly, 1981), which is feasible but again reduces the efficiency of the process (lower yields and heterogeneity in cell numbers between carriers).
7. Enzymic harvesting (e.g. trypsin) of some cells from microcarriers can be difficult or damaging. Consideration can be given to digesting some microcarriers (e.g. by dextranase, collagenase) because this gives a far quicker and less damaging means of getting a single-cell suspension.
8. Microcarrier culture requires more critical preparation than non-dynamic culture systems. To ensure success it is very important that all experimental details are carried out optimally. Of particular importance is the quality of the cell inoculum. This should be rapidly dividing, not stationary, and cells should be in a good condition (i.e. not trypsin-damaged) and as near a single-cell suspension as possible. It is good practice to feed the seed culture 24 h before use. Also ensure that medium is prewarmed and pre-equilibrated for pH (shifts in pH during attachment are extremely damaging). Due to the number of manipulations in the process, extreme care with aseptic technique should be taken, and as many as possible of the steps carried out in a tissue culture cabinet. Do not inoculate below the minimum number; the culture may not initiate or, if it does, will not reach maximum cell density.

PRELIMINARY PROCEDURE: SILICONIZATION

Materials and equipment

- Dimethyldichlorosilane (Merck or Hopkin and Williams (as Repelcote))

1. Add a small volume of siliconizing fluid to clean glassware (spinner flask, bottles and pipettes used for handling microcarriers) and wet all surfaces.
2. Drain off excess fluid and allow glassware to dry.
3. Wash glassware in distilled water (combination of three washes/prolonged immersion).
4. Autoclave.

This siliconization process will allow glassware to be used through many repeat processes but glassware should be re-treated at least annually.

PROCEDURE: GROWTH OF CELLS ON MICROCARRIERS

Materials and equipment

- Eagle's minimum essential medium (MEM) (or equivalent alternative) with 10% foetal bovine serum (FBS), Eagle's non-essential amino acids and other supplements as necessary (see 'General Principles', point 4).
- Anchorage-dependent cell line (e.g. MRC-5)
- Microcarrier, e.g. Cytodex 3 (Pharmacia)
- Ca^{2+}/Mg^{2+}-free PBS
- Spinner vessel adapted for microcarrier culture (e.g. Bellco or Techne)

1. Add complete medium to spinner flask (200 ml in 1-l flask), gas with 5% CO_2 and allow to equilibrate.
2. Decant PBS from sterilized stock solution of Cytodex 3 and replace with growth medium (1 g to 30–50 ml). Add Cytodex 3 to spinner vessel to give a final concentration of 2 g l^{-1} (use in range 1–3 g l^{-1}).
3. Put spinner on magnetic stirrer at 37°C and allow temperature and physiological conditions to equilibrate (minimum 1 h).
4. Add cell inoculum obtained by trypsinization of late log phase cells (pre-stationary phase) at over five cells per bead (optimum to ensure cells on all beads is seven); inoculate at the same density per square centimetre as with other culture types, e.g. $5-10 \times 10^4$ cm^{-2}. Cytodex 3 at 2 g l^{-1} inoculated at six cells per bead gives:

$$8 \times 10^6 \text{ microcarriers} = 4.8 \times 10^7 \text{ cells } l^{-1}$$

$$9500 \text{ cm}^2 = 5 \times 10^4 \text{ cells cm}^{-2}$$

$$200 \text{ ml} = 2.5 \times 10^5 \text{ cells ml}^{-1}$$

5. Place spinner flask on magnetic stirrer and stir at the minimum speed to ensure that all cells and carriers are in suspension (usually 20–30 rpm). It is advantageous if cells and carriers are limited to the lower 60–70% of the culture for this purpose. Alternatively either:
   - Inoculate in 50% of the final medium volume and add the rest of the medium after 4–8 h, or
   - Stir intermittently (for 1 min every 20 min) for the first 4–8 h. However, only use this option for cells with very poor plating efficiency because it causes clumping and uneven distribution of cells per bead
6. When the cells have attached (expect 90% attachment), the stirring speed can be increased to allow complete homogeneity (40–60 rpm).
7. Monitor progress of culture by taking 1-ml samples at least daily; observe microscopically and carry out cell (nuclei) counts.
8. As the cell density increases there is often a tendency for microcarriers to begin clumping. This can be avoided by increasing the stirring speed to 75–90 rpm.
9. After 3–4 days the culture will become acid. Remove the spinner flask and re-gas the headspace and/or add sodium bicarbonate (5.5% stock solution).

With some cells, or at microcarrier densities of 3 g l^{-1} or more, partial medium changes should be carried out. Allow culture to settle (5–10 min), siphon off at least 50% (usually 70%) of the medium and replace with prewarmed fresh medium (serum can be reduced or omitted at this stage). Replace spinner flask on stirrer.

10. After 4–5 days cells reach a maximum cell density (confluency) at the same level as in static cultures, although multilayering is not so prevalent. Thus a cell yield of $1–2 \times 10^6$ ml^{-1} ($2–4 \times 10^5$ cm^{-2}) can be expected.
11. Cells can be harvested in the following way:
    - Allow culture to settle (10 min)
    - Decant off as much medium as possible (>90%)
    - Add warm Ca^{2+}/Mg^{2+}-free PBS (or EDTA in PBS) and mix
    - Allow culture to settle and decant off as much PBS as possible
    - Add 0.25% trypsin (30 ml) and stir at 75–100 rpm for 10–20 min at 37°C
    - Allow the beads to settle out (2 min). Either decant trypsin plus cells or pour mixture through a sterile, coarse-sinter, glass filter or a specially designed filter such as the Cellector (Bellco)
    - Centrifuge cells (800 g for 5 min) and resuspend in fresh medium (with serum or trypsin inhibitor, e.g. soybean inhibitor at 0.5 mg ml^{-1})

As a guide to calculating settled volume and medium entrapment by microcarriers, 1 g of Cytodex, for example, has a volume of 15–18 ml

DISCUSSION

The basic principles for using microcarrier culture are described, together with many notes on how to avoid problems and get the most out of this very powerful and useful technology. The description is suited to small-scale processes based on spinner flasks (i.e. 200 ml to 5 l) at levels that do not need special adaptations for perfusion, etc. This does not mean that scale-up is not possible; in fact, it has been volumetrically scaled up to 4000 l for the production of interferon and viral vaccines. It has also been scaled up in density by the use of spin-filters (Griffiths et al., 1987) to allow continuous perfusion of the culture and thus operation at microcarrier concentrations up to 15 g l^{-1} and cell densities over 10^7 ml^{-1}.

REFERENCES

Crespi CL & Thilly WG (1981) Continuous cell propagation using low-charge microcarriers. *Biotechnology and Bioengineering* 23: 983–993.

Griffiths JB (1991) Cultural revolutions. *Chemistry and Industry* 18: 682–684.

Griffiths JB, Cameron DR & Looby D (1987) A comparison of unit process systems for anchorage dependent cells. *Developments in Biological Standardization* 66: 331–338.

Meignier B, Mougeot H & Favre H (1980) Foot and mouth disease virus production on microcarrier-grown cells. *Developments in Biological Standardization* 46: 249–256.

Sanford KK, Earle WR & Evans VJ (1951) The measurement of proliferation in tissue cultures by enumeration of cell nuclei. *Journal of the National Cancer Institute* 11: 773–795.

van Wezel AL (1967) The growth of cell strains and primary cells on microcarriers in homogeneous culture. *Nature (London)* 216: 64–65.

BACKGROUND READING

Pharmacia (1981) *Microcarrier cell culture: Principles and methods*. Pharmacia LKB Biotechnology, 5-75182 Uppsala, Sweden.

Reuveny S (1990) Microcarrier culture systems. In: Lubiniecki AS (ed.) *Large-scale Mammalian Cell Culture Technology*, pp. 271–341. Marcel Dekker, New York.

5.9 POROUS MICROCARRIER AND FIXED-BED CULTURES

Porous microcarrier technology is currently the most successful scale-up method for high-density perfused cultures. The technology was pioneered by the Verax Corporation (now Cellex Biosciences Inc.) and systems are available from 16 ml to 24-l fluidized-bed bioreactors (Runstadler *et al.*, 1989; Ray *et al.*, 1990). The smallest system in the range, Verax System One, which is a benchtop continuous perfusion fluidized-bed reactor, was designed for process assessment and development. A great advantage is that the results achieved in the System One will scale up directly to the System 2000. The process is based on the immobilization of cells (both anchorage-dependent and suspension) in porous collagen microspheres. The spheres are weighted (specific gravity 1.6) so that they can be used in fluidized beds at high recycle flow rates (typically 75 cm min^{-1}). The microspheres (Figure 5.9.1) have a sponge-like structure with a pore size of 20–40 μm and pore volume of 85%, allowing the immobilization of cells to high density ($1\text{--}4 \times 10^8$ cells ml^{-1}).

The culture system is based on a fluidized-bed bioreactor containing the microspheres, through which the culture fluid flows upward at a velocity sufficient to suspend the microspheres in the form of a slurry. For oxygenation the medium is recycled through a membrane oxygenator. The system is run for long culture periods (typically over 100 days) by continuously removing the harvest and replacing it with fresh medium (Figure 5.9.2).

The protocols for using the Verax System One are supplied with the culture unit and have been published previously (Looby, 1993). The culture productivity data presented in Tables 5.9.1 and 5.9.2 demonstrate its capability in comparison with other systems and during scale-up.

An alternative commercial system that is now available is the Cytopilot (Pharmacia) (Reiter *et al.*, 1990, 1991), which is a 25-l system using polyethylene carriers (Cytoline) that supports up to 1.2×10^8 CHO K1 cells ml^{-1} carrier. The range of porous microcarriers available is given in Table 5.9.3.

Porous microcarriers can be used in stirred tank (Mignot *et al.*, 1990) and fixed-bed cultures (Looby & Griffiths, 1989; Looby *et al.*, 1990; Kratje *et al.*, 1991) as well as the more commonly used fluidized beds. As stirred tank procedures are still very developmental, even though the design of specific porous microcarriers such as ImmobaSil (Table 5.9.3) is ensuring rapid progress, a procedure for fixed-bed culture is given in detail.

Cell and Tissue Culture: Laboratory Procedures in Biotechnology, edited by A. Doyle and J.B. Griffiths.
© 1998 John Wiley & Sons Ltd.

POROUS MICROCARRIER AND FIXED-BED CULTURES

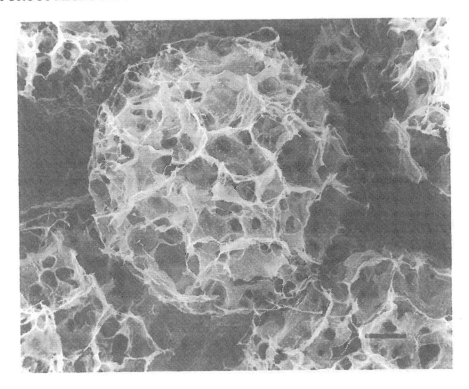

Figure 5.9.1 Scanning electron micrograph of Verax microsphere without cells. Reproduced by permission of Cellex Biosciences Inc.

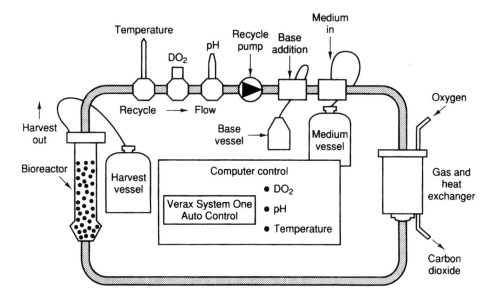

Figure 5.9.2 Schematic diagram of Verax culture system.

Table 5.9.1 Comparison of hybridoma culture productivity in different culture systems[a]

System	Max. cell number viable ($\times 10^{10}$ l^{-1})[b]	Monoclonal antibody productivity average (mg l^{-1} day^{-1})[b]
Verax System One	15.00	540.0
Chemostat	0.12	24.5
Airlift	0.12	18.5
Stirred reactor	0.19	25.5

[a]Data from Looby et al. (1992)
[b]Results are normalized to per litre reactor active volume (i.e. volume containing most of the cells).

Table 5.9.2 Predicted scale-up of hybridoma growth and antibody productivity of Verax fluidized beds[a]

System	Cell numbers ($\times 10^{10}$ viable)	Steady-state productivity[b] (g day^{-1})	Bed volume fluidized (l)	Medium feed rate[b] (l day^{-1})
System One	0.30	0.04	0.02	1.10
System Ten	2.25	0.30	0.15	8.25
System 20	4.50	0.60	0.30	16.50
System 200	24.00	3.20	1.60	88.00
System 2000	360.00	48.00	24.00	1320.00

[a]Data from Looby et al. (1992).
[b]Assuming linear scale-up.

Table 5.9.3 Porous microcarriers

Porous microcarriers	Manufacturer	Type	Diameter (μm)
Cellsnow	Kirin Ltd	Cellulose	800–1000
Cytocell	Pharmacia	Cellulose	180–210
Cultispher	Hyclone	Gelatin	170–270
Cytoline 1, 2	Pharmacia	Polyethylene	1200–1500
ImmobaSil	Ashby Scientific	Silicone rubber	1000
Siran	Schott Glaswerke	Glass	400–5000

Fixed-bed culture

The fixed-bed, porous-glass-sphere culture system was designed for the production of secreted cell products and lytic virus. The system is based on the immobilization of cells (anchorage-dependent or suspension) to high cell densities in porous glass spheres (supplied by Schott Glaswerke).

Large spheres (5 mm diameter) are used in fixed beds because they give an open bed structure (approximately 1 cm^2 channel cross-sectional area), which minimizes blockages due to biomass build-up, uneven distribution of the inoculum and media channelling within the bed (Looby & Griffiths, 1988, 1989, 1990; Griffiths & Looby, 1991). The characteristics of the spheres are given in Table 5.9.4.

POROUS MICROCARRIER AND FIXED-BED CULTURES

Table 5.9.4 Characteristics of Siran porous glass spheres suitable for fixed-bed reactors

Material	Borosilicate glass
Average diameter	3–5 mm
Pore size	60–300 μm
Pore volume	60% open
Total surface area	75 m^2 l^{-1} [a]
Biocompatible	Yes
Steam sterilizable/autoclavable	Yes
Re-usable	Yes

[a] Surface area based on fixed-bed volume.

The system consists of a reactor vessel containing the carriers for cell growth and a reservoir vessel for medium (Figure 5.9.3). A recycle flow of medium from 40 ml, increasing with cell growth to 450 ml (per litre packed bed volume per minute), is pumped up through the bed and returned to the reservoir for oxygenation. The system can be operated in batch, repeated batch feed and harvest, or continuous mode.

The operation of a 1.0 l reactor will be described here. The reactor vessel is custom made from borosilicate glass with a water-jacket for temperature control.

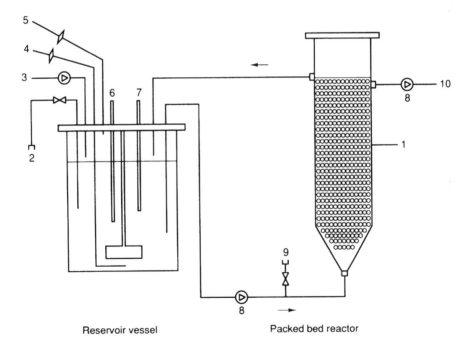

Figure 5.9.3 Schematic diagram of fixed-bed porous-glass-sphere culture system: (1) packed bed of porous glass spheres; (2) sampling port; (3) medium fed in; (4) air/oxygen sparge; (5) off-gas filters; (6) pH probe; (7) dissolved oxygen probe; (8) peristaltic pump; (9) inoculation port; (10) harvest.

A 15 l (11 l working volume) Applikon stirred tank, or equivalent, reactor is used as a medium reservoir.

PRELIMINARY PROCEDURE: INITIAL PREPARATION AND CALIBRATION OF EQUIPMENT

Materials and equipment

- 4% Decon
- 1.0 l reactor with water-jacket
- 15 l (11 l working volume) Applikon stirred tank or equivalent
- pH probes (Ingold)
- Polarographic dissolved oxygen (DO) probes (Ingold)

Bead preparation

1. Boil the beads in distilled water for 2 h.
2. Rinse thoroughly in distilled water.
3. Place beads in an oven to dry (approximately 4 h at 100°C).
4. Rinse thoroughly in distilled water.
5. Place beads in an oven to dry (approximately 4 h at 100°C).
6. Repeat Steps 4 and 5 three times.
7. Place dry beads in a beaker, cover with foil and seal with autoclave tape.
8. Sterilize the beads with dry heat at 180°C for 2 h.

Preparation of culture vessels

1. Wash the reservoir and reactor in 4% Decon.
2. Rinse thoroughly in distilled water.

Calibration of probes

1. Calibrate the pH probes (Ingold) according to manufacturer's instructions.
2. Calibrate the polarographic DO probes (Ingold) with N_2/air-saturated medium (at 37°C) according to manufacturer's instructions.

PROCEDURE: ASSEMBLY OF CULTURE VESSELS

Materials

- Silicon tubing
- Air filter and vent filter
- Marprene tubine
- Peristaltic pump
- Male and female stericonnectors
- Porous glass spheres (Schott Glaswerke)

Note: Maintenance of sterility during all steps involving additions or connections to sterilized equipment is essential.

Reservoir

1. Attach air filter with silicon tubing to sparge tube.
2. Attach air filter with silicon tubing to top air tube.
3. Attach vent filter with silicon tubing to air condenser.
4. Insert Ingold polarographic DO probe into DO holder.
5. Insert Ingold pH probe into pH holder.
6. Prepare the medium-out recycle tube, i.e. silicon tubing (5 mm diameter) with a short piece of marprene tubing inserted for the peristaltic pump, and a male stericonnector with blanking plug attached to the end of the tube.
7. Attach the medium-out recycle tube to the medium-out tube (long tube).
8. Attach short piece (50 cm) with a female stericonnector and blanking plug to the medium recycle return tube (short tube).
9. Attach three short pieces of silicon tubing with female stericonnectors and blanking plugs, i.e. for base and medium addition, with one spare entry port.
10. Place a Universal bottle on the sampling plot.

Reactor

1. Attach two pieces of silicon tubing with female stericonnectors and blanking pieces to Y-piece with single piece of tubing to inlet at base of reactor, i.e. for the medium-out recycle tube from reservoirs and inoculation vessel.
2. Attach silicon tube with male stericonnector and blanking piece to the medium-out recycle port on the reactor (cut to sufficient length to reach the medium-in recycle tube on the reservoir vessel).
3. Attach silicon tubing with male stericonnector and blanking plug to harvest port.
4. Insert probes (DO polarographic and pH) into probe holders.
5. Attach vent filter.

Sterilization

1. Add 200 ml of distilled water to reservoir and 100 ml of distilled water to culture vessel.
2. Cover stericonnectors with foil.
3. Autoclave at 121°C for 30 min.

Bead addition

1. Place the reactor in a tissue culture cabinet.
2. Spray the headplate with 70% ethanol.
3. Remove the headplate.
4. Remove the foil from beaker.
5. Pour beads into the reactor.
6. Replace the headplate.
7. Autoclave at 121°C for 30 min.

PROCEDURE: SYSTEM SET-UP

Materials

- Thermocirculator
- pH controller
- DO controller

1. Attach thermocirculator with silicon tubing to reactor and reservoir, fill with water and set temperature to 37°C.
2. Connect medium recycle stericonnectors.
3. Place marprene tubing in the peristaltic pump head.
4. Add prewarmed medium to the reservoir vessel.
5. Leave for 4 h to equilibrate.
6. Take a sample from the reservoir and measure pH using an independently calibrated pH meter.
7. Adjust pH controller to read the pH of the medium.
8. Check calibration of DO probe in reservoir according to manufacturer's instructions and adjust if necessary.
9. Adjust pH and DO controller set points to desired values.

PROCEDURE: INOCULATION AND MAINTENANCE OF CULTURE SYSTEM

Inoculation

1. Resuspend inoculum (5.0×10^9 cells l^{-1} bed volume) in 800 ml (bed void volume) fresh medium.
2. Add to inoculation vessel.
3. Connect to reactor vessel.
4. Gently swirl the inoculation vessel to ensure an even cell suspension.
5. Slowly pump the inoculum into the bed with air pressure (hand pump), being careful not to introduce air bubbles into the system.
6. Drain and refill the bed twice.
7. For suspension cells, start perfusing immediately at 40 ml min^{-1} or at a linear flow velocity of 2 cm min^{-1}.
8. For anchorage-dependent cells, leave stationary for 2–4 h for the cells to attach before perfusing at 40 ml min^{-1}.

Batch operation

Run the culture until the maximum product concentration is reached.

Repeated feed and harvest operation

The reservoir medium volume (11 l) is changed when the glucose concentration in the medium falls below 2 mg ml^{-1}, i.e. daily in steady/pseudo-steady-state culture.

Continuous operation

The dilution rate (medium feed rate) is adjusted to give a medium glucose concentration of approximately 2 mg l^{-1}.

SUPPLEMENTARY PROCEDURE: ANALYSIS OF CONSUMPTION AND PRODUCTION RATES IN THE FIXED-BED POROUS-GLASS-SPHERE CULTURE SYSTEM

Take a 10-ml sample daily from the reservoir, and perform the following:

- Viable cell count (free cells)
- Glucose assay
- Lactate assay
- Product assay

Process calculations

Batch, repeated batch feed and harvest modes

Glucose consumption rate (GCR):

$$\text{GCR} = \frac{(G_1 - G_2)}{(t_1 - t_2)} V$$

Lactate production rate (LPR):

$$\text{LPR} = \frac{(L_1 - L_2)}{(t_1 - t_2)} V$$

Product production rate (PPR):

$$\text{PPR} = \frac{(P_1 - P_2)}{(t_1 - t_2)} V$$

Continuous perfusion

Glucose consumption rate (GCR):

$$\text{GCR} = \text{MFR}\,(G_5 - G_3) - \frac{(\Delta C)}{(\Delta t)} V$$

when:

$$\frac{\Delta C}{\Delta t} = \frac{(G_3 - G_4)}{(t_2 - t_1)}$$

Lactate production rate (LPR):

$$\text{LPR} = \text{MFR}(L_2) + \frac{(\Delta L)}{(\Delta t)} V$$

when:

$$\frac{\Delta L}{\Delta t} = \frac{(L_3 - L_1)}{(t_2 - t_1)}$$

Product production rate (PPR):

$$\text{PPR} = \text{MFR}(P_4) + \frac{(\Delta P)}{(\Delta t)} V$$

when:

$$\frac{\Delta P}{\Delta t} = \frac{(P_3 - P_4)}{(t_2 - t_1)}$$

See Table 5.9.5 for definitions of the above terms.

Table 5.9.5 List of symbols

Symbol	Definition
G_1	Glucose concentration before medium change – current (mg ml^{-1})
G_2	Glucose concentration after medium change – previous (mg ml^{-1})
G_3	Glucose concentration – current (mg ml^{-1})
G_4	Glucose concentration – previous (mg ml^{-1})
G_5	Glucose concentration in feed medium (mg ml^{-1})
V	Total medium volume in reactor and reservoir (ml)
t_1	Time of previous sample (days)
t_2	Time of current sample (days)
L_1	Lactate concentration before medium change – current (mg ml^{-1})
L_2	Lactate concentration after medium change – previous (mg ml^{-1})
L_3	Lactate concentration – current (mg ml^{-1})
L_4	Lactate concentration – previous (mg ml^{-1})
F	Recycle flow rate (ml min^{-1})
P_1	Product concentration before medium change – current (mg ml^{-1})
P_2	Product concentration after medium change – previous (mg ml^{-1})
P_3	Product concentration – previous (mg ml^{-1})
P_4	Product concentration – current (mg ml^{-1})
MFR	Medium feed rate (ml day^{-1})
GCR	Glucose consumption rate (mg day^{-1})
LPR	Lactate production rate (mg day^{-1})
PPR	Product production rate (mg day^{-1})
ΔC	Change in concentration of glucose (mg ml^{-1})
ΔL	Change in concentration of lactate (mg ml^{-1})
ΔP	Change in concentration of product (mg ml^{-1})

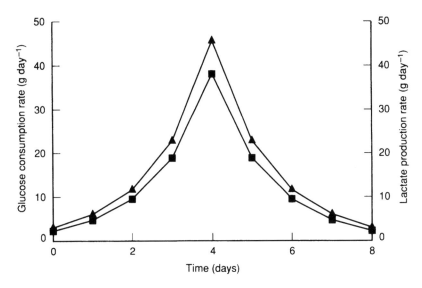

Figure 5.9.4 Glucose consumption rate (■) and lactate production rate (▲) in batch mode.

Example of fixed-bed culture system in different modes of operation

A graphic presentation of a murine hybridoma grown in a fixed-bed bioreactor under batch mode is given in Figures 5.9.4 and 5.9.5. The medium used in these examples was Dulbecco's modified Eagle's medium (DMEM) with 4.5 g l^{-1} glucose, 4 mM glutamine and 5% foetal calf serum (FCS).

SUPPLEMENTARY PROCEDURE: TERMINATION OF CULTURE AND DETERMINATION OF CELL NUMBERS

Run termination

1. Switch off DO controllers.
2. Switch off pH controllers.
3. Switch off thermocirculator.
4. Drain remaining medium/harvest from the system.
5. Drain water from the water-jacket.
6. Remove beads from the bed.
7. Take a 20-ml sample of beads.
8. Autoclave vessels and medium bottles.
9. Do a cell count on the 20-ml sample of beads (as below).

Determination of final cell numbers (total)

1. Place 20-ml sample of beads into a 250-ml conical flask.
2. Add 100 ml of 0.1% citric acid and 0.2% Triton X-100.

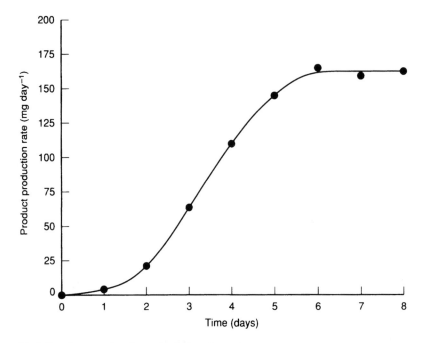

Figure 5.9.5 Product production rate in batch mode.

3. Cover with Parafilm.
4. Place the flask in an orbital shaker at 37°C and shake at 150 rpm for 2 h.
5. Remove all supernatant and measure volume.
6. Wash the beads with phosphate-buffered saline (PBS).
7. Add crystal violet (10%) to the supernatant to give a final concentration of 0.1%.
8. Count the nuclei in a haemocytometer.

Productivity and scale-up

The productivities of the different culture systems are compared in Table 5.9.6, with reference to viable cell numbers and monoclonal antibody productivity. Table 5.9.7 gives a prediction of the scale-up of hybridoma growth and productivity in fixed-bed Siran porous-glass-sphere reactors.

DISCUSSION

Fluidized bed technology is ideally suited for the production of secreted cell products in long-term culture. It can also be used for lytic virus production; however, because this is a batch process, it is not ideally suited to fluidized beds. It must also be emphasized that this technology is not suitable for cell-associated products, e.g. some viruses.

Table 5.9.6 Comparison of culture productivity in different culture systems[a]

System	Max. cell numbers viable ($\times 10^{10}$ l^{-1})[b]	Monoclonal antibody productivity average (mg l^{-1} day^{-1})[b]
Fixed bed	2.75	166.0
Chemostat	0.12	24.5
Airlift	0.12	18.5
Stirred reactor	0.19	25.5

[a]Data from Looby et al. (1992).
[b]Results are normalized to per litre reactor active volume (i.e. volume containing most of the cells).

Table 5.9.7 Predicted scale-up of hybridoma growth and productivity in fixed-bed Siran porous glass-sphere reactors

System bed volume (l)	Total viable cell no. ($\times 10^7$)	Harvest (l day^{-1})	Monoclonal antibody productivity (mg day^{-1})
0.1	275.0	1.0	15.0
1.0[a]	2750.0	10.0	150.0
2.5	6875.0	25.0	375.0
100.0	275 000.0	1000.0	15 000.0

[a]Data from Looby et al. (1992).

A great deal of effort has gone into the development of high-density immobilized perfusion culture systems, which can be operated continuously. The limitation with many of these systems (e.g. hollow fibre) is that, whilst they achieve very high unit cell density (typically 10^8 ml^{-1}), they do not scale up well volumetrically. This limitation has been overcome by the use of porous carrier immobilization techniques, where high unit cell density can be combined with good volumetric scale-up potential and long-term continuous operation.

Fixed-bed solid-glass-sphere culture systems have been in existence for many years (Spier & Whiteside, 1976; Robinson et al., 1980; Whiteside & Spier, 1981; Looby & Griffiths, 1987; Brown et al., 1988); however, they have not achieved widespread use, mainly because they have a low surface area per unit volume (0.7 m^2 l^{-1} for 5-mm spheres) and they are not suitable for the immobilization of suspension cells. These disadvantages were overcome by the introduction of the fixed-bed porous-glass-sphere (Siran) culture system, which has a very large surface area per unit volume (74 m^2 l^{-1}) and is suitable for suspension as well as anchorage-dependent cells (Brown et al., 1988; Bohmann et al., 1992). This has led to an increased interest in the potential of fixed-bed glass-sphere culture systems for the production of animal cell products (Bohmann et al., 1992).

REFERENCES

Bohmann A, Portner R, Schnmieding J, Kusche V & Manke H (1992) The membrane dialysis bioreactor with integrated radial flow fixed bed – a new approach for continuous cultivation of animal cells. *Cytotechnology* 9: 51–57.

Brown PC, Figueroa C, Costello MAC & Maciukas SM (1988) Protein production from mammalian cells grown on glass beads. In: Spier RE & Griffiths JB (eds) *Animal Cell Technology*, Vol. 3, pp. 251–262. Academic Press, London.

Griffiths JB & Looby D (1991) Fixed immobilised beds for the cultivation of animal cells. In: Ho CS & Wang DIC (eds) *Animal Cell Bioreactors*, pp. 165–188. Butterworth–Heinemann, NY.

Kratje R, Jager V & Wagner R (1991) Comparison of the production efficiency of cells grown in a fluidised and in a stirred tank bioreactor. In: Spier RE, Griffiths JB & Meignier B (eds) *Production of Biologicals from Animal Cells in Culture*, pp. 528–532. Butterworth, Guildford.

Looby D (1993) Fluidized bed culture. In: Doyle A, Griffiths JB & Newall D (eds) *Cell and Tissue Culture: Laboratory Procedures*, pp. 28D: 1.1–1.13. Wiley, Chichester.

Looby D & Griffiths JB (1987) Optimisation of glass sphere immobilised bed cultures. In: Spier RE & Griffiths JB (eds) *Modern Approaches to Animal Cell Technology*, pp. 342–452. Butterworth, Guildford.

Looby D & Griffiths JB (1988) Fixed bed porous glass sphere (porosphere) bioreactors for animal cells. *Cytotechnology* 1: 339–346.

Looby D & Griffiths JB (1989) Immobilisation of animal cells in fixed and fluidised porous glass sphere reactors. In: Spier RE, Griffiths JB, Stephenne J & Crooy PJ (eds) *Advances in Animal Cell Biology and Technology for Bioprocessors*, pp. 336–344. Butterworth, Guildford.

Looby D & Griffiths JB (1990) Immobilisation and animal cells in porous carrier culture. *Trends in Biotechnology* 8: 204–209.

Looby D, Racher AJ, Griffiths JB & Dowsett AB (1990) The immobilisation of animal cells in fixed and fluidised porous glass sphere reactors. In: de Bont JAM, Visser J, Mattiasson B & Tramper J (eds) *Physiology of Immobilised Cells*, pp. 255–264. Elsevier Science, Amsterdam.

Looby D, Griffiths JB & Racher AJ (1992) Productivity of a hybridoma cell line in a range of suspension and immobilised culture systems. In: Spier RE, Griffiths JB & MacDonald C (eds) *Animal Cell Technology: Developments, Processes and Products*, pp. 331–335. Butterworth–Heinemann, Oxford.

Mignot G, Faure T, Ganne V, Arbeille B, Pavirani A & Romet-Lemonne JL (1990) Production of recombinant Von Willebrand factor by CHO cells cultured in macroporous microcarriers. *Cytotechnology* 4: 163–171.

Racher AJ, Looby D & Griffiths JB (1990) Studies on monoclonal antibody production of a hybridoma cell line (C1E3) immobilised in a fixed bed porosphere culture system. *Journal of Biotechnology* 15: 129–146.

Ray NG, Tung AS, Hayman EG, Vournakis JN & Runstadler PW Jr (1990) Continuous cell culture in fluidized bed reactors: cultivation of hybridomas and recombinant CHO cells immobilised in collagen microspheres. *Annals of the New York Academy of Sciences, Biochemical Engineering VI* 589: 443–457.

Reiter M, Hohenwarter O, Gaida T, Zach N, Schmatz C, Blüml G, Weigang F, Nilsson K & Katinger H (1990) The use of macroporous carriers for the cultivation of mammalian cells in fluidised bed reactors. *Cytotechnology* 3: 271–277.

Reiter M, Blüml G, Gaida T, Zach N, Unterluggaaer F, Dublhoff-Dier M, Noe R, Placl R, Huss S & Katinger H (1991) Modular integrated fluidised bed bioreactor technology. *Biotechnology* 9: 1100–1102.

Robinson NH, Butlin PM & Imrie RC (1980) Growth characteristics of human diploid fibroblasts in packed beds of glass beads. *Developments in Biological Standardization* 46: 173–181.

Runstadler PW Jr, Tung AS, Hayman EG, Ray NG, Sample JvG & DeLucia DE (1989) Production of tissue plasminogen activator and monoclonal antibodies in continuous culture employing serum free media. In: Lubiniecki AS (ed.) *Large Scale Mammalian Cell Culture Technology*, vol. 3, pp. 363–381. Marcel Dekker, New York.

Spier RE & Whiteside JP (1976) The production of foot-and-mouth disease virus from BHK 21 cells grown on the surface of glass spheres. *Biotechnology and Bioengineering* 18: 649–657.

Whiteside JP & Spier RE (1981) The scale-up from 0.1 to 100 liter of unit process system based on 3 mm diameter glass spheres for the production of four strains of FMDV from BHK monolayer cells. *Biotechnology and Bioengineering* 23: 551–565.

5.10 CONTROL PROCESSES

The classical cell culture method (which is still used) is very simple but lacks efficient process control strategies. There is an incubator, providing reasonably good temperature control, and a CO_2-enriched atmosphere that interacts with a carbonate-buffered system to keep the pH within an acceptable range. All parameters for the regulation of the culture are taken off-line, mainly relying upon microscope observation by experienced operators. This strategy is appropriate for laboratory use, where a fully controlled, probably automated, process and good economy are not the main issues. For production processes, and for many investigative purposes where high reproducibility, high efficiency or high safety level, or all three, are needed, then more control and regulation of the environment is necessary.

Depending on the features needed, there is a wide range of possibilities available from commercial suppliers. The minimum configuration is a ready-to-use control cabinet measuring (and controlling) temperature, pH and pO_2 by set point devices. This may be sufficient for cell mass or product generation purposes using laboratory-scale, low-cell-density cultures. There should be no intention of creating a high-efficiency production process from such a set-up, because the very high cell densities are very sensitive to a balance between the physiological status of the cells and the environmental parameters. This cannot be achieved by set point control alone. Additional measurements and automatic, very quick and accurate control loops are needed for such processes, which are mainly dedicated to industrial production.

Defining the needs for the different levels of process sophistication is the first prerequisite for creating the appropriate set-up. To facilitate this, some of the principal features of processes and their control are discussed. For a detailed insight into process design, sensors and electronic devices, the following articles are recommended: Merten (1988a), Fleischacker *et al.* (1981), Bliem & Katinger (1988a,b), Einsele *et al.* (1985), Schügerl (1988), Scheirer & Merten (1991), Harris & Spier (1985), Webb & Mavituna (1987), Werner & Nöe (1993 a,b,c) and Buetemeyer *et al.*, 1994).

BASIC PROCESS CONTROL

Process control for animal cell fermentation should at least include a set point control of temperature, pH, pO_2 and agitation rate.

Cell and Tissue Culture: Laboratory Procedures in Biotechnology, edited by A. Doyle and J.B. Griffiths.
© 1998 John Wiley & Sons Ltd.

CONTROL PROCESSES

Temperature control

As a consequence of low stirring rates, and the long mixing time inherent in animal cell fermentation, there is a high risk of localized overheating when using standard microbiological fermentation devices. Therefore, the best solution is a water-jacketed vessel fitted with warm water circulation, with the temperature limited to a value just 1–3°C higher than that of the reactor by a control loop overriding the reactor loop. Use hot water or low-energy electric heating elements rather than steam injection as the energy source to ensure minimum temperature variation.

Alternatively, for vessels up to 10 l, an electric heating band (barrel heater) may be used, having a power of approximately 30 W l^{-1} of reactor volume. In no case should heating rods or tubes inside the reactor be used, because localized high temperatures, aggravated by the low stirring rate, will cause unfavourable medium alterations and cell damage.

The control unit of the reactor should be a quick-response sensor inside the reactor (preferably Pt100 with a three-wire connection) and a proportional controller to achieve optimal temperature constancy. In no case should there be a deviation of more than 0.2°C during operation (during initial warming a little more has to be accepted).

pH control

The use of pressurized, sterilizable combined glass electrodes mounted through the reactor vessel wall is widely established. A high degree of accuracy and stability is needed to provide a good basis for the control loop. However, a recalibration is necessary both after sterilization and at regular intervals. This is easily done by external measurement after sampling the reactor, because the pH of the sample drifts very slowly.

There are system amplifiers/controllers supplied by the equipment manufacturers, but a standard laboratory titration instrument is also very useful for experimental fermentations.

With glass membrane pH probes there is some fouling of the membrane after several weeks of operation in protein-containing media. Therefore, for continuous processes that need accurate pH control, a changeable electrode mounting device should be considered (e.g. Bioengineering or Ingold).

Because all cell culture media are carbonate-buffered systems, the pH is dependent on the CO_2 in solution, which in turn is in equilibrium with the CO_2 in the gas phase. This offers the opportunity of using the CO_2 concentration within the gas phase for gentle, very efficient pH control (Figure 5.10.1). Alternatively, or in addition, pH control is possible by using base titration (acid titration is given by CO_2 anyway). When using hydroxide, one should use a 1 M mixture of NaOH and KOH (95:5) to avoid too great a shift in Na/K ratio and osmolarity.

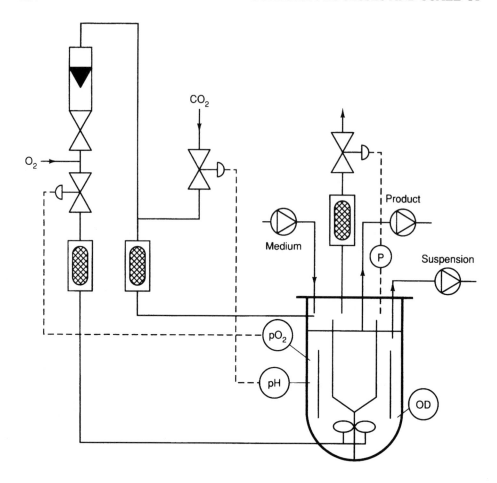

Figure 5.10.1 pH control by CO_2. OD = optical density.

Control of dissolved oxygen

The availability of polarimetric electrodes means that there is now a stable and accurate sensor suitable for long-term animal cell fermentations. However, there are still several shortcomings:

1. The working cycle between the regular electrode services is limited to 6–8 weeks, depending on the oxygen tension. This could be overcome by the use of interchangeable housings or by the use of a second electrode, which is installed at the beginning of the run but connected electrically several weeks after the first electrode to save measuring capacity but to allow tuning while the first electrode is still in operation.
2. Recalibration after the start of the fermentation is difficult.
3. The response time for these electrodes is in the range of 1 min for 98% response.

To overcome these drawbacks with high-density cultures, a well-balanced design of the oxygenation system is necessary. The gas volume of the system from the control valve to the sparger outlet should be as low as possible to avoid over-titration. The proportional controller should have a self-optimization program that allows adaptation of the control characteristics to the electrode, pneumatics and sparger characteristics. Additionally, an overriding limitation of gas flow should be installed to minimize oscillations and restrict foaming. This can be done in a simple way by gas flow meters and needle valves or by using electronic mass flow measurement and control.

Control of agitation

'Agitation rate' is synonymous with 'stirrer speed'. Because the stirrer speed with animal cell fermentations is much lower than for bacterial fermentations (20–200 rpm versus 1000–3000 rpm), there is a need for a slower drive, even in the case of a nominal control range starting from zero. This is because there is too much energy loss with wide-range electronic down-regulation, resulting in poor uniformity of speed.

With the proper mechanical drive range, all standard-speed controllers can be used successfully, but a tachometric control may be the most favourable solution. Additionally, this offers a convenient signal for speed recording.

The consideration of agitation rate/shear stress is a particular problem with scaling-up. Some systems show a favourable behaviour (e.g. airlift, circular loop), while others are quite difficult (e.g. Rushton/Microcarrier). For a detailed analysis of these problems, please consult the specialized literature (Katinger *et al.*, 1979; Bliem & Katinger, 1988a,b; Handa-Corrigan, 1990; Kretzmer *et al.*, 1991; Merchuk, 1991; Van t'Riet & Tamper, 1991; Chapter 4, section 4.7).

Suppliers of standard cell culture fermentation equipment are:

- Applicon BV, Schiedam AC, the Netherlands
- Bioengineering AG, Wald, Switzerland
- B. Braun, Melsungen, Germany
- LH Fermentation Ltd, Maidenhead, Berks, UK
- LSL Biolafitte, St Germain en Laye, France
- Marubishi Bioengineering Co., Ltd, Tokyo 101, Japan
- New Brunswick Scientific Co., Inc., Edison, NJ, USA
- Setric Genie Industriel, Toulouse, France

ENHANCED CONTROL OF PHYSICAL/CHEMICAL PARAMETERS

The cultivation of cells at high densities and/or for long periods of time may be very attractive with respect to productivity and economy. Therefore, these methods will be used mainly for processes that are intended for industrial production. Such methods need not only a more detailed and safer process control, but also additional measures for meeting the requirements of good manufacturing practice.

Process control at high cell density is much more critical with regard to oxygen (see above) and nutrient supply as compared with standard batch and continuous cultures. The more intense use of medium contributes to the economy of the process, but leaves a deficit in nutritional buffer capacity. This causes rapid starvation of cells after relatively small alterations in specific nutrient supply. Therefore, cell number estimation, liquid flow/level control and the measurement of specific medium components and osmotic pressure will be dealt with here.

Cell number determination (see also Chapter 2)

Cell number and viability determination is not only an excellent direct process parameter, but also the basis for all specific and calculated parameters (growth rate, specific consumption and production rates). Therefore, the application of a sophisticated process control system is largely dependent on a reliable cell number measurement. Most laboratories determine the cell number within the reactor by sampling and off-line analysis. This is not satisfactory because of the necessity of frequent handling, overnight attendance and problems with reproducibility. On the other hand, automatic devices are either very complicated (like sampling photometers or sampling cell counters) or are relatively new and not validated.

The use of an infrared sensor (Aquasant Messtechnik AG, Bubendorf, Switzerland) may be a solution for many applications, particularly in combination with a control of viability (Merten *et al.*, 1987). Other alternatives are sensors measuring conductance/capacitance (ABER Instruments, Cefnllan, Aberystwyth, UK) and software sensors (Pelletier *et al.*, 1994). For further reading see De Gouys *et al.* (1996). A recently published method based on real-time imaging opens up new possibilities by real cell counting (Ożturk *et al.*, 1997).

Liquid flow/level control

As outlined before, particularly with high-cell-density continuous processes, there is the need for accurate control of medium flow and liquid level. Wide variations in the feeding rate will lower viability, and variations in the liquid level will alter hydraulic behaviour and the oxygen transfer rate.

Liquid flow

The liquid flow cannot be controlled properly by peristaltic or other non-metering pumps because there are huge variations that are dependent on temperature, differential pressure, age of the tubing/membranes, air bubbles, etc. This can be compensated for by the introduction of an electronic flow meter together with a set point control loop (Figure 5.10.2). The flow may be measured using either magneto-inductive or thermoelectric principles. Very few companies can supply devices with a flow range as low as $1-5\ l\ h^{-1}$:

CONTROL PROCESSES

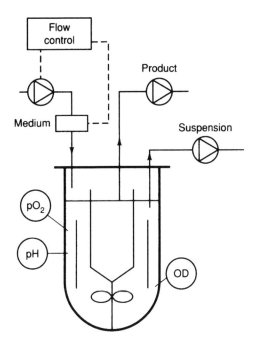

Figure 5.10.2 Level control by electronic flow meter with a set point control loop. OD = optical density.

- Magneto-inductive principle: Turbo Messtechnik, Cologne, Germany
- Thermoelectric principle: Fluid Components, Inc., San Marcos, CA, USA

An alternative is provided by sterilizable metering pumps, but there are drawbacks: they are sensitive to gas bubbles and, for documentation purposes, an additional flow measurement device is necessary. A special device is made by Bioengineering Co., using a membrane valve combination for pumping (Kobio pump).

A third possibility is the use of an automatic valve combination together with a small intermediate vessel that is placed on an electronic balance coupled to a computer for control of the valves. This is a very accurate method that can be adjusted to very small feeding steps down to a minute range, resembling an almost continuous medium supply.

Liquid level

Liquid level measurement may be accomplished in several ways. The classical way is the use of a liquid level sensor, either based on conductivity or on electrical capacity. They can be supplied by all fermenter companies. Both are affected by foam.

Another well-established method is to place the fermenter on an electronic balance. Most manufacturers offer such systems. They are very accurate as long as there are no changes with the peripheral installation (tubing, stoppers, etc.).

Two other useful methods should also be noted here:

1. Measurement of the differential hydrostatic pressure by two piezoelectric sensors mounted at the top and at the bottom of the reactor, respectively. The pressure difference corresponds to the liquid height (Hottinger-Baldwin Messtechnik, Vienna, Austria; Bioengineering AG, Wald, Switzerland).
2. Measurement by depth sounding using an ultrasonic device is also possible, but its use is dependent on a special mounting plate for the sonar head on the bottom of the fermenter (X'SONAR by Moore Electronics Inc., Klamath Falls, OR, USA).

Measurement of medium components

With high-efficiency processes it may be beneficial or even necessary to control the process by a distinct, critical medium parameter, either for adjusting the optimum feeding rate or for defining a distinct state of the process (harvest time, switching point, etc.). The parameter may be sugar level, concentration of a distinct amino acid, free ammonia, etc. For the state of the art of measuring techniques, see Chapter 3, section 3.4.

As the metabolic behaviour of cell lines changes with many parameters, such as growth rate, cell density, pH, pO_2, glucose concentration and probably other physicochemical parameters, it may be useful to monitor more than one of the medium components. This will allow correct modification of the medium composition/environmental conditions. The goal may be directed to cost-effectiveness or to favour the enhanced generation of a distinct isoproduct (Konstantinov et al., 1996; Ożturk et al., 1997).

Measurement of osmotic pressure

The osmolarity of the cell suspension increases during a fermentation process as a result of metabolic events (Øyaas, 1989). The influence on the cells and the product is still unclear. Therefore, with each development of a new cell culture process, a study on the importance of this parameter must be performed. Unfortunately, there is no on-line measurement system commercially available, although sterile installation of a membrane osmometer should be possible.

Currently, the standard method is off-line measurement using a freezing-point osmometer (Fiske, Needham Heights, MA, USA; Gonotec, Berlin, Germany; Knauer, Bad Homburg, Germany). For maintaining constant osmolarity, a second medium inlet system for distilled water must be installed and adjustments made occasionally according to sample readings.

ENHANCED CONTROL OF CELL METABOLISM

A rather more futuristic aspect of fermentation control is the use of complex parameters, calculated from basic and biochemical parameters (Fenge et al., 1991;

Glacken, 1991). These include growth rate, uptake and production rates as well as intracellular metabolites. Such parameters directly monitor the physiological state of the cells and would allow very reliable and fast control of the process (Grammatikos, 1997).

The automatic calculation of growth rate from two density measurements and the flow rate is not very useful because of the long time intervals necessary for a reliable calculation; this inherently leads to a retrospective value. A better possibility in this respect is the measurement of specific intracellular ATP, which correlates with the growth rate. This can be done by automatic off-vessel, on-line analysis of total ATP using commercial kits (Packard Chemicals, Groningen, The Netherlands), the actual cell number from an on-line sensor and a calculation model (Figure 5.10.3). This value allows continuous fermentation at a constant growth rate by nutrient manipulation or temperature regulation. Productivity may be a function of growth rate (Merten, 1988b), so this may lead to an improvement in process economy.

Our own unpublished results show a correlation between glucose uptake rate and monoclonal antibody production rate with a hybridoma. However, due to limited technical resources, it has only become possible recently to use this phenomenon within a production process. Nevertheless, the automatic off-vessel, on-line measurement of glucose, twice per hour, can be established (Ożturk, 1997). An in-line glucose measurement system has been developed recently, based on

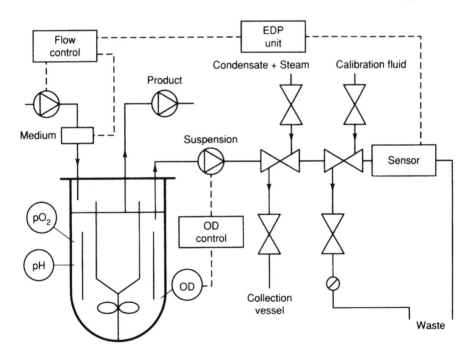

Figure 5.10.3 Calculation of cell mass by on-line measurement of ATP, allowing a constant growth rate to be maintained. OD = optical density; EDP = electronic data processing.

near-infrared measurement (Ed Baars, Hogeschool West-Brabant, Etten-Leur, The Netherlands).

A very interesting task would be automatic measurement of the product and the calculation of production rates. This would allow automatic optimization of the process by a computer program that varies all relevant parameters, probably by multiparameter analysis, to find the best production conditions. As long as there are no product sensors available, the main problem may be the time necessary for the measurement of an automatically taken sample. However, the use of HPLC methods can give accurate results within 20 min, and high-performance capillary electrophoresis (HPCE), with an analysis time of 5 min, could be introduced (Beckman PIACE 2000, E. Wasserbauer, personal communication; and James et al., 1994). Nonetheless, further development is necessary before these methods can be used routinely for automatic fermentation analysis.

Other parameters may include specific amino acid uptake or production rate, or the specific release rate of intracellular enzymes such as lactate dehydrogenase or glutamate oxaloacetate transaminase, both instant parameters for changing viability and particularly useful for alarm settings. Some of these possibilities are feasible today by using an automatic sampling device, either on a filtration base or free suspension sampling with an automatic analyser (Figure 5.10.3). For amino acids, a system called Biotronic LC5001 has been described (Buzsaky et al., 1989; Duval et al., 1989). For enzymes, a modified clinical analyser system could be used; and for products (e.g. monoclonal antibodies), HPLC methods can be useful (Holmberg et al., 1991; Ożturk et al., 1995).

DISCUSSION

In summary, it can be seen that basic process control may be satisfactory for laboratory fermentations, where the feasibility of the process should be investigated and a laboratory amount of material has to be generated without too much attention to economy. Enhanced process control will be necessary where either high product concentration or distinct product qualities are the goal and the manufacturing costs for the product are a major issue. Direct control on the chemical/physical level will probably give good results in many cases, but there is a clear advantage for process optimization in introducing control at the metabolic level. However, this may be a labour-intensive and costly task, which may pay back only with products achieving full production capacity for long periods of time.

REFERENCES

Blackie J, Dean A, Konstantinov K & Naveh D (1998) Real-time imaging for monitoring of mammalian cell culture processes. In: Merten OW & Griffiths B (eds) *Animal Cell Technology: New Developments and New Applications*. Kluwer, Dordrecht, in press.

Bliem R & Katinger H (1988a) Scale-up engineering in animal cell technology: Part I. *Tibtech* 6: 190–195.

Bliem R & Katinger H (1988b) Scale-up engineering in animal cell technology: Part II. *Tibtech* 6: 224–230.

Buetemeyer H, Marzah R & Lehmann J (1994) A direct computer control concept for mammalian cell fermentation. *Cytotechnology* 15: 271–279.

Buzsaky F, Lie K & Lindner-Olsson E (1989) Amino acid utilization by hamster cells during continuous perfusion. In: Spier RE, Griffiths JB, Stephenne J & Crooy PJ (eds) *Advances in Animal Cell Biology and Technology for Bioprocesses*, pp. 252–256. Butterworth, Oxford.

De Gouys V, Menozzi FD, Harfield J, Fabry L & Miller AOA (1996) Electronic estimation of the biomass in animal cell cultures. *Chimica Oggi* 14: 33–36.

Duval D, Geahel A, Dufau AF & Hache J (1989) Effect of amino acids on the growth and productivity of hybridoma cell cultures. In: Spier RE, Griffiths JB, Stephenne J & Crooy PJ (eds) *Advances in Animal Cell Biology and Technology for Bioprocesses*, pp. 257–259. Butterworth, Oxford.

Einsele A, Finn RK & Samhaber W (1985) *Mikrobiologische und Biochemische Verfahrenstechnik*. VCH Verlagsgesellschaft, Weinheim.

Fenge C, Fraune E & Schügerl K (1991) Physiological investigations in high density perfusion culture of free suspended animal cells. In: Spier RE, Griffiths JB & Meignier B (eds) *Production of Biologicals from Animal Cells in Culture*, pp. 262–265. Butterworth–Heinemann, Oxford.

Fleischacker RJ, Weaver JC & Sinskey AJ (1981) Instrumentation for process control in cell culture. *Advances in Applied Microbiology* 27: 137–167.

Glacken MW (1991) Bioreactor control and optimization. In: Ho CS & Wang DIC (eds) *Animal Cell Bioreactors*, pp. 373–404. Butterworth–Heinemann, Oxford.

Grammatikos SI, Tobien K, Noé W & Werner R (1998) From applied research to industrial applications: the success story of monitoring intracellular nibonucleotide pools. In: Griffiths B & Merten OW (eds) *Animal Cell Technology: New Developments and New Applications*. Kluwer, in press.

Handa-Corrigan A (1990) Oxygenating animal cell cultures: the remaining problems. In: Spier RE & Griffiths JB (eds) *Animal Cell Biotechnology*, vol. 4, pp. 123–132. Academic Press, London.

Harris JL & Spier RE (1985) Physical and chemical parameters: measurement and control. In: Spier RE & Griffiths JB (eds) *Animal Cell Biotechnology*, vol. 1, pp. 283–319. Academic Press, London.

Holmberg A, Ohlson S & Lundgren T (1991) Rapid monitoring of monoclonal antibodies in cell culture media by high performance liquid affinity chromatography (HPLAC). In: Spier RE, Griffiths JB & Meignier B (eds) *Production of Biologicals from Animal Cells in Culture*, pp. 594–596. Butterworth–Heinemann, Oxford.

James DC, Freedman RB, Hoare M & Jenkins N (1994) High resolution separation of recombinant human interferon-gamma glycoforms by unicellar electrokinetic capillary chromatography. *Analytical Biochemistry* 222: 315–322.

Katinger HWD, Scheirer W & Krömer E (1979) Bubble column reactor for mass propagation of animal cells in suspension culture. *German Chemical Engineering* 2: 31.

Konstantinov KB, Tsai Y, Moles D & Matanguihan R (1996) Control of long-term perfusion Chinese hamster ovary cell culture by glucose autostat. *Biotechnology Progress* 12: 100–109.

Kretzmer G, Ludwig A & Schügerl K (1991) Determination of the 'critical shear stress level' for adherent BHK cells. In: Spier RE, Griffiths JB & Meignier B (eds) *Production of Biologicals from Animal Cells in Culture*, pp. 244–246. Butterworth–Heinemann, Oxford.

Merchuk JC (1991) Shear effect on suspended cells. *Advances in Biochemical Engineering Biotechnology* 44: 65–95.

Merten OW (1988a) Sensors for the control of mammalian cell processes. In: Spier RE & Griffiths JB (eds) *Animal Cell Biotechnology*, vol. 3, pp. 75–140. Academic Press, London.

Merten OW (1988b) Batch production and growth kinetics of hybridomas. *Cytotechnology* 1: 113–121.

Merten OW, Palfi GE, Stäheli J & Steiner J (1987) Invasive infrared sensor for the determination of the cell number in a continuous fermentation of hybridomas. *Developments in Biological Standardization* 66: 357–363.

Øyaas K, Berg TM, Bakke O & Levine DW (1989) Hybridoma growth and antibody production under conditions of hyperosmotic stress. In: Spier RE, Griffiths JB, Stephenne J & Crooy PJ (eds) *Advances in Animal Cell Biology and Technology for Bioprocesses*, pp. 212–220. Butterworth, Oxford.

Ożturk SS, Thrift JC, Blackie JD & Naveh D (1995) Real-time monitoring of protein secretion in mammalian cell fermentation: measurement of monoclonal antibodies using a computer-controlled HPLC-system (BioCad/RPN). Biotechnology and Bioengineering 48: 201–206.

Ożturk SS, Thrift JC, Blackie JD & Naveh D (1997). Real-time monitoring and control of glucose and lactate concentrations in a mammalian cell perfusion reactor. *Biotechnology and Bioengineering* 53: 372–378.

Pelletier, F, Fonteix C, Lourenco-Da-Silva A, Marc A & Engasser JM (1994) Software sensors for the monitoring of perfusion cultures: evolution of the hybridoma density and the medium composition from glucose concentration measurements. *Cytotechnology* 15: 291–299.

Scheirer W & Merten OW (1991) Instrumentation of animal cell culture reactors. In: Ho CS & Wang DIC (eds) *Animal Cell Bioreactors*, pp. 405–443. Butterworth-Heinemann, Oxford.

Schügerl K (1988) Measurement and bioreactor control. In: Durand G, Bobichon L & Florent J (eds), *Proceedings of 8th International Biotechnology Symposium*, Paris, 1988, pp. 547–562. Societé Française de Microbiologie, Paris, Cedex 15.

Van t'Riet K & Tramper J (eds) (1991) *Basic Bioreactor Design*. Marcel Dekker, New York.

Webb C & Mavituna F (eds) (1987) *Plant and Animal Cells: Process Possibilities*. Ellis Horwood, Chichester.

Werner RG & Nöe W (1993a) Mammalian cell cultures. Part I: Characterization morphology and metabolism. *Arzneimittel Forschung* 43: 1134–1139.

Werner RG & Nöe W (1993b) Mammalian cell cultures. Part II: Genetic engineering, protein glycosylation, fermentation and process control. *Arzneimittel Forschung* 43: 1242–1249.

Werner RG & Nöe W (1993c) Mammalian cell cultures. Part III: Safety and future aspects. *Arzneimittel Forschung* 43: 1388–1390.

CHAPTER 6

REGULATORY ISSUES

6.1 REGULATORY ASPECTS OF CELLS UTILIZED IN BIOTECHNOLOGICAL PROCESSES

There is a range of regulatory guidelines relevant to the utilization of cell lines or their products in the manufacture of both biotherapeutic and diagnostics.

Quality assurance by means of strict quality control of all aspects of a cell culture process has always been of prime importance, given the sensitivity of cells to suboptimal medium and environmental factors and the ease with which cells can become contaminated with viruses and other microorganisms. The potential for biological changes in scale-up of cell culture processes demands even greater standardization and testing of the system.

The first *in vitro* development processes for the manufacture of vaccines utilized primary primate cells as the virus substrate. However, contamination of these cells with viruses, notably SV40, highlighted the need for stocks of the cell substrate that could be standardized. Standardized procedures only became reality when the human diploid cell (HDC) line WI-38, derived from foetal lung tissue, was accepted for the production of human vaccines (World Health Organization (WHO), 1987; Wood & Minor, 1990). This cell line was shown to be free of all known contaminating viruses (endogenous and adventitious), to be genetically stable within defined population doubling levels (which have been increased in more recent times) (Wood & Minor, 1990) and to retain a diploid karyotype. These factors, which can be delineated under the headings of identity, senescence and purity, enabled a set of sensible and rationalized quality assurance parameters to be established and a truly standardized process, from seed banks to final product, to be introduced (WHO, 1989).

The use of HDCs confined the culture system to a unit process in either multiple flasks or roller bottles, which is a labour-intensive and inefficient production methodology. In some ways it was unfortunate that the parameters set by the characterization of HDCs then appeared to become the 'norm' by which other cell types were assessed. As a result it was highly unlikely that in the development of cell-based biotherapeutics more *production amenable* cell lines would receive ready acceptance by the regulatory agencies, given that genetic instability and the presence of potentially oncogenic DNA in transformed cells could give the regulators cause for anxiety.

This situation has now altered with increasing acceptance of alternative cell substrates, including both genetically manipulated and spontaneously transformed (but stable) cell lines for production processes. A good example is the WHO Vero

---

Cell and Tissue Culture: Laboratory Procedures in Biotechnology, edited by A. Doyle and J.B. Griffiths.
© 1998 John Wiley & Sons Ltd.

distribution cell bank held at The European Collection of Cell Cultures (ECACC) and the American Type Culture Collection (ATCC), which is fully authenticated for vaccine manufacture (WHO, 1989).

CELL LINE DERIVATION

Nine times out of ten, cell line derivation occurs during a research programme and many properties of the cell may well be different after 2–6 months of continuous cultivation. It is important therefore to re-start cultures from the original cell bank every 2–3 months in order to avoid this problem. Also, cell lines and clones are usually selected on the basis of their growth ability, rather than as a targeted attempt to address particular requirements for protein products. The difficulties that this can present are gradually being redressed in an attempt to avoid the problems encountered in the past due to provision of inappropriately prepared cell substrates from such research programmes. The introduction of good laboratory practice (Department of Health, 1996) and associated diligent record keeping, enhanced by accreditation to international standards (e.g. ISO 9001, ISO25/EN4500), ensure that the risks of cross-contamination between cell lines or the potential for introduction of adventitious agents are kept to a minimum. Even so, in practice, tissue culture work at the research level is often performed under less than ideal conditions in cramped laboratories with multiply-used facilities that can lead to difficulties at a later date.

In many cases insufficient thought is given to the eventual regulatory implications concerning the choice of starting materials, which are commonly chosen simply because they are readily available in the organization concerned. With HDC strains and heteroploid cells there is no element of 'choice' – the starting tissue must be derived from normal healthy individuals (or foetal material) and is thus in some way seen to be free from contaminating organisms. Where heteroploid tumour-derived cells are under consideration, there are obvious parameters concerning the status of the patient or animal providing the tissue (e.g. HIV/HTLV, hepatitis) and in human material this must be reflected in the testing regime for the final master cell bank (Centres for Biologics Evaluation and Research (CBER), 1993).

In the derivation of monoclonal antibody-secreting hybridomas and the generation of genetic recombinant cell lines there is an element of 'choice' of starting material and valued judgments regarding safety and reliability can be made on the final selection of cells. Mouse myelomas developed for hybridoma preparation must be obtained from reputable sources and undergo routine quality control procedures prior to their use. Many laboratories have decided to use rat myelomas in preference to mouse because they lack the endogenous C-type virus particles frequently observed in mouse myeloma cells. However, both mouse and rat cells from laboratory strains may represent an infection hazard. These are important considerations if the protein product is destined for *in vitro* human use. Identification of certain viruses, designated Group 1 (Table 6.1.1) (Minor, 1994) would contraindicate the use of the associated cells whilst viruses designated to Group 2 (Table 6.1.1) may be satisfactorily eliminated in downstream purification of the product.

At present there is no human myeloma available that can be used to prepare hybridomas with the same efficiency as the mouse or rat system. This has led to the development of several heteromyelomas (mouse × human, e.g. HF2 × 653) as immortalizing fusion partners, although their use may lead to concerns regarding the possible generation of recombinant or reactivated viruses in the resulting hybrid cells. An alternative is to use Epstein-Barr virus (EBV)-transformed B lymphoblastoid cell lines secreting antibody, but in this case questions are raised concerning the oncogenic potential of the cellular and viral DNA that may remain in the final product.

Table 6.1.1 Categorization of viral contaminants of murine hybridoma cells[a]

Group 1	Hantavirus
	Lymphocytic choriomeningitis virus
	Reovirus type 3
	Sendai virus
Group 2	Ectromelia virus
	K virus
	Lactate dehydrogenase virus
	Minute virus of mice
	Mouse adenovirus
	Mouse cytomegalovirus
	Thielers virus
	Mouse hepatitis virus
	Mouse rotavirus
	Pneumonia virus of mice
	Polyoma virus
	Retroviruses
	Thymic virus

[a]See Minor (1994)

RECOMBINANT CELLS

Several host/vector systems are available for the preparation of recombinant cell lines. The choice of host system is based on the ability to insert genes into a host cell, which will permit integration of the foreign DNA. This very permissiveness reveals a lot about the cell and therefore the long-term stability of the cell line constructs generated. For example, the generation of CHO DHFR-stains (Urlaub et al., 1983), which permit transfection and amplification of vectors carrying inserts encoding protein products, was directly enabled by the high level of genomic instability of CHO cells. This means that there is still a question mark over the long-term stability of such recombinant cell lines, which must be considered when assessing their safety and regulatory acceptability.

CELL CHARACTERIZATION STUDIES

Development of testing strategies for cell banks

The ground rules for cellular characterization were set with human diploid cells and these particular substrates for vaccines are subject to guidelines developed by the WHO (1997). In turn, WHO guidelines have provided the parameters for characterization of heteroploid cells (WHO, 1987). It is therefore a natural progression that this level of characterization should also provide the basis for the characterization of master and working cell banks (see Chapter 1, section 1.3), even though this does present problems.

In particular, the relevance of some of the procedures is questionable (e.g. karyology) because a lot of the cells are inherently unstable and a certain amount of controversy surrounds *in vivo* tests involving animals that are now considered 'redundant'. For example, if it is accepted that every hybridoma is capable of producing tumours in model systems, what is the point of repeating the exercise with newly derived hybridomas? Unfortunately, what has tended to happen is that each new test that is added as technology advances is *additional* rather than a *replacement*. This leads to an ever-increasing number of quality control tests to be performed and thus adds to the expense of validation studies and is a growth industry in itself.

Characterization techniques

Identity testing

Ensuring that the cells in culture are in fact what they are meant to be is a fairly important validation step, however obvious it may seem. The analysis of the chromosomal make-up of a cell line by karyology remains a key tool in cell line identity testing and for determining the stability of cell cultures (CBER, 1993). However, the use of this technique for establishing stability is questionable in the light of inherent drift in the chromosomal complement of animal cells in culture (Rutzky *et al.*, 1980). Isoenzyme analysis is also a key regulatory requirement in the testing of cell banks (CBER, 1993). In general this technique is used to exclude the possibility of cross-contamination of a cell line with cells from another species (O'Brien *et al.*, 1997; Halton *et al.*, 1983). Recently, however, significant progress has been made in the refinement of molecular techniques for cell line identity testing, such as multilocus DNA fingerprinting (Gilbert *et al.*, 1990; Stacey *et al.*, 1991, 1992), DNA profiling (Masters *et al.*, 1988; Stacey *et al.*, 1997) and random amplified polymorphic DNA (RAPD) methods (Williams *et al.*, 1990). Multilocus DNA fingerprinting methods provide a unique identification pattern for each human individual (Jeffreys *et al.*, 1985) and in many other animal and plant species (Vassart *et al.*, 1987; Ryskov *et al.*, 1988) using Southern blot analysis. Whilst fingerprinting and profiling techniques have received intensive validation in human and animal populations, RAPD methods have yet to come through rigorous scrutiny, although microsatellite and mini-satellite anchored methodologies are promising in terms of reproducibility (Meyer & Mitchell, 1995). Provided that convincing validation data are produced for these new technologies, it may be that other test procedures (karyology and isoenzyme analysis) will be gradually replaced (see also Chapter 1, section 1.3).

Adventitious agents

In practice, most animal cell-derived biotherapeutic products are evaluated by a series of meetings with experts from the regulatory bodies to discuss results and thus build up an overall picture of the cell line and its product. In particular, virus testing procedures are important to exclude contamination of therapeutic products, and therefore have to be exhaustive. Regulatory authorities may request an extended range of tests, including retrovirus testing.

While viruses are the major concern, other common organisms that may contaminate and proliferate within animal cells should not be disregarded from the point of view of safety: in particular mycoplasmas, which can proliferate within cell cultures without their presence being suspected (see chapter 1, sections 1.6 and 1.7). The fastidious nature of these organisms – which require cholesterol from animal cell membranes for growth – and their ability to persist at low levels under antibiotic treatment necessitate careful testing regimes, which have been a focus of the regulatory authorities (Council of Europe, 1991: US Food and Drugs Administration, 1993), although this is primarily to ensure the purity and safety of therapeutic products.

If adventitious agents *are* found, how does this affect the final acceptability of the product? The answer to this question rests on the product itself and the efficiency of the purification procedures in removing contaminants. In such cases the amount of residual DNA in the product is an important factor. Recent guidelines state that each dose must contain less than 100 pg DNA (WHO, 1987). Validation of purification procedures must also include the establishment of 'spiking' experiments to provide that viruses representative of pathogens (i.e. in terms of their characteristics relevant to the purification process) are removed (Tao et al., 1994).

Undoubtedly, bringing a cell line-derived product into therapeutic use is a long and expensive process for which regulatory approval is only one – but fundamental – facet. There seems little likelihood of any relaxation in the stringent testing procedures required by the regulatory authorities before approval for a product is given. This is becoming a consideration for diagnostic products as well as therapeutics, particularly with respect to the new EC *In Vitro* Medical Devices Directive.

REFERENCES

Centres for Biologics Evaluation and Research (CBER) (1993) *Points to Consider in the Characterisation of Cell Lines Used to Produce Biologicals*. CBER, US Food and Drug Administration, Bethesda, MD.

Council of Europe (1991) Biological tests. In *European Pharmacopoeia*, suppl. to 3rd edn, vol. 2.6 (sterility) and vol. 2.6.5 (mycoplasmas) European Pharmacopoeia Secretariat, Strasbourg.

Department of Health (1996) *Good Laboratory Practice*. The United Kingdom Compliance Programme, Department of Health, London, UK.

Gilbert DA, Reid YA, Gail MH, Pee D,

White C, Hay RJ & O'Brien SJ (1990) Application of DNA fingerprints for cell line individualisation. *American Journal of Human Genetics* 47: 499–514.

Halton, DM, Peterson WD & Hukku B (1983) Cell culture quality control by isoenzymatic characterisation. *In Vitro* 19: 16–24.

Jeffreys AJ, Wilson V & Thein S-L (1985) Hypervariable 'minisatellite' regions in human DNA. *Nature (London)* 314: 67–73.

Masters JR, Bedford P, Kearney A, Povey S & Franks LM (1988) Bladder cancer cell line cross-contamination: identification using a locus specific minisatellite probe. *British Journal of Cancer* 57: 284–286.

Meyer W & Mitchell TG (1995) Polymerase chain reaction fingerprinting in fungi using single primers specific to minisatellites and simple repetitive DNA sequences: strain variation in *Cryptococcum neoformans*. *Electrophoresis* 16: 1648–1656.

Minor, PD (1994) Significance of contamination with viruses of cell lines used in the production of biological medicinal products. In: Spier RE, Griffiths JB & Berthold W (eds) *Animal Cell Technology: Products for Today, Prospects for Tomorrow*. Butterworths-Heinemann, Oxford, pp. 741–750.

O'Brien SJ, Cliener G, Olson R & Shannon JE (1977) Enzyme polymorphisms as genetic signatures in human cell cultures. *Science* 195: 1345–1348.

Rutzky CP, Kaye CJ, Siciliano, MJ, Chao M & Kahan BD (1980) Longitudinal karyotype and genetic signature analysis of cultured human colon adenocarcinoma cell lines LS180 and LS174T. *Cancer Research* 40: 1443–1448.

Ryskov AP, Jincharadze AG, Prosnyak MI, Livanov PL & Limborska SA (1988) M13 phage DNA as a universal marker for DNA fingerprinting of animals, plants and micro-organisms. *FEBS Letters* 233: 388–392.

Stacey GN, Bolton BJ & Doyle A (1991) The quality control of cell banks using DNA fingerprinting. In: Burke T, Jeffreys AJ, Dolf G & Wolf R (eds) *DNA Fingerprinting: Approaches and Applications*, Birkhauser, Berlin.

Stacey GN, Bolton BJ & Doyle A (1992). DNA fingerprinting transforms the art of cell authentication. *Nature (London)* 391: 261–262.

Stacey GN, Hoelzl H, Stephenson JR & Doyle A (1997) Authentication of animal cell cultures by direct visualization of repetitive DNA, aldolase gene PCR and isoenzyme analysis. *Biologicals* 25: 75–85.

Tao CZ, Cameron R, Harbour C & Barford JP (1994) The development of appropriate viral models for the validation of viral inactivation procedures. In: Spier RE, Griffiths JB & Berthold W (eds) *Animal Cell Technology: Products for Today Prospects for Tomorrow*, pp. 754–756. Butterworth-Heinemann, Oxford.

Urlaub G, Kas E, Carothers AM & Chasin LA (1983) Detection of the diploid dihydrofolate reductase locus from cultured mammalian cells. *Cell* 33: 405–412.

US Food and Drugs Administration (1993) *Code of Federal Regulations (Sterility) 21CFR 610.12 (1993)* and *Code of Federal Regulations (Mycoplasma) 21 CFR 610.30 (1993)*. US FDA, Bethesda, MD.

Vassart G, Georges M, Monsieur R, Brocas H, Lequarre AS & Christophne D (1987) A sequence in M13 phage detects hypervariable minisatellites in human and animal DNA. *Science* 235: 683–684.

WHO Expert Committee on Biological Standardisation. Requirements for Use of Animal Cells as *In Vitro* Substrates for the Production of Biologicals. Technical Report Series No. 50 (in press). World Health Organisation, Geneva.

Williams JGK, Kubelik AR, Livak KL, Rafalski JA & Tingey SV (1990) DNA polymorphisms amplified by arbitrary primers are useful as genetic markers. *Nucleic Acid Research* 18: 6531–6535.

Wood DT & Minor PD (1990) Meeting report: use of human diploid cells in vaccine production. *Biologicals* 18: 143.

World Health Organization (1987) *Acceptability of Cell Substrates for Production of Biologicals*, Technical Report Series 747. WHO, Geneva. World Health Organization (1989) *WHO Cell Banks of Continuous Cell Lines for the Production of Biologicals*, Technical Report Series 756. WHO, Geneva.

CONCLUDING REMARKS

The first animal cell product prepared *in vitro* was Salk polio vaccine (1954), produced in a pot of magnetically stirred primary cells. Now there are over 200 licensed (or in the process of being licensed) cell products, including vaccines, immunoregulators and growth factors, therapeutic monoclonal antibodies and a range of therapeutic proteins (e.g. tPA, EPO, blood factor VIII, hGH and FSH) (see Chapter 5, section 5.1). Furthermore, these are produced in a range of highly controlled and sophisticated bioreactors (at scales up to 20 000 l or at tissue-like cell densities) in optimized and defined media. The engineered cell lines that have been developed represent the cutting-edge of molecular biology. Additionally, the degree of regulation required by the licensing authorities to ensure the safety of the product (as well as the process) means that a safe and effective biological is produced utilizing these advanced molecular and engineering technologies. Taking a long-term perspective of such phenomenal progress, and considering the vast number and range of products, one can derive a degree of satisfaction that a job has been well done and successfully completed. However, there are still many improvements that can be made to give a more economical and efficient process, and in fact to allow some products to be manufactured at all. The following discussion highlights the approaches currently being made to give these improvements and develop new products.

PRODUCTIVITY

In terms of both volumetric and cell specific productivity, yields are well below those obtained using competing systems such as bacteria and yeasts. A number of core parameters, such as a combination of slow growth rate, low unit cell density, critical environmental (particularly physical) factors and the complexity (and expense) of culture media, conspire to make cell culture a low-productivity, high-cost process. Fortunately for animal cell technologists, because the quality of the product is very much higher than that from competing systems, and cell products are of low concentration/high activity, then cell culture is the preferred, if not the only, option for many products. However, costs should be put in perspective because the disproportionately high costs of quality control and assurance help to reduce the impact of the actual production costs.

How can the productivity problem be solved? There are three possible approaches

1. *The system: developing more efficient bioreactors and processes, i.e. the chemical engineering approach with the ultimate goal of reducing costs yet further.* This approach has been very effective over the past 20 years with the development of microcarriers, scale-up of stirred tanks to over 10 000 l and airlift to 2000 l,

and the introduction of high-density (100 million cells ml^{-1}) perfusion systems. In particular, the difficult problem of oxygen limitation has been alleviated by many innovative techniques. However, apart from improving process control, reliability and integrated downstream processing (to improve on low recovery of purified product), this approach in future will probably be less rewarding than others in terms of the order of magnitude of expected return.

2. *The cell: cell development via the cell engineering approach.* Increasing product expression levels to *in vivo* values that can be maintained over extended periods of time by introducing new genetic materials is still to be achieved, but progress is bringing worthwhile improvements in productivity. New developments in this area should allow very significant increases in productivity to be achieved. In addition, cells are being modified genetically to produce their own growth factors (reduce media costs) and to become resistant to apoptotic death. A move from ubiquitous cell lines such as BHK and CHO to more specialized cell types (e.g. hepatic and lymphocytic) is also likely to occur, albeit slowly. This will be partly due to the reluctance of regulatory authorities to accept new engineered lines.

3. *Metabolism/physiology: controlling cell metabolism via the metabolic engineering approach.* This is a combination of developing an increased range of sensitive sensors for metabolites together with a better understanding of cell biochemistry and control of metabolic pathways. The increasing knowledge on cell regulation from surface receptors, through cytoplasmic cascades to gene function as modulated by growth factors and other regulatory molecules, is becoming a key element in increasing cell productivity. It is believed by many that this approach will become the most rewarding in terms of significantly increasing the efficiency of cell processes.

The use of animal cells to produce biologicals in large-scale industrial processes is well established and there is enormous potential to expand. This is despite the many process disadvantages compared with the competing technologies offered by bacteria, fungi and transgenic animals. Although transgenics offers a cheaper process for large quantity production of biologicals, public opinion on the use of animals may severely limit this technology. The complexity of many of the products in terms of protein folding, three-dimensional configuration and glycosylation guarantees that they can only be expressed from eukaryotic cells. Additionally, new products are emerging and that is in the use of the cell itself rather than an expressed product. Tissue grown *in vitro* has been used for many years in the treatment of burns but new horizons have widened, from skin to fully functional replacement organs and for transplantation. The most advanced model provides temporary cover for kidney dysfunction, but systems for liver are also well advanced. As well as tissue engineering application, sights are set on utilizing cells in gene therapy, with diabetes and a range of neuronal diseases, including Parkinson's, being targeted. This new area of animal cell biotechnology depends upon the derivation of suitable engineered cell lines (often with a 'suicide' gene for safety reasons) and encapsulation technology for implantation. The implant requires long-term (months) activity utilizing the body's conventional systems for nutrition and oxygenation, and perhaps even more importantly protection from immune 'rejection'. Test systems are now so well advanced that this is undoubtedly going to be the most dynamic area of cell biotechnology in the next few years.

History has proved that advanced technologies have a habit of cropping up in areas where only science fiction writers once held sway; cell technology is by no means an exception to this rule.

A. Doyle and J. B. Griffiths
Managing Editors

APPENDIX 1: TERMINOLOGY

Many people wonder why one should spend time worrying about scientific terminology when, it is said, people will continue to use terms with which they are comfortable in spite of any communication difficulties that occur or action taken by standardization committees.They often point out that major discoveries have been, and continue to be, made in spite of any problems with communication. On the contrary, it should be a major concern of scientists that their ideas be properly conveyed to others. According to the dictionary, to communicate is to 'impart or give information'. Many feel that scientists have not done particularly well in communicating with the lay community (Schaeffer, 1984; Iglewski, 1989). That this has become a major problem is exemplified by the difficulties that the scientific community encounters when attempting to influence legislation dealing with funding for research. There has been a continual erosion of the money allocated to biomedical research, on a worldwide scale, and this has created a virtual crisis in research and training of future scientists. It may also be argued that a significant aspect of the poor communication with the lay community stems from the fact that scientists do not communicate as well as they should with each other. This can be seen in the microcosm of the larger scientific community constituted by researchers using the techniques of cell and tissue culture.

SOME ASPECTS OF THE PROBLEM

Communication within scientific disciplines has long been efficient via a system of jargon. Individuals within particular fields communicate efficiently but individuals outside the field or discipline find that jargon is simply, as defined in the dictionary, 'unintelligible talk'. Some feel that this is not a major problem in fairly circumscribed fields because there is little necessity for 'outsiders' to understand technical presentations. This may be true for such fields of research. However, difficulties arise when areas of research become interdisciplinary. Then the jargon, techniques and body of information must be understood widely and in diverse disciplines. Terminology adopted by individuals who have not previously used it is often changed and misused, causing confusion. That the above is so is confirmed by the confusion in the fields of vertebrate, invertebrate and plant cell culture (hereafter referred to only as cell culture), molecular biology and molecular genetics over the use and abuse of terminology.

There is hardly a field of biological investigation in which cell cultures or the techniques of molecular biology and molecular genetics are not used. These technologies are being employed by an ever-widening group of researchers who are, therefore, communicating more globally via scientific presentations, publications and research proposals. Unfortunately, often the communicator and recipient represent different areas of specialization and have been brought together by the common technology used in their work. In such situations, misuse of terminology can prove unfortunate indeed, leading to inability to repeat a piece of research, or the notion of relative incompetence in completing a task proposed for a research project, or problems in publishing a paper describing a completed research project. There needs to be a major standardization of the terminology associated with cell culture, molecular biology and molecular genetics lest the situation becomes even worse.

SOLUTIONS TO THE PROBLEM

Because our goal is effective communication, we should be striving to make our communications as lucid as possible. We should make every effort to describe phenomena so that others can repeat and, hopefully, reproduce our findings.

It is important for a commission, such as the Society for *In Vitro* Biology (formerly the Tissue Culture Association Terminology Committee), wherein collective wisdom is brought to bear on an issue, to come forth with definitions that can be adopted by the scientific community. Because these experts come to agreements on terms and phrases after much deliberation, their definitions should carry greater weight than any used in an individual publication. Misuse of terms is widespread and is probably a result of *individual* vis-à-vis *collective* use of terms. In oral communication, it is easy to explain ambiguities. However, with written communication, where there is no immediate dialogue, it is more difficult to do so. For a manuscript, a grant proposal or a report to a special interest group, it is important not to be misunderstood. Therefore, only precisely defined terms, which are universally accepted, should be used.

In the interest of helping to alleviate the problems mentioned above, the following compendium of terms, compiled and defined by the Tissue Culture Association Terminology Committee (Mueller *et al.*, 1990), is reprinted. The glossary is reprinted with the permission of the Society for *In Vitro* Biology (formerly the Tissue Culture Association), Largo, Maryland, USA.

TERMINOLOGY ASSOCIATED WITH CELL, TISSUE AND ORGAN CULTURE, MOLECULAR BIOLOGY AND MOLECULAR GENETICS

Adventitious: Developing from unusual points of origin, such as shoots or root tissues from callus or embryos from sources other than zygotes. This term can also be used to describe agents that contaminate cell cultures.

Anchorage-dependent cells or cultures: Cells, or cultures derived from them, that will grow, survive or maintain function only when attached to a surface such as glass or plastic. The use of this term does not imply that the cells are normal or that they are or are not transformed neoplastically.

Aneuploid: The situation that exists when the nucleus of a cell does not contain an exact multiple of the haploid number of chromosomes; one or more chromosomes being present in greater or lesser number than the rest. The chromosomes may or may not show rearrangements.

Apoptosis: A naturally occurring process of cell death that plays a complementary, but opposite, role to mitosis in regulation of animal cell populations. This is an active cellular process that provides a means for precisely regulating cell numbers and, therefore, biological activity. Unlike simple degeneration, cellular death that is dependent on active participation of cellular components can, potentially, be suppressed. Inhibition of cell death may contribute to oncogenesis. Characteristics of apoptosis may include: volume reduction through blebbing of the cell surface (membranes remain intact; there is no cell lysis but there is a selective loss of intracellular fluids); chromatin condensation (activation of nuclear endonucleases that cleave chromatin at internucleosomal sites); and cell surface changes that allow recognition and disposition by phagocytic cells before they autolyse.

Attachment efficiency: The percentage of cells plated (seeded, inoculated) that attach to the surface of the culture vessel within a specified period of time. The conditions under which such a determination is made should always be stated.

Autocrine cell: In animals, a cell that produces hormones, growth factors or other signalling substances for which it also expresses the corresponding receptors. See also 'Endocrine' and 'Paracrine'.

Axenic culture: A culture without foreign or undesired life forms. An axenic culture may include the purposeful co-cultivation of different types of cells, tissues or organisms.

Cell culture: Term used to denote the maintenance or cultivation of cells *in vitro*, including the culture of single cells. In cell cultures, the cells are no longer organized into tissues.

Cell generation time: The interval between consecutive divisions of a cell. This interval can best be determined, at present, with the aid of cinephotomicrography. *This term is not synonymous with 'population doubling time'.*

Cell hybridization: The fusion of two or more dissimilar cells leading to the formation of a synkaryon.

Cell line: A cell line arises from a primary culture at the time of the first successful subculture. The term cell line implies that cultures from it consist of lineages of cells originally present in the primary culture. The terms **finite** or **continuous** are used as prefixes if the status of the culture is known. If not, the term **line** will suffice. The term **continuous line** replaces the term **established line**. In any published description of a culture, one must make every attempt to publish the characterization or history of the culture. If such has already been published, a reference to the original publication must be made. In obtaining a culture from another laboratory, the proper designation of the culture, as originally named and described, must be maintained and any deviations in cultivation from the original must be reported in any publication.

Cell strain: A cell strain is derived either from a primary culture or a cell line by the selection or cloning of cells having specific properties or markers. In describing a cell strain, its specific features must be defined. The terms **finite** or **continuous** are to be used as prefixes if the status of the culture is known. If not, the term **strain** will suffice. In any published description of a cell strain, one must make every attempt to publish the characterization or history of the strain. If such has already been published, a reference to the original publication must be made. In obtaining a culture from another laboratory, the proper designation of the culture, as originally named and described, must be maintained and any deviations in cultivation from the original must be reported in any publication.

Chemically defined medium: A nutritive solution for culturing cells in which each component is specifiable and, ideally, of known chemical structure.

Clone: In animal cell culture terminology, a population of cells derived from a single cell by mitoses. A clone is not necessarily homogeneous and, therefore, the terms **clone** and **cloned** do not indicate homogeneity in a cell population, genetic or otherwise. In plant culture terminology, the term may refer to a culture derived as above or it may refer to a group of plants propagated only by vegetative and asexual means, all members of which have been derived by repeated propagation from a single individual.

Cloning efficiency: The percentage of cells plated (seeded, inoculated) that form a clone. One must be certain that the colonies formed arose from single cells in order to use this term properly. See 'Colony forming efficiency'.

Colony forming efficiency: The percentage of cells plated (seeded, inoculated) that form a colony.

Complementation: The ability of two different genetic defects to compensate for one another.

Contact inhibition of locomotion: A phenomenon characterizing certain cells in which two cells meet, locomotory activity diminishes and the forward motion of one cell over the surface of the other is stopped.

Continuous cell culture: A culture that is apparently capable of an unlimited number of population doublings; often referred to as an immortal cell culture. Such cells may or may not express the characteristics of *in vitro* neoplastic or malignant transformation. See also 'Immortalization'.

Crisis: A stage of the *in vitro* transformation of cells. It is characterized by reduced proliferation of the culture, abnormal mitotic figures, detachment of cells from the culture substrate and the formation of multinucleated or giant cells. During this massive cultural degeneration, a small number of colonies usually, but not always, survive and give rise to a culture with an apparent unlimited *in vitro* lifespan. This process was first described in human cells following infection with an oncogenic virus (SV40). See also 'Cell line', '*In vitro* transformation' and '*In vitro* senescence'.

Cryopreservation: Ultra-low temperature storage of cells, tissues, embryos or seeds. This storage is usually carried out using temperatures below $-100°C$.

Cumulative population doublings: See 'Population doubling level'.

Cybrid: The viable cell resulting from the fusion of a cytoplast with a whole cell, thus creating a cytoplasmic hybrid.

Cytoplast: The intact cytoplasm remaining following the enucleation of a cell.

Cytoplasmic hybrid: Synonymous with 'cybrid'.

Cytoplasmic inheritance: Inheritance attributable to extranuclear genes, e.g. genes in cytoplasmic organelles such as mitchondria or chloroplasts, or in plasmids, etc.

Density-dependent inhibition of growth: Mitotic inhibition correlated with increased cell density.

Differentiated: Cells that maintain, in culture, all or much of the specialized structure and function typical of the cell type *in vivo*.

Diploid: The state of the cell in which all chromosomes, except sex chromosomes, are two in number and are structurally identical with those of the species from which the culture was derived. Where there is a Commission Report available, the experimenter should adhere to the convention for reporting the karyotype of the donor. Commission Reports have been published for mouse (Committee on Standardized Genetic Nomenclature for Mice, 1972), human (Paris Conference, 1971) and rat (Committee for a Standardized Karyotype of *Rattus norvegicus*, 1973). In defining a diploid culture, one should present a graph depicting the chromosome number distribution leading to the modal number determination along with representative karyotypes.

Electroporation: Creation, by means of an electrical current, of transient pores in the plasmalemma usually for the purpose of introducing exogenous material, especially DNA, from the medium.

Embryo culture: *In vitro* development or maintenance of isolated mature or immature embryos.

Embryogenesis: The process of embryo initiation and development.

Endocrine cell: In animals, a cell that produces hormones, growth factors or other signalling substances for which the target cells, expressing the corresponding receptors, are located at a distance. See also 'Autocrine' and 'Paracrine'.

Epigenetic event: Any change in a phenotype that does not result from an alteration in DNA sequence. This change may be stable and heritable and includes alteration in DNA methylation, transcriptional activation, translational control and post-translational modifications.

Epigenetic variation: Phenotypic variability that has a non-genetic basis.

Epithelial-like: Resembling or characteristic of, having the form or appearance of epithelial cells. In order to define a cell as an epithelial cell, it must possess characteristics typical of epithelial cells. Often one can be certain of the histological origin and/or function of the cells placed into culture and, under these conditions, one can be reasonably confident in designating the cells as epithelial. It is incumbent upon the individual reporting on such cells to use as many parameters as possible in assigning this term to a culture. Until such time as a rigorous definition is possible, it would be most correct to use the term epithelial-like.

Euploid: The situation that exists when the nucleus of a cell contains exact multiples of the haploid number of chromosomes.

Explant: Tissue taken from its original site and transferred to an artificial medium for growth or maintenance.

Explant culture: The maintenance or growth of an explant in culture.

Feeder layer: A layer of cells (usually lethally irradiated for animal cell culture) upon which are cultured a fastidious cell type.

Fibroblast-like: Resembling or characteristic of, having the form or appearance of fibroblast cells. In order to define a cell as a fibroblast cell, it must possess characteristics typical of fibroblast cells. Often one can be certain of the histological origin and/or function of the cells placed into culture and, under these conditions, one can be reasonably confident in designating the cells as fibroblast. It is incumbent upon the individual reporting on such cells to use as many parameters as possible in assigning this term to a culture. Until such time as a rigorous definition is possible, it would be most correct to use the term **fibroblast-like**.

Finite cell culture: A culture that is capable of only a limited number of population doublings after which the culture ceases proliferation. See also '*In vitro* senescence'.

Habituation: The acquired ability of a population of cells to grow and divide independently of exogenously supplied growth regulators.

Heterokaryon: A cell possessing two or more genetically different nuclei in a common cytoplasm, usually derived as a result of cell-to-cell fusion.

Heteroploid: The term given to a cell culture when the cells comprising the culture possess nuclei containing chromosome numbers other than the diploid number. This is a term used only to describe a culture and is not used to describe individual cells. Thus, a heteroploid culture would be one that contains aneuploid cells.

Histiotypic: The *in vitro* resemblance, of cells in culture, to a tissue in form or function or both. For example, a suspension of fibroblast-like cells may secrete a glycosaminoglycan–collagen matrix and the result is a structure resembling fibrous connective tissue, which is, therefore, histiotypic. This term is not meant to be used along with the word 'culture'. Thus, a tissue culture system demonstrating form and function typical of cells *in vivo* would be said to be histiotypic.

Homokaryon: A cell possessing two or more genetically identical nuclei in a common cytoplasm, derived as a result of cell-to-cell fusion.

Hybrid cell: The term used to describe the mononucleate cell that results from the fusion of two different cells, leading to the formation of a synkaryon.

Hybridoma: The cell that results from the fusion of an antibody-producing tumour cell (myeloma) and an antigenically-stimulated normal plasma cell. Such cells are constructed because they produce a single antibody directed against the antigen epitope that stimulated the plasma cell. This antibody is referred to as a monoclonal antibody.

Immortalization: The attainment by a finite cell culture, whether by perturbation or intrinsically, of the attributes of a continuous cell line. An immortalized cell is not necessarily one that is neoplastically or malignantly transformed.

Immortal cell culture: See 'Continuous cell culture'.

Induction: Initiation of a structure, organ or process *in vitro*.

***In vitro* neoplastic transformation:** The acquisition, by cultured cells, of the property to form neoplasms, benign or malignant, when inoculated into animals. Many transformed cell populations that arise *in vitro* intrinsically or through deliberate manipulation by the investigator produce only benign tumours that show no local invasion or metastasis following animal inoculation. If there is supporting evidence, the term ***in vitro* malignant neoplastic transformation** or ***in vitro* malignant transformation** can be used to indicate that an injected cell line does, indeed, invade or metastasize.

***In vitro* senescence:** In vertebrate cell cultures, the property attributable to finite cell cultures; namely, their inability to grow beyond a finite number of population doublings. Neither invertebrate nor plant cell cultures exhibit this property.

***In vitro* transformation:** A heritable change, occurring in cells in culture, either intrinsically or from treatment with chemical carcinogens, oncogenic viruses, irradiation, transfection with oncogenes, etc., and leading to the acquisition of altered morphological, antigenic, neoplastic, proliferative or other properties. This expression is distinguished from the term ***in vitro* neoplastic transformation** in that the alterations occurring in the cell population may not always include the ability of the cells to produce tumours in appropriate hosts. The type of transformation should always be specified in any description.

Karyoplast: A cell nucleus, obtained from the cell by enucleation, surrounded by a narrow rim of cytoplasm and a plasma membrane.

Liposome: A closed lipid vesicle surrounding an aqueous interior; it may be used to encapsulate exogenous materials for their ultimate delivery into cells by fusion with the cell.

Microcell: A cell fragment, containing one to a few chromosomes, formed by the enucleation or disruption of a micronucleated cell.

Micronucleated cell: A cell that has been mitotically arrested and in which small groups of chromosomes function as foci for the reassembly of the nuclear membrane, thus forming micronuclei, the maximum of which would be equal to the total number of chromosomes.

Morphogenesis: (a) The evolution of a structure from an undifferentiated to a differentiated state. (b) The process of growth and development of differentiated structures.

Mutant: A phenotypic variant resulting from a changed or new gene.

Organ culture: The maintenance or growth of organ primordia or the whole or parts of an organ *in vitro* in a way that may allow differentiation and preservation of the architecture and/or function.

Organogenesis: In animal cell cultures, the evolution, from dissociated cells, of a structure that shows natural organ form or function or both. In plant tissue culture, a process of differentiation by which plant organs are formed *de novo* or from pre-existing structures. In developmental biology, this term refers to differentiation of an organ system from stem or precursor cells.

Organotypic: Resembling an organ *in vivo* in three-dimensional form or function or both. For example, a rudimentary organ in culture may differentiate in an *organotypic* manner, or a population of dispersed cells may become rearranged into an *organotypic* structure and may also function in an *organotypic* manner. This term is not meant to be used along with the word 'culture' but is meant to be used as a descriptive term.

Paracrine: In animals, a cell that produces hormones, growth factors or other signalling substances for which the target cells, expressing the corresponding receptors, are located in its vicinity or in a group adjacent to it. See also 'Autocrine' and 'Endocrine'.

Passage: The transfer or transplantation of cells, with or without dilution, from one culture vessel to another. It is understood that any time cells are transferred from one vessel to another, a certain portion of the cells may be lost and, therefore, dilution of cells, whether deliberate or not, may occur. This term is synonymous with the term **subculture**.

Passage number: The number of times the cells in the culture have been subcultured or passaged. In descriptions of this process, the ratio or dilution of the cells should be stated so that the relative cultural age can be ascertained.

Pathogen free: Free from specific organisms based on specific tests for the designated organisms.

Plating efficiency: This is a term that originally encompassed the terms **Attachment (Seeding) efficiency, Cloning efficiency** and **Colony forming efficiency** but is now better described by using one or more of them in its place because the term **plating** is not sufficiently descriptive of what is taking place. See also 'Attachment', 'Seeding', 'Cloning' and 'Colony forming efficiency'.

Population density: The number of cells per unit area or volume of a culture vessel; also, the number of cells per unit volume of medium in a suspension culture.

Population doubling level: The total number of population doublings of a cell line or strain since its initiation *in vitro*. A formula to use for the calculation of 'population doublings' in a single passage is:

$$\textit{Number of population doublings} = \frac{\log}{10 \ (N/N_o) \times 3.33}$$

where N = number of cells in the growth vessel at the end of a period of growth and N_o = number of cells plates in the growth vessel. *It is best to use the number of viable cells or number of attached cells for this determination.* Population doubling level is synonymous with the term **cumulative population doublings**.

Population doubling time: The interval calculated during the logarithmic phase of growth in which, for example, 1.0×10^6 cells increase to 2.0×10^6 cells. This term is not synonymous with the term **cell generation time**.

Primary culture: A culture started from cells, tissues or organs taken directly from organisms. A primary culture may be regarded as such until it is successfully subcultured for the first time. It then becomes a **cell line**.

Pseudodiploid: This describes the condition where the number of chromosomes in a cell is diploid but, as a result of chromosomal rearrangements, the karyotype is abnormal and linkage relationships may be disrupted.

Recon: The viable cell reconstructed by the fusion of a karyoplast with a cytoplast.

Saturation density: The maximum cell number attainable, under specified culture conditions, in a culture vessel. This term is usually expressed as the number of cells per square centimetre in a monolayer culture or the number of cells per cubic centimetre in a suspension culture.

Seeding efficiency: See 'Attachment efficiency'.

Senescence: See '*In vitro* senescence'.

Somatic cell genetics: The study of genetic phenomena of somatic cells. The cells under study are most often cells grown in culture.

Somatic cell hybrid: The cell resulting from the fusion of animal cells derived from somatic cells that differ genetically.

Somatic cell hybridization: The *in vitro* fusion of animal cells derived from somatic cells that differ genetically.

Subculture: See 'Passage'. With plant cultures, this is the process by which the tissue or explant is first subdivided and then transferred into fresh culture medium.

Substrain: A substrain can be derived from a strain by isolating a single cell or groups of cells having properties or markers not shared by all cells of the parent strain.

Suspension culture: A type of culture in which cells, or aggregates of cells, multiply while suspended in liquid medium.

Synkaryon: A hybrid cell that results from the fusion of the nuclei it carries.

Tissue culture: The maintenance or growth of tissues, *in vitro*, in a way that may allow differentiation and preservation of their architecture and/or function.

Totipotency: A cell characteristic in which the potential for forming all the cell types in the adult organism is retained.

Transfection: The transfer, for the purposes of genomic integration, of naked, foreign DNA into cells in culture. The traditional *microbiological* usage of this term implied that the DNA being transferred was derived from a virus. The definition as stated here is that which is in use to describe the general transfer of DNA irrespective of its source. See also 'Transformation'.

Transformation: In plant cell culture, the introduction and stable genomic integration of foreign DNA into a plant cell by any means, resulting in a genetic modification. This definition is the traditional microbiological definition. For animal cell culture, see '*In vitro* transformation', '*In vitro* neoplastic transformation' and 'Transfection'.

Variant: A culture exhibiting a stable phenotypic change, whether genetic or epigenetic in origin.

Virus-free: Free from specified viruses based on tests designed to detect the presence of the organisms in question.

REFERENCES

Committee for a Standardized Karyotype of *Rattus norvegicus* (1973) Standard karyotype of the Norway rat, *Rattus norvegicus*. *Cytogenetics and Cell Genetics* 12: 199–205.

Committee on Standardized Genetic Nomenclature for Mice (1972) Standard karyotype of the mouse, *Mus musculus*. *Journal of Heredity* 63: 69–72.

Iglewski B (1989) Communicating with the public: a new scientific imperative. *ASM News* 55(6): 306.

Mueller S, Renfroe M, Schaeffer WI, Shay JW, Vaughn J & Wright M (Tissue Culture Association Terminology Committee Members) (1990) Terminology associated with cell, tissue and organ culture, molecular biology and molecular genetics. *In Vitro Cellular and Developmental Biology* 26: 97–101.

Paris Conference (1971), Supplement (1975) Standardization in Human Cytogenetics. Birth Defects: Original Article Series, XI, 9, 1975. The National Foundation, New York. Reprinted in: *Cytogenetics and Cell Genetics* (1975) 15: 201–238.

Schaeffer WI (1984) In the interest of clear communication. *In Vitro Cellular and Developmental Biology* 25: 389–390.

APPENDIX 2: COMPANY ADDRESSES

Aldrich Chemical Co.
1001 West St Paul Avenue
PO Box 355
Milwaukee
WI 53201
USA

American Type Culture Collection (ATCC)
10801 University Blvd
Manassas
VA 20110
USA

Amersham International
White Lion Road
Aylesbury
Buckinghamshire HP20 2TP
UK

Amicon Products
Millipore Corp.
80 Ashby Road
Bedford
MA 01730
USA

Amicon Products
Millipore (UK) Ltd
The Boulevard
Blackmoor Lane
Watford
Herts WD1 8YW

Anachem Scientific Ltd
Whatman House
St Leonard's Road
20/20 Maidstone
Kent ME16 0LS
UK

Applikon BV
De Brauwweg 13
PO Box 149
NL-3100 Schiedam AC
The Netherlands

Baker Co., Inc.
Sanford Airport
PO Drawer E
Sanford ME 04073
USA

BASF Corp
8 Campus Drive
Parsipanny
NJ 07054
USA

Baxter International Inc.
1 Baxter Parkway
Deerfield
IL 60015
USA

Bayer Diagnostics
Stoke Court
Stoke Poges
Slough SL2 4LY
UK

BDH Laboratory Supplies
Merck Ltd
Merck House
Poole
Dorset BH15 1TD
UK

Beckman Instruments Inc.
2500 Harbor Blvd
Fullerton
CA 92634
USA

Beckman UK
Progress Road
Sands Industrial Estate
High Wycombe
Buckinghamshire HP12 4LJ
UK

Becton Dickinson Labware (Research & Development)
1 Becton Drive
Franklin Lakes
NJ 07417-1886
USA

Becton Dickinson UK Ltd
Between Towns Road
Cowley
Oxford OX4 3LY
UK

Bellco Glass Inc./Bellco Biotechnology
340 Elrudo Road
Vineland
NJ 08360
USA

Bethesda Research Laboratoris
See Gibco BRL

Bibby J. Science Products Ltd
Stone
Staffordshire ST15 0SA
UK

Bioengineering AG
Sagenrainstrasse 7
CH-8636 Wald/ZH
Switzerland

Biological Industries
Kibbutz Bet Haemek 25115
Israel

Bio-Rad Laboratories
32nd & Griffin Avenue
Richmond
California 94804
USA

Bio-Rad Laboratories (UK) Ltd
Bio-Rad House
Marylands Avenue
Hemel Hempstead
Hertfordshire HP2 7TD
UK

Biotech Instruments Ltd
183 Camford Way
Luton
Bedfordshire LU3 3AN
UK

Bio-Whittaker UK Ltd
The Atrium Court
Apex Plaza
Reading RG1 1AX
UK

BOC Ltd
The Priestley Centre
10 Priestley Road
The Surrey Research Park
Guildford
Surrey GU2 5XY
UK

Boehringer Mannheim GmbH
Sandhofer Strasse 116
PO Box 310120
D6800 Mannheim 31
Germany

Roche Diagnostics (UK) Ltd
Bell Lane
Lewes
East Sussex BN7 1LG
UK

B. Braun Biotech International GmbH
PO Box 1120
D-34209 Melsungen
Germany

APPENDIX 2: COMPANY ADDRESSES

B. Braun Biotech Inc.
999 Postal Road
Allentown
PA 18103
USA

Calbiochem-Behring Corp.
PO Box 12087
San Diego
California 2112
USA

Cambridge Instruments
Kent Industrial Instruments
Howart Road
Eaton-Socon
Huntingdon
Cambridgeshire PE19 3EU
UK

Camlab Limited
Nuffield Road
Cambridge CB4 1TH
UK

Cellex Biosciences
8500 Evergreen Blvd
Minneapolis
MN 55433
USA

Cellon
22 rue Dernier
Sol L-2543
Luxembourg

Chemap, Inc.
See B. Brawn Biotech

Clonetics Corp.
1800 30th Street
Boulder
CO 80301
USA

Collaborative Res. Inc.
Two Oak Park
Bedford
Massachusetts 01730
USA

Collagen Corp.
2500 Faber Place
Palo Alto
California 94303
USA

Corning Ltd
Stone
Staffordshire ST15 0BG
UK

Corning Medical
Corning Glass Works
Medfield
MA 02052
USA

Costar Europe Ltd
Sloterweg 305a 1171 VC
Badhoevedorp
The Netherlands

Coulter Corp
601 West 20th Street
West Hialeah
FL 33012
USA

Coulter Electronics Ltd
Northwell Drive
Luton
Bedfordshire LU3 3RH
UK

Cryomed
51529 Birch Street
New Baltimore
MI 48047
USA

Dako Corp.
6392 Via Real
Carpenteria
CA 93013
USA

Dextran Products Ltd
PO Box 1360
Princeton
NJ 08542
USA

Difco Laboratories
PO Box 1058A
Detroit MI 48232
USA

Dow Corning Corp.
Dow Corning Center
Box 0994
Midland
MI 48686-0994
USA

Du Pont UK Ltd
Coal Road
Leeds LS14 2AL
UK

Du Pont (UK) Ltd
Wedgwood Way
Stevenage
Hertfordshire SG1 4QN
UK

Dynal A.S.
PO Box 158
Skoyen N-0212
Oslo 2
Norway

Elga Ltd
Lane End
High Wycombe
Buckinghamshire HP14 3JH
UK

Envair Ltd
York Avenue
Haslingden
Rossendale
Lancashire BB4 4HX
UK

Eppendorf North America, Inc.
545 Science Drive
Madison
WI 53711
USA

Eppendorf-Netheler-Hinz GmbH
Postfach 650670
D-2000 Hamburg 65
Germany

European Collection of Cell Cultures
(ECACC)
CAMR (Centre for Applied
Microbiology and Research)
Salisbury
Wiltshire SP4 0JG
UK

European Society for Animal Cell
Technology (ESACT)
ECACC
CAMR
Salisbury
Wiltshire SP4 0JG
UK

European Tissue Culture Society
(ETCS)
ECACC
CAMR
Salisbury
Wiltshire SP4 0JG
UK

Fisher Scientific Co.
711 Forbes Avenue
Pittsburgh
PA 15219
USA

APPENDIX 2: COMPANY ADDRESSES

Fisons Scientific Equipment
Bishop Meadow Road
Loughborough
Leicestershire LE11 0RG
UK

Forma Scientific, Inc.
PO Box 649, Marietta
OH 45750
USA

Pall Gelman Sciences
Brackmills Business Park
Caswell Road
Northampton NN4 0EZ
UK

Pall Gelman Sciences Inc.
600 South Wagner Road
Ann Arbor
MI 48106
USA

Genzyme Corporation
1 Kendall Square
Cambridge
MA 02139-1562
USA

Gibco BRL
Division of Life Technologies, Inc.
PO Box 9418
Gaithersburg
MD 20898
USA

Gibco/BRL/Life Technologies
(Gibco Europe Ltd)
PO Box 35
Trident House
Renfrew Road
Paisley PA3 4EF
UK

Gibco Laboratoires
Grand Island Biological Co.
3175 Staley Road
Grand Island
NY 14072
USA

Gilson Medical Electronics (France)
S.A.
72 Rue Gambetta
BP 45
F-95400 Villiers le Bel
France

Gilson Medical Electronics, Inc.
3000 West Beltline Highway
PO Box 27
Middleton
WI 53562
USA

Grant Instruments (Cambridge) Ltd
Barrington
Cambridge CB2 5QZ
UK

Hazelton Research Products
See JRH Biosciences

Hepaire Manufacturing Ltd
Aire Cool House
Spring Gardens
London Road
Romford
Essex RM7 9LY
UK

Heraeus Equipment
111A Corporate Blvd
South Plainfield
NJ 07080
USA

Heraeus Equipment Ltd
Unit 9
Wates Way
Ongar Road
Brentwood
Essex
UK

Hitachi Scientific Instruments
Nissei Sangyo America Ltd
460 East Middlefield Road
Mountain View
CA 94043
USA

Hoechst (UK) Ltd
Hoechst House
Salisbury Road
Hounslow
Middlesex TW4 6HJ
UK

Horwell AR Ltd
Laboratory & Clinical Supplies
73 Maygrove Road
West Hampstead
London NW6 2BP
UK

HyClone Laboratories Inc
1725 South Road HyClone
Logan UT 84321
USA

N.V. HyClone Europe S.A.
Industriezone III
B-9320 Erembodegem-Aalst
Belgium

N.V. HyClone Europe S.A.
Unit 9, Atley Way
North Nelson Industrial Estate
Cramlington
Northumberland NE23 9WA
UK

ICI Cellmark Diagnostics
Blacklands Way
Abingdon Business Park
Abingdon
Oxfordshire OX14 1DY
UK

ICN Biomedicals
PO Box 28050
Cleveland
OH 44128
USA

ICN/Flow Laboratories
PO Box 17
Second Avenue Industrial Estate
Irvine KA12 8NB
UK

Imperial Laboratories (Europe) Ltd
West Portway
Andover
Hampshire SP10 3LF
UK

Ingold Messtechnik AG
Industrie Nord
Urdorf
Switzerland
CH-8902

Innovative Chemistry Inc.
PO Box 90
Marshfield
MA 02050
USA

Irvine Scientific
2511 Daimler Street
Santa Ana
CA 92705
USA

Jencons (Scientific) Ltd
Cherry Court Way Industrial Estate
Stanbridge Road
Leighton Buzzard
Bedfordshire LU7 8UA
UK

APPENDIX 2: COMPANY ADDRESSES

Jouan Inc.
170 Marcel Drive
Winchester
VA 22602
USA

JRH Biosciences
13804 West 107th Street
Lenexa
KS 66215
USA

L'Air Liquide
57 av Carnot
BP13 94503 Champigny Cedex
France

L'Aire Liquide
Cryogenic Equipment Division
PO Box 395
Allentown
PA 18105
USA

Life Science Laboratories
Sedgewick Road
Luton
Bedfordshire LU4 9DT
UK

Life Technologies (Gibco BRL), Inc.
PO Box 68
Grand Island
NY 14072-0068
USA

MDH Ltd
Walworth Road
Andover
Hampshire SP10 5AA
UK

Media-Cult A/s
Symbion Science Park
Haraldsgade 68
DK2100 Copenhagen
Denmark

Merck Ltd (BDH Laboratory Supplies)
Merck House
Poole
Dorset BH15 1TD
UK

Microbiological Associates
Life Sciences Center
9900 Blackwell Road
Rockville
MD 20850
USA

Millipore Corp.
80 Ashby Road
Bedford
MA 01730
USA

Millipore (UK) Ltd
The Boulevard
Blackmoor Lane
Watford
Hertfordshire WD1 8YW
UK

Nalge Company
Nalgene Labware Department
PO Box 20365
Rochester
NY 14602-0365
USA

National Institutes of Health
Small Animal Section
Veterinary Resources Branch
Division of Research Services
Building 14A
Room 103
Bethesda MD 20205
USA

New Brunswick Scientific Co., Inc.
44 Talmadge Road
Box 4005
Edison
NJ 08818-4005
USA

New Brunswick Scientific (UK) Ltd
Edison House
163 Dixons Hill Road
North Mimms
Hatfield
Hertfordshire AL9 7JE
UK

Nikon Corpopration
Fuji Building
2–3 Marunouchi 3-chome
Chiyoda-ku
Tokyo 100
Japan

Nuaire, Inc.
2100 Fernbrook Lane
Plymouth
MN 55447
USA

Nuclepore Corp.
7035 Commerce Circus
Pleasanton
CA 94566-3294
USA

Nune A/s
PO Box 280
Kamstrup
DK-4000
Roskilde
Denmark

Nycomed Pharma
Box 4284
Torshov
N-0401 Oslo 4
Norway

Olympus Corp.
Clinical Instruments Division
Precision Instrument Division
4 Nevada Drive
Lake Success
NY 11042-1179
USA

Olympus Optical Co., UK, Ltd
2–8 Honduras Street
London EC1Y 0TX
UK

Ortho Pharmaceuticals Inc.
410 University Avenue
Westwood
MA 02090
USA

Ortho Pharmaceuticals Ltd
Enterprise House
Station Road
Loudwater
High Wycombe
Buckinghamshire
UK

Oxoid USA Inc.
Wade Road
Basingstoke
Hampshire RG24 0PW
UK

Oxoid USA Inc.
PO Box 691
Ogdensburg
NY 13669
USA

PAA Labor-und
Forchungsgesellschaft mBh
Weiner Strasse 131
A-4020 Linz
Austria

Pall Europe Ltd
Europa House
Havant Street
Portsmouth PO1 3PD
UK

Pall Ultrafine Filtration Co
2200 Northern Blvd
East Hills
NY 11548
USA

APPENDIX 2: COMPANY ADDRESSES

Pel-Freez Biologicals
PO Box 68
Rogers
AR 72756
USA

Pfeifer & Langen Dormagen
Frankenstrasse 25
D-4047 Dormagen
Germany

Pharmacia LKB Biotechnology
800 Centennial Avenue
PO Box 1327
Piscataway
NJ 08855-1327
USA

Pharmacia LKB Biotechnology
Bjorkgatan 30
S-751 82 Uppsala
Sweden

Pharmacia LKB Biotechnology
23 Grosvenor Road
St Albans
Hertfordshire AL1 3AW
UK

Planer Products Ltd
Windmill Road
Sunbury-on-Thames
Middlesex TW16 7HD
UK

Revco Scientific, Inc.
275 Aiken Road
Asheville
NC 28804
USA

Revco Scientific International
PO Box 321
NL-8600 AH Sneek
The Netherlands

Sartorius AG
PO Box 32 43
Weender Landstrasse 94-108
3400 Goettingen
Germany

Sartorius Corp.
140 Wilbur Place
Bohemia
NY 11716
USA

Schott Glaswerke
Chemical Division
Laboratory Product Group
Hattenbergstrasse
10 Postfach 2480
D-6500 Mainz 1
Germany

Serotec
22 Bankside Station Approach
Kidlington
Oxford OX5 1JE
UK

Sigma Chemical Co. Ltd
Fancy Road
Poole
Dorset BH17 1NH
UK

Squibb & Sons Ltd
Regal House
Twickenham
Middlesex TW1 3QT
UK

Stratagene
11011 North Torrey Pines Road
La Jolla
CA 92037
USA

Taylor-Wharton Cryogenics
PO Box 568
Theodore
AL 36590
USA

Techne (Cambridge) Ltd
Hinxton Road
Duxford
Cambridge CB2 4PZ
UK

Tissue Culture Association
19110 Montgomery Village Avenue
Suite 300
Gaithersburg
MD 20879
USA

Tissue Culture Services Ltd
Botolph
Claydon
Buckingham MK18 2LR
UK

Ventrex Laboratories Inc.
217 Read Street
Portland
ME 04103
USA

Volac
77–93 Tanner Street
Barking
Essex IG11 8QD
UK

Watson-Marlow Ltd
Falmouth
Cornwall TR11 4RU
UK

Whatman LabSales
PO Box 1359
Hillsboro
OR 97123
USA

Wheaton Scientific
1000 North Tenth Street
Millville
NJ 08332
USA

Whittaker Bioproducts
8830 Biggs Ford Road
Walkersville
MD 21793-0127
USA

Worthington Biomedical Corporation
Halls Mill Road
Freehold
NJ 07728
USA

Zeiss (Carl) Germany
PO Box 1380
D-7082 Oberkochen
Germany

Zeiss (Carl) Inc.
Microscope Division
1 Zeiss Drive
Thornwood
NY 10594
USA

APPENDIX 3: RESOURCE CENTRES FOR BIOTECHNOLOGISTS

The concept of a culture collection acting as a resource centre for the benefit of the scientific community is by no means a recent innovation. The first recognized culture collection was established in 1880 in Prague by Dr Franz Kral. In more recent times collections have been created that follow the developing needs of biologists, and the American Type Culture Collection (ATCC) was the first animal cell culture bank established in the early 1960s. Since then, the requirement to have readily available authenticated cultures has become widely recognized. Today, there are many culture collections available to serve the needs of both academic and industrial users, and potential depositors can select a safe repository for their cultures from a range of institutions, some specialized in nature and others with more generalized activities (Table A3.1).

Table A3.1

Collection	Address	Holdings
American Type Culture Collection (ATCC)	PO Box 1549, Manassas, Virginia, 20108 1549, USA http://www.atcc.org/	General
European Collection of Cell Cultures (ECACC)	Centre for Applied Microbiology & Research Salisbury, Wiltshire SP4 0JG, UK http://ww.camr.org.uk	General/human/ genetic/HLA defined
DSM-Deutsche Sammlung Von Mikro-organismen und Zellkulturen GmbH (DSMZ)	Mascheroder Weg 1b D-38124 Braunschweig, Germany http//www.gbf.braunschweig. de.bl/DSMZ/	General
Interlab Cell Line Collection Servizio Biotechnologie, Istituto Nazationale per la Ricerca sul Cancro	CBA: Centro di Biotecnologie Avanzate L.go R Benzi 10, 16132 Genova, Italy http://arginine.umdnj.edu/home.html	General/human genetic/HLA defined
Coriell Institute for Medical Research	401 Haddon Avenue, Camden, NJ 08103, USA http://locus.umdnj.edu\nigms	Human genetic
Institution for Fermentation Osaka	17-85 Juso-Honmachi 2-Chome, Yodogawa-Ku, Osaka 532, Japan	General
Riken Gene Bank	3-1-1 Koyadai, Tsukuba Science City 305 Iboraki, Japan	General

Each collection has an accession procedure that will follow the general pattern. This is designed to provide users with some guarantee of authenticity and purity of the material supplied. A major concern with animal cell cultures is the potential for contamination with mycoplasmas, which often have deleterious effects on cells. Accordingly, responsible cell banks refuse to distribute material contaminated with mycoplasmas or indeed potentially pathogenic viruses. Some viruses (e.g. bovine viral diarrhoea virus) are widespread in tissue culture due to contamination of source materials but it is nevertheless important for responsible culture collections to test for such organisms: to prevent wider spread and to give commercial users of the collections confidence in the quality of material that they may ultimately use for biological production purposes.

Whilst quality control and safety data for cell lines are continuously being improved, the fundamental standards associated with the running of repositories are now established and provide reassurance to the user community that a minimum quality and safety standard is in place. In addition, many individuals and groups deposit new cell lines with collections as a matter of routine because they provide a valuable backup safe repository, give independent validation and relieve laboratories of the drain on time and resources required to maintain and distribute cells. A culture collection will also enhance the value of a culture by collating data on it and collections are increasingly efficient at making these data available to potential users.

It is usual for each collection to be registered with the World Data Centre established by the World Federation for Culture Collections (WFCC). In addition, there are numerous access points for data on the holdings of repositories available on the internet, e.g. WFCC home page http//www.wdcm.riken.go.jp/wfcc/wfcc.html. However, when investigating these sources of information a major consideration is to establish the quality and authenticity of the available data and biological material in the physical collections. As an example, there is little quality control of the DNA sequence data supplied by the international databases, and bacterial plasmid sequences from cloning vectors are known to occur in recorded gene sequences. For this reason cell culture groups coordinate quality initiatives to

Table A3.2 Suppliers of cell stocks

Culture collection	Address
ECACC	European Collection of Cell Cultures CAMR (Centre for Applied Microbiology & Research), Salisbury, Wiltshire SP4 0JG, UK
ATCC	American Type Culture Collection PO Box 1549, Manassas, Virginia, 20108 1549, USA
DSMZ	Deutsche Sammlung von Mikroorganismen und Zeilkuituren GmbH (German Collection of Microorganisms & Cell Cultures), Mascheroder Weg 1B, D-3300 Braunschweig, Germany
Riken	Riken Cell Bank, 3-1-1 Koyadai, Tsukuba Science City, 305 Iboraki, Japan

APPENDIX 3: RESOURCE CENTRES FOR BIOTECHNOLOGISTS

ensure that data on international electronic networks are backed up by quality-controlled cultures. The major resource centres are listed in Table A3.2 Each has a supply capability, usually there are related services and also training courses are available.

INDEX

Adaptation to serum-free, 5, 89, 92–98
Adenovirus, 8, 12–13
Adhesion, 116–117
Adventitious agents, 3, 19, 85, 299
Aeration systems, 195
Aeration, membrane, 198, 236
Aeration, surface, 194, 204
Aggregates, 5
Aging, 18
Agitation, 195, 202–204, 208, 210, 236, 244, 285
Airlift culture, 10, 12, 205, 222, 226, 270, 279, 301
Albumin, 87, 95
Amino acid analysis, 101
Amino acid, specific uptake rate, 290
Amino acids, essential, 87, 100, 160
Amino acids, non-essential, 100, 263
Ammonia, 131, 133, 162, 238
Ammonia production, specific rate of, 161, 166, 170, 172
Anchorage dependent cells, 11, 111, 119, 222, 265
Antibiotics, 38, 50–52, 85
Antifoam, 193, 197, 232
Apoptosis, 132, 179, 181–185
Apoptotic bodies, 179–180
ATCC, 3, 9, 49, 296, 325–326
ATP, 56, 151–152, 154–155, 194, 289
Attachment factors, 95
Authentication, 25–27, 296

B cells, 10
Bacteria, 19, 47–52
Baculovirus, 6
Batch culture, 133, 142, 163, 171, 225, 237, 242
BHK-21 cells, 12, 14, 192, 262, 302
Binding proteins, 95
Biochemistry, cell, 131, 302
Bioreactor, stirred, 133–134, 164, 203, 224, 236, 241–242, 244, 270, 279, 285, 301

Bovine serum albumin, 216
Bubble column reactor, 196, 215 *see also* airlift culture

Carboxymethylcellulose (CMC), 215, 232
CD7, 8
Cell bank (*see also* MCB, WCB), 19–20, 298
Cell culture suppliers, 326
Cell death kinetics, 179–186
Cell death, 11, 162, 210 (*see also* Apoptosis and Necrosis)
Cell factory (multitray), 119, 226, 230, 254–261
Cell lysis, 55, 140, 175, 180, 202
Cell nuclei count, 55, 264
CellCube, 226, 230
Cell number count, see Haemocytometer
Center for Biologics Evaluation & Research (CBER), 18, 296, 298
Chemostat, 135, 142, 163, 225, 237, 242–244, 246–252, 270, 279
CHO cells, 5, 14, 95, 192, 197, 213, 250, 297, 302
Chromosomes, 25–26, 35, 298
Clonogenic assay, 76
CMRL 1969 medium, 11
Collagen, 124
Co-cultivation, 123
Collagenase, 264
Colorimetric methods, 76–81
Company addresses, 315–324, 326
Contact, cell, 131
Contaminants (*see* Bacteria, Fungi, Mycoplasma, Adventitious agents, Viruses)
Contamination, elimination, 50–52
Continuous culture, *see* Chemostat
Coomassie blue stain, 68
COS 1/COS 7 cells, 7
Coulter counter (see Electronic counting)
ψCre/ψCrip cells, 15
Cryopreservation, 3, 20
Crystal violet stain, 76, 78–79

Culture collections, 1, 325–326
Culture control processes, 282–291
Culture systems, comparison, 226
Cytodex microcarrier, 262–263, 265–266
Cytogenetic analysis, 25, 27, 31
Cytoline microcarrier, 268, 270
Cytopathology, 35, 65
Cytotoxicity, 65, 68, 80

Detoxification, 123, 187–189
Dextran, 112, 207, 211, 215–216
Dextranase, 264
Dienes stain, 40
Dihydrofolate reductase, 6, 297
Dilution rate, 136, 139, 246–247, 249–250, 275
DMSO, 23, 78, 259
DNA fingerprinting, 19, 23, 29–34, 298
DNA polymerase, 42
DNA stain (see Hoechst)
DNA transfection, 8
DON cells, 106
Doubling time, 151, 244
Dry weight, 55
Dulbecco's modified MEM, 8, 100, 277, 254

Eagle's basal medium, 11, 12
Eagle's MEM, 9, 11, 94, 100, 164
EBV, 297
ECACC, 3, 6, 9, 25, 35, 296, 325–326
EDTA, 76
Electronic counting, 56, 77, 78, 93
Endotoxin, 93
Ethanolamine, 95
Exponential growth phase, 187
Extended cell bank, 20
Extracellular matrices, 112, 121

F-12 medium, 5, 90
Fed-batch culture, 133, 225, 237
Fetuin, 89, 95
Fibronectin, 87, 112, 117, 263
Filterwell, 123
Fixed bed culture, 268–281
Flow cytometric analysis, 26
Flow rate, 246
Fluidized bed culture, 268, 278–279
FMDV, 262
Fungi, 19, 47–52

Gene therapy, 13, 302
Genomic DNA, 31–32
GH3 cells, 15
GLP, 21, 296

Glucose, 131, 146–152, 153, 160–164, 251, 275
Glucose consumption, specific rate, 55, 133–4, 161, 170, 172, 275, 289
Glucose-6–phosphate dehydrogenase, 25–26
Glutamine, 131, 133, 160–164, 250
Glutamine, rate of degradation, 161, 163
Glutamine consumption, Specific rate of, 161,166, 170, 172
Glycolysis, 151–152
Glycoprotein, 117
Glycosylation, 5, 302
GMP, 21
Good laboratory practice (see GLP)
Good manufacturing practice (see GMP)
Gram stain, 52
Growth rate, specific, 134–137, 144, 154, 161–163, 166, 170, 172, 174, 176, 246–247, 249, 289
Growth factors, 87–88, 93, 131
Growth limiting substrate, 246–247, 250
Growth yields, 55

Haematoxylin stain, 264
Haemocytometer, 55, 57–61, 77–78, 93, 164
HAT medium, 10
HeLa cells, 8, 25
Henry's law constant, 151
Hepatocytes, 122–123, 192
HEPES buffer, 85, 263
Heteromyeloma cells, 297
Hoechst stain, 4, 19, 36, 38, 40
Hollow fibre culture, 10, 226, 235
Hormones, 95
HPLC, 105–106, 290
Human diploid cells (HDC), 11, 18, 20, 254–262, 296, 295–298 (see also MRC-5, WI-38)
Hybridization, 30, 33, 42, 225
Hybridoma cells, 10, 14, 19, 57, 95, 156, 164, 172, 192, 203, 213, 235–245, 250, 270, 289, 297

Identity testing, 25–28, 298
ImmobaSil microcarrier, 268, 270
Immobilized culture, 10, 133, 144, 155–157, 222, 234, 268–281
Immortalization, 10, 297
Impeller, 195, 197, 203–205, 208, 211, 224, 236
Indicator cell lines, 39
Insulin, 89, 94–95, 98
Integrated shear factor, 203
Integrity tests, 21

Interferon, 6, 14, 221, 259, 266
ISO 9001, 21, 296
Isoenzyme analysis, 19, 25–27, 29, 31, 299
ITS, 95

J558L cells, 9, 14

Karyology, 19, 26, 298–299
Karyotype, 25,
Kinetic model, 160–178
K_La, 134, 191, 193, 195
Kolmogorov eddy, 203–204, 207–208

Lactate, 71, 148, 150, 152–153, 162–164, 275
Lactate dehydrogenase (LDH), 55, 140–142, 202
Lactate production, specific rate of, 161, 166, 170, 172
LDH assay, 71–74, 175, 202
Laminin, 112
LETS, 117
Lipids, 95
Liquid flow control, 286
Liquid level control, 286–288
Lysosomes, 65–66

M3 medium, 7
Maintenance energy, 154–155
Mass balance equation, 171
Master cell bank (see MCB)
Matrigel, 124
Matrix culture, 124, 235
MCB, 18, 296, 298
MCDB32 medium, 95
MDCK cells, 13–14
Media, defined, 87
Medium 199, 9
Medium residence time, 141, 224
β-Mercaptoethanol, 95
Metabolic quotients, 134, 144–146, 150, 155–156
Metabolic yield ratios, 152
Metabolism, amino acids, 100–108
Metabolism, cell, 144–155
Metabolites, toxic, 189
Methocel, 215
Methylcellulose, 214
Microcarrier, aggregates, 208
Microcarrier culture, 55, 85, 112–113, 119, 121, 123, 142–144, 197, 206–208, 211, 216, 222, 225–226, 231, 254, 262–267, 301
Microcontainer, 124
Microgravity, 125
Mitotic index, 55

Mixing, 202–209, 210, 222, 231
Modelling, see Kinetic model
Moloney murine leukaemia virus, 15
Monoclonal antibodies, 10, 14, 156, 165, 187, 237, 325
MRC-5 cells, 11, 14, 111, 254, 265
MTT, 56, 62–64, 76, 79, 80
Multilocus DNA, 26, 34, 298
Mycoplasma, 3–4, 19, 35–46, 50–52, 299
Myeloma cells, 9, 89, 296

Namalwa cells, 12, 14
Necrosis, 132, 179, 181–182
Neutral red, 56, 65–70, 76, 78–79
NIH/3T3 cells, 8, 14
NS0 cells, 10, 14
Nuclear magnetic resonance, 153
Nutrient uptake, specific rate, 162

Oligonucleotide probes, 32
Osmotic pressure, 162, 288
Oxygen, 12, 131, 151, 153, 190–199, 222, 226, 243, 286
Oxygenation, 190–199, 205, 236
Oxygen consumption rate, 55, 151, 190–192
Oxygen, dissolved (DO), 190, 194, 196, 242, 284–285
Oxygen tension, 190–191, 242, 282
Oxygen transfer, 194, 198, 210

Patents, 21
PCR, 4, 27, 42–46
PCR amplification, 43
PDGF, 89
Perfusion rate, 139
Perfusion systems, 9, 133, 222, 225, 233, 238, 242–243
pH, 138, 190, 162, 226, 242, 264, 282
pH control, 283
Phenylisothiocyanate (PITC), 101
Pituitary tumor cells, 13
Plasticware, 117
Plating efficiency, 56
Pluronic F-68, 12, 98, 196, 198, 206, 212–214, 232
Polio virus, 8–9, 11, 12, 221, 259, 301
Polyethylene glycol (PEG), 12, 198, 212–214
Poly-lysine, 112, 118
Polymorphism, 31
Polystyrene, 109, 116
Polyvinylpyrrolidones, 213
Population doublings (PD), 11, 18, 20, 255, 359

Porous membrane culture, 113–114
Porous microcarrier, 119, 123, 206, 226, 268–281
Process control, 282–286
Productivity, 278–279, 301–303

Quality control, 3, 18–19, 23, 29, 31, 40, 295

Rabies virus, 12
Radiochromium, 56
RAPD, 298
Rate limiting factors, 174
Recombinant cell lines, 296–297
Recombinant proteins, 5–8, 14, 106, 121, 187, 221
Redox, 56
Regulatory guidelines, 295–299
Residual DNA, 299
Reynold's number, 204
Robotics, 230
Roller bottle culture, 10, 118–119, 222, 228–230, 262
Rotating wall vessel, 125
Roux bottle, 116, 164
RPMI 1640 medium, 10, 12, 94, 100, 164

Sarcoma virus, 8
Scale-up, 119, 221–227, 242, 278
Schneider-2 cells, 7
Seed lot system, 18, 22
Selenium, 94–95, 98
Semi-continuous culture, 139
Sendai virus, 12
Sensors, 282
Serum, 85, 87, 213
Serum factors, 117
Serum-free medium, 7, 87–99
Serum, substitution, 88–90, 94–95
Sf9 cells, 7
Shear forces, 196, 236, 243–244
Shear stress, 98, 125, 285
Siliconization, 233, 263–264
Siran glass spheres, 270–279
SOP, 22, 23
Soybean trypsin inhibitor, 93, 266
Sparging, 12, 194, 196–197, 203, 205–206, 222, 236, 244
Specific death rate, 134, 162, 166, 170, 172, 175–176, 247
Specific product formation rate, 145, 156, 275
Specific rate of antibody production, 161, 166, 170, 172, 289

Specific utilization rate, 55
Spheroids, 121–123, 144
Spin filter, 223, 226, 238, 266
Spinner culture, 7, 205, 222, 231–234
Spinner flasks, 118, 208, 231–234, 241, 263, 265
Standard operating procedure (*see* SOP)
Steady-state, 140, 238, 247, 250–251
Sterility tests, 19
Stirrer speed, 234, 243–244, 265, 285
Stoichometric coefficient, 172
Subculture, 23
Sulphorhodamine B stain, 76, 78–79
Suppliers, *see* Company Addresses
Surfaces, tissue culture, 109–120
SV40, 8, 295

Temperature control, 283
Terminology, 305–313
TGF-β, 89
Three-dimensional culture, 121–125
Tissue engineering, 302
tPA, 6, 221
Trace elements, 95
Transfected cell lines, 5
Transferrin, 94–95, 98, 263
Transformation, 13, 35
Trypan blue, 55, 57–61, 77–78, 80, 179, 202
Trypsin, 77
Tryptose phosphate broth, 12, 263

UV radiation, 69

Verax system, 268–270
Vero cells, 9, 39, 134, 192, 295
Viability, 19, 55–62, 65, 67, 76, 136, 140, 210, 286
Viral contaminants, 92, 297
Viral vaccines, 9, 11, 14, 221, 254–262, 266, 295
Viruses, 49
Vital stains, 55
VNTR's, 27

WCB, 18, 20, 22, 245, 254–255, 259, 298
WFCC, 326
WHO, 18, 295–296, 298–299
WI-38 cells, 11, 14, 254, 295
Working cell bank (*see* WCB)

Yield, cell, 154, 247
Yield, product, 157

Zinc deficiency, 185